Telecommunication Networks and Computer Systems

Series Editors

Mario Gerla
Aurel Lazar
Paul Kühn
Hideaki Takagi

Springer
London
Berlin
Heidelberg
New York
Barcelona
Budapest
Hong Kong
Milan
Paris
Santa Clara
Singapore
Tokyo

Hisao Kameda, Jie Li, Chonggun Kim and
Yongbing Zhang

Optimal Load Balancing in Distributed Computer Systems

With 49 Figures

 Springer

Hisao Kameda
Jie Li
Institute of Information Sciences
and Electronics,
University of Tsukuba,
1-1-1 Tennoudai, Tsukuba-shi,
Ibaraki 305, Japan

Chonggun Kim
Department of Computer Engineering,
Yeungnam University,
Gyongsan, Kyongbuk, 712-749, Korea

Yongbing Zhang
Institute of Policy and Planning Sciences,
University of Tsukuba,
1-1-1 Tennoudai, Tsukuba-shi,
Ibaraki 305, Japan

Series Editors

Mario Gerla
Department of Computer Science,
University of California,
Los Angeles,
CA 90024, USA

Paul Kühn
Institute of Communications
Switching and Data Technics,
University of Stuttgart,
D-70174 Stuttgart, Germany

Aurel Lazar
Department of Electrical Engineering and
Center for Telecommunications Research,
Columbia University,
New York, NY 10027, USA

Hideaki Takagi
Institute of Policy and Planning Sciences,
University of Tsukuba,
1-1-1 Tennoudai, Tsukuba-shi,
Ibaraki 305, Japan

British Library Cataloguing in Publication Data
Optimal load balancing in distributed computer systems.
    (Telecommunication networks and computer systems)
    1.Electronic data processing - Distributed processing
    I.Kameda, Hisao
    004.3'6

ISBN-13: 978-1-4471-1246-4   e-ISBN-13: 978-1-4471-0969-3
DOI: 10.1007/978-1-4471-0969-3

Library of Congress Cataloging-in-Publication Data
A catalog record for this book is available from the Library of Congress

Typesetting: Camera ready by authors

69/3830-543210 Printed on acid-free paper

Preface

A recent trend in computer systems is to distribute computation among several physical processors. There are basically two schemes for building such systems. One is a *tightly coupled* system and the other is a *loosely coupled* system. In the tightly coupled system, the processors share memory and a clock. In these *multiprocessor* systems, communication usually takes place through the shared memory.

In the loosely coupled system, the processors do not share memory or a clock. Instead, each processor has its own local memory. The processors communicate with each other through various communication lines, such as high-speed buses or telephone lines. These systems are usually referred to as *distributed* computer systems.

The processors in distributed computer systems may vary in size and function. They may include small microprocessors, workstations, minicomputers, and large general-purpose computer systems. These processors are referred to by a number of different names such as *sites*, *nodes*, *computers*, and so on, depending on the context in which they are mentioned. The term *node* is mainly used through this book.

Distributed computer system research includes many areas, including: communications networks, distributed operating systems, distributed databases, concurrent and distributed programming languages, theory of parallel and distributed algorithms, interconnection structures, fault tolerant and ultrareliable systems, distributed real-time systems, distributed debugging, and distributed applications [Sta84]. Therefore, a distributed computer system consists of the physical networks, nodes, and all the controlling software.

There are five major reasons for building distributed computer systems: *resource sharing*, *performance improvement*, *reliability*, *communication*, and *extensibility*. We will briefly elaborate on each of them.

(1) Resource Sharing: If a number of different nodes are connected to each other, then a user at one node may be able to use the resources available at another. For example, a user at node A may be using a laser printer that is provided by node B. Meanwhile, a user in B may access a file that resides at A. In general, resource sharing in a distributed computer system provides mechanisms for sharing files at remote nodes, processing information in a distributed database, printing files at remote nodes, using

remote specialized hardware devices, and other operations.

(2) Performance Improvement: If a particular computation can be partitioned into a number of subcomputations that can run concurrently, then the availability of a distributed computer system may allow us to distribute the computation among the various nodes, to run it concurrently. In addition, if a particular node is currently overloaded with jobs, some of them may be moved to other, lightly loaded nodes. This movement of jobs is called *load balancing* .

(3) Reliability: If one node fails in a distributed computer system, the remaining nodes can potentially continue operating. If the system is composed of a number of large general-purpose computers, the failure of one of them should not affect the rest. In general, if enough redundancy exists in the system in both hardware and software, the system can continue with its operation, even if some of its nodes have failed.

(4) Communication: When a number of nodes are connected to each other via a communications network, the users at different nodes have the opportunity to exchange information. In a distributed system we refer to such activity as *electronic mail*. The similarity between an electronic mail system, which is an application visible to the users, and a message system, which is intended for communication between processes, should be apparent.

(5) Extensibility: This is the ability to easily adapt to both short and long term changes without significant disruption of the system. Short term changes include varying workloads and subnet traffic, and host or subnet failures or additions. Long term changes are associated with major modifications to the requirements of the system.

One of the most interesting problems in distributed computer systems is improving the performance of the system by balancing the load among nodes. In this book, we study load balancing problems.

Load balancing may be either static or dynamic. Static load balancing strategies are generally based on the information about the average behavior of system; transfer decisions are independent of the actual current system state. Dynamic strategies, on the other hand, react to the actual current system state in making transfer decisions. It seems that dynamic strategies are more effective than static strategies whereas the former may have more overhead than the latter. Furthermore, it seems that there currently exists no optimal dynamic strategy that is sophisticated enough to be applicable generally to various environments and analytically tractable.

This book investigates mainly the optimal static load balancing in distributed computer systems. The results of optimal static load balancing might be helpful in designing a distributed computer system or making parametric adjustments to improve the system performance.

In this book, the optimal load balancing in five important network configurations are considered. These network configurations are single channel network configurations, star network configurations with one-way traffic, tree hierarchy network configurations, star network configurations with two-way traffic, and tree network configurations with two-way traffic.

The single channel network configuration is common in distributed computer systems. In the network configurations, host computers share the same communication media (e.g., bus) for communication.

In the star network configuration with one-way traffic, there are many satellite nodes (stations) and a central node (hub). All satellite nodes are connected to the central node by communication lines. A job externally arriving at a satellite node can be processed locally or be forwarded to the central node for processing. A job arriving at a central node, however, can only be processed locally. That is, there is only one way traffic from satellite nodes to the central node in this network configuration.

The tree hierarchy network configuration is an extension of the star network with one-way traffic. Notice that many of distributed computer communication networks naturally have the tree hierarchy network configurations. That is, we often have tree organizations of computing machines. For example, a super–power center machine is connected with remote station mainframe computers each of which is, in turn, connected with server-type workstations each of which is connected with client-type workstations and personal computers, etc. In a tree hierarchy network, all nodes are connected in a hierarchy structure. We assume that a job arriving at a node can only be processed locally or be forwarded to a node in a higher layer for processing in a tree hierarchy network. This assumption is reasonable if the processing capacities of the nodes in the higher layer of a node i are larger than that of node i, and the processing capacities of the nodes in the lower layer are smaller than that of node i.

In the same as the star network configuration with one-way traffic, there are many satellite nodes (stations) and a central node in a star network with two-way traffic. A job externally arriving at the central node, however, can be forwarded to a satellite node for processing. In this network configuration, the central node can be taken as a switch so that a job arriving at a node (a satellite node or the central node) can be sent to any nodes in the network for processing.

In tree network configurations with two-way traffic, there are no hierarchical structures. Every node is connected to one of its neighbor nodes by one and only one communication line. Since there is two-way traffic between any two nodes (host computers) in this configuration, a job arriving at a node can be processed locally or be forwarded to any other nodes in a network for processing.

Furthermore, two types of load balancing policies each of which has distinct performance objective are considered. One is the *overall* optimal policy and the other is the *individually* optimal policy. The overall optimal policy whose goal is to minimize the system-wide mean job response time. The solution is referred as the *optimum*. The individually optimal load balancing, on the other hand, is the policy whereby each job is scheduled so as to minimize its own expected job response time, given the expected node and communication delays. The solution of the individually optimal load balancing is referred as the *equilibrium*.

Chapter 1 studies the optimal load balancing in single channel and one-way traffic star network configurations. At first, two effective load balancing algorithms are presented for the single channel and one-way traffic star network configurations one for each. For the multi-class job environment of the single channel network configurations, the model of the distributed system is also presented. On the basis of the model, the load balancing problem is formulated as a nonlinear optimization problem. An effective load balancing for the multi-class job environment is also presented.

Chapter 2 provides comparative study on the overall optimal load balancing and individually optimal load balancing policies in single channel and one-way traffic star network configurations by parameter analysis in detail.

Three important parameters in distributed computer systems can be considered: the communication time of the network, the processing capacity of each node, and the job arrival rate of each node. The effects of the three parameters on the behavior of the systems in the overall optimal policy and in the individually optimal policy is studied for the multi-class job environment through numerical experimentation.

Chapter 3 presents optimal load balancing in tree hierarchy network configurations. We derive theorems which give the necessary and sufficient conditions for the optimal load balancing to the tree hierarchy problem. It is proven that the tree hierarchy optimization problem can be solved by solving much simpler star sub–optimization problem iteratively. A decomposition algorithm to solve the optimal static load balancing problem in tree hierarchy network configurations is presented. Furthermore, the effects of link communication time and node processing time on optimal load balancing in tree hierarchy network configurations are studied. Clear and simple analytical results are presented.

In chapter 4, optimal load balancing in star network configurations with two-way traffic is presented. It is proven that in the optimal solution the satellite nodes of the star network are classified into the following categories: the idle source nodes, the active source nodes, the neutral nodes, and the sink nodes. The necessary and sufficient conditions for optimal solution are studied, and an effective $O(n)$ algorithm is proposed to solve the optimization problem for an n-satellite system. Furthermore, the effects of link communication time on optimal load balancing in a star network are studied analytically.

As an important extension the work on load balancing in star network configurations with two-way traffic, chapter 5 contains the study of optimal load balancing in tree networks with two-way traffic. It is demonstrated that the optimization problem of the particular structure can be solved effectively by a decomposition technique. An effective algorithm is developed for obtaining the optimal solution.

In finding the optimal load balancing policies, the study of the uniqueness of the solution is important. Chapter 6 provides the necessary and sufficient conditions of the uniqueness for a certain class of the static optimal load balancing problems.

As we have noted above, there are two kinds of load balancing: static and dynamic. The dynamic load balancing is also an active research field. Although this book mainly studies the static load balancing problems in distributed computer systems, a survey of studies on dynamic load balancing is provided in chapter 7.

Chapter 8 provides a comparative study of static and dynamic load balancing. Dynamic load balancing policies offer the possibility of improving load distribution at the expense of additional communication and processing overhead. The overhead of dynamic load balancing may be large, in particular for a large heterogeneous distributed system. Comparing the performance provided by static load balancing policies and dynamic load balancing policies, it shows that the static load balancing policies are preferable when the system loads are light and moderate or when the overhead is not negligible high.

Acknowledgements

We have been fortunate in having the help of many people during our work in The University of Electro-Communications and University of Tsukuba. Without their help, it would be impossible to finish this work. We would like to express our special gratitude to Professor Hideaki Takagi for his kindly introducing us the opportunity to publish our work in the Springer-Verlag London Limited series, Telecommunications Networks and Computer Systems (TNCS). We are greatly indebted to the support of our publisher, Spring-Verlag London Limited.

Credits

We are grateful to the copyright holders of the following articles for permitting us to use material contained within them.

C. Kim and H. Kameda. Optimal static load balancing of multi-class jobs in a distributed computer system. In *10th IEEE International Conference on Distributed Computing Systems*, pages 562–569, Paris, France, June 1990. Reprinted with permission of IEEE.

C. Kim and H. Kameda.   Optimal static load balancing of multi-class jobs in a distributed computer system. *The Transactions of the IEICE*, E73(7):1207–1214, July 1990. Reprinted with permission of IEICE.

C. Kim and H. Kameda.   An algorithm for optimal static load balancing in distributed computer systems. *IEEE Trans. on Computers*, 41(3):381–384, March 1992. permission of IEEE.

C. Kim and H. Kameda.   Parametric analysis of static load balancing of multi-class jobs in a distributed computer system. *The Transactions of the IEICE*, E75-D(4):527–534, July 1992. Reprinted with permission of IEICE.

H. Kameda and Y. Zhang.   Uniqueness of the solution for optimal static routing in open bcmp queueing networks. *Mathl. Comput. Modelling*, 22:119–130, 1995. Reprinted with permission of Elsevier Science Ltd.

J. Li and H. Kameda.   Optimal static load balancing in tree hierarchy network configurations. In *Proc. of the 10th IASTED International Conference, Applied Informatics*, pages 13–16, Innsbruck, Austria, February 10 - 12 1992. Acta Press. Reprinted with permission of IASTED.

J. Li and H. Kameda.   Effects of node processing time on optimal load balancing in tree hierarchy network configurations. *Computer Communications*, 16(12):781–793, 1993. Reprinted with permission of Elesevier Science B.V.

J. Li and H. Kameda.   Optimal load balancing in tree networks with two-way traffic. *Computer Networks and ISDN Systems*, 25(12):1335–1348, 1993. Reprinted with permission of Elsevier Science Publishers B.V.

J. Li and H. Kameda.   A decomposition algorithm for optimal static load balancing in tree hierarchy network configurations. *IEEE Trans. on Parallel and Distributed Systems*, 5(5):540–548, 12 1994. Reprinted with permission of IEEE.

J. Li and H. Kameda.   Optimal load balancing in star network configurations with two-way traffic. *Journal of Parallel and Distributed Computing*, 23(3):364–375, 12 1994. Reprinted with permission of Academic Press, Inc.

J. Li, H. Kameda, and K. Shimizu.   Effects of link communication time on optimal load balancing in tree hierarchy network configurations. *IEICE Trans. on Information and Systems*, E76-D(2):199–209, 1993. Reprinted with permission of IEICE.

Y. Zhang, K. Hakozaki, H. Kameda, and K. Shimizu.   A performance comparison of adaptive and static load balancing in heterogeneous distributed systems. In *Proc. 28th Annual Simulation Symposium, IEEE Computer Society Press*, pages 332–340, Phoenix, A.Z., 1995. Reprinted with permission of IEEE.

Y. Zhang, H. Kameda, and K. Shimizu.   Parametric analysis of static load balancing in star network configurations. *Transactions IEICE*, J74-D-I(9):644–655, 1991. Reprinted with permission of IEICE.

Y. Zhang, H. Kameda, and K. Shimizu.   A comparison of adaptive and static load balancing strategies by using simulation methods. In *Proc. 1992 Singapore Int. Conf. Intelligent Control and Instrumentation, IEEE Singapore Section*, pages 1162–1167, Singapore, February 1992. Reprinted with permission of IEEE.

Y. Zhang, H. Kameda, and K. Shimizu.   Parametric analysis of optimal static load balancing in distributed computer systems. *Journal of Information Processing*, 14(4):433–441, 1992. Reprinted with permission of the Informatin Processing Society of Japan.

Y. Zhang, H. Kameda, and K. Shimizu.   Adaptive bidding job scheduling in heterogeneous distributed systems. In *Proc. 1994 Int. Workshop Modeling, Analysis, and Simulation of Computer and Telecommunication Systems, IEEE Computer Society Press*, pages 250–254, Durham, N.C., 1994. Reprinted with permission of IEEE.

Contents

List of Tables

List of Figures

Chapter 1

Load Balancing in Single Channel and Star Network Configurations

1.1 Introduction

In this chapter, the static load balancing in single channel and star network configurations is studied. Tantawi and Towsley [TT84, TT85] studied load balancing of single class jobs in a distributed computer system that consists of a set of heterogeneous host computers connected by single channel and star communications networks. They considered an optimal static load balancing strategy which determines the optimal load at each host so as to minimize the mean job response time, and derived two algorithms (called single-point algorithms) that determines the optimal load at each host for given system parameters. Static load balancing may be useful for system sizing (e.g., allocation of resources, identification of bottleneck, sensitivity studies, etc.). The solution of optimal static load balancing may help us design the system.

First, the same model for sigle channel network configurations as Tantawi and Towsley is considered. Some additional properties that the optimal solution satisfies are derived. On the basis of these properties, another single-point algorithm that seems more easily understandable and more straightforward than that of Tantawi and Towsley is derived. The performance of the derived algorithm is compared with that of Tantawi and Towsley (T & T). The number of program steps for implementing the new algorithm is about a third of that of the T & T algorithm. During the course of numerical experiment, the new algorithm required about two thirds of the computation time that is required by the T & T algorithm [KK92a]. For star network configuration system, an efficient load balancing algorithm than that of Tantawi and Towsley [TT84] is also derived [KK92a].

Second, the single job class model is extended to a multiple job class environment of single channel network configurations [KK90a] [KK90b]. An efficient algorithm for optimal load balancing of multi-class jobs is proposed. The performance of the proposed algorithm for multi-class jobs is favorably compared with those of other well-known algorithms: the Flow Deviation (FD) algorithm [FGK73] and the Dafermos algorithm [DS69]. Both the proposed algorithm and the FD algorithm require a comparable amount of storage that is far less than that required by the Dafermos algorithm. During the course of numerical experimentation, the proposed algorithm and the Dafermos algorithm required comparable computation times for obtaining the optimal solution that are far less than that of the FD algorithm [KK90a] [KK90b].

1.2 Load Balancing in the Single Job Class Environment

1.2.1 Introduction

Tantawi and Towsley [TT85] have considered an optimal static load balancing strategy which determines the optimal load at each host so as to minimize the mean job response time. They have proposed a model of a distributed computer system that consists of a set of heterogeneous host computers connected by a single channel communications network. A key assumption of theirs was that the communication delay does not depend on the source-destination pair. This assumption may apply to single channel networks such as satellite networks and some LAN's. Given this assumption, they determined the requirement that the optimal load at each host satisfies, and derived an algorithm that determines the optimal load at each host for given system parameters. It is this algorithm that they call a single-point algorithm.

The Tantawi and Towsley single-point algorithm [TT85] is surprising in the sense that it does not calculate the load at each node iteratively. Note that previous algorithms on related models such as flow-deviation type algorithms (see, e.g., Fratta, Gerla and Kleinrock [FGK73]) and Gauss-Seidel type algorithms (see, e.g., Dafermos and Sparrow [DS69] and Magnanti [Mag84]) require iterative calculation of loads. However, the algorithm appears to be complicated and difficult to understand.

In this section, the same model as Tantawi and Towsley [TT85] under the same assumptions concerning the communication delay is considered. Additionally, some properties that the optimal solution satisfies are derived. On the basis of these properties, another single-point algorithm that is more easily understandable and more straightforward than that of Tantawi and Towsley is offered [TT85]. Furthermore, several properties relating to the convergence of our algorithm are identified and its performance is demonstrated. A more efficient load balancing algorithm than that of Tantawi and Towsley [TT84] for a distributed computer system with star network configurations is also de-

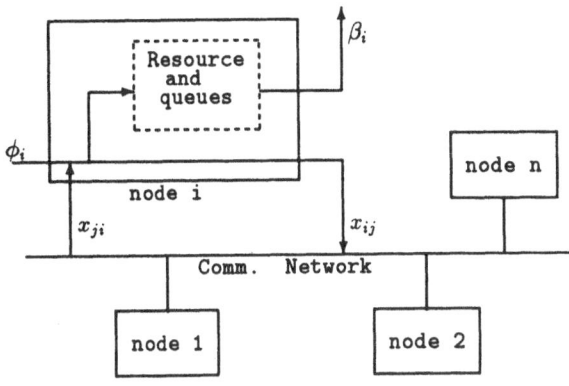

Figure 1.1: A distributed computer system.

rived.

1.2.2 Model description

The same model as that of Tantawi and Towsley is assumed [TT85]. That is, the system consists of n nodes (hosts) connected by a single channel communications network (Figure 1.1).

Jobs arrive at node i, $i = 1, 2, ..., n$, according to time-invariant Poisson process. A job arriving at node i may be either processed at node i or transferred through the communications network to another node j. We assume that the decision of transferring a job does not depend on the state of the system, and hence is *static* in nature. Also, we assume that a transferred job from node i to node j receives its service at node j and is not transferred to other nodes. It is assumed that the expected communication delay from node i to node j is independent of the source-destination pair (i, j).

For reference, a portion of the notation and assumptions contained in [TT85] are repeated here (see Appendix C of [TT85]).

n Number of nodes.

β_i Job processing rate (load) at node i.

β $[\beta_1, \beta_2, .., \beta_n]$.

x_{ij} Job flow rate from node i to node j.

ϕ_i External job arrival rate to node i.

Φ Total external job arrival rate ($\Phi = \sum_{i=1}^{n} \phi_i$).

λ Network traffic.

$F_i(\beta_i)$ Mean node delay of a job processed at node i — an increasing positive function.

$G(\lambda)$ Source-destination-independent mean communication delay — a nondecreasing positive function.

Tantawi and Towsley classified nodes in the following [TT85].

(1) idle source (R_d) : The node does not process any jobs, that is , $\beta_i = 0$.

(2) active source (R_a) : The node sends jobs and does not receive any jobs. But the node processes a part of the jobs that arrive at the node.

(3) neutral (N) : The node processes jobs locally without sending or receiving jobs.

(4) sink (S) : The node receives jobs from other nodes but does not send out any jobs.

The problem of minimizing the mean response time of a job is expressed in the following formulations, as stated by Tantawi and Towsley [TT85].

$$\text{minimize}\quad D(\boldsymbol{\beta}) = \frac{1}{\Phi}[\sum_{i=1}^{n} \beta_i F_i(\beta_i) + \lambda G(\lambda)], \tag{1.1}$$

$$\text{subject to}\quad \sum_{i=1}^{n} \beta_i = \Phi,$$

$$\beta_i \geq 0, \quad i = 1, 2, .., n,$$

where the network traffic λ may be expressed in terms of the variable β_i as

$$\lambda = \frac{1}{2}\sum_{i=1}^{n} |\phi_i - \beta_i|. \tag{1.2}$$

We note that the job flow rates x_{ij} through the communications network may be determined arbitrarily as long as the job flow balance constraint is satisfied. In particular, we may assign the flow rate from node i to node j proportionate to the contribution of node j to the network traffic; that is,

$$x_{ij} = \begin{cases} \frac{(\beta_j - \phi_j)}{\lambda}(\phi_i - \beta_i), & i \in R_d, R_a \ j \in S, \\ 0, & \text{otherwise.} \end{cases}$$

Since node j is a sink, the quantity $\beta_j - \phi_j (> 0)$ represents the traffic sent to node j. Node i, on the other hand, is a source and therefore $\phi_i - \beta_i (> 0)$ represents the traffic sent from node i.

1.2.3 Properties of the optimal solution and an optimal load balancing algorithm

Define the following two functions.

$$f_i(\beta_i) = \frac{\partial}{\partial \beta_i}(\beta_i F_i(\beta_i)),$$

$$g(\lambda) = \frac{\partial}{\partial \lambda}(\lambda G(\lambda)).$$

The inverse of the marginal node delay is defined as

$$f_i^{-1}(x) = \begin{cases} a, & f_i(a) = x, \\ 0, & f_i(0) \geq x. \end{cases}$$

Tantawi and Towsley [TT85] derived the following theorem by using the Kuhn-Tucker theorem (see Theorem 2 of [TT85]):

[Tantawi-Towsley Theorem] *The optimal solution to problem (1.1) satisfies the relations,*

$$\begin{array}{llll} f_i(\beta_i) \geq \alpha + g(\lambda), & \beta_i = 0 & (i \in R_d), \\ f_i(\beta_i) = \alpha + g(\lambda), & 0 < \beta_i < \phi_i & (i \in R_a), \\ \alpha \leq f_i(\beta_i) \leq \alpha + g(\lambda), & \beta_i = \phi_i & (i \in N), \\ \alpha = f_i(\beta_i), & \beta_i > \phi_i & (i \in S), \end{array} \quad (1.3)$$

subject to the total flow constraint,

$$\sum_{i \in R_a} f_i^{-1}(\alpha + g(\lambda)) + \sum_{i \in N} \phi_i + \sum_{i \in S} f_i^{-1}(\alpha) = \Phi, \quad (1.4)$$

where α is the Lagrange multiplier.

Tantawi and Towsley's single-point algorithm [TT85] first determines the node partitions (see steps 2 ~ 5 [TT85]). Then it solves eq.(1.4) for α, and obtains the values of β_i determined by that value of α (see step 6 [TT85]). In practical situations, we can rarely obtain the closed form giving α from eq.(1.4). Therefore some kind of iterative calculation of α is necessary. Since load balancing does not usually need very high accuracy, a simple method such as a binary search may be practical in solving eq.(1.4) iteratively for α.

For the purpose of developing our algorithm, the following properties are derived directly from the Tantawi-Towsley theorem. Let β be an optimal solution to problem (1.1) and let

$$\alpha = \min_i f_i(\beta_i), \quad (1.5)$$

$$\lambda = \frac{1}{2} \sum_{i=1}^{n} |\phi_i - \beta_i|. \quad (1.6)$$

It is now shown from the above theorem that the following three properties hold true in the optimal solution.

Property 1.1

$$\begin{array}{ll}
f_i(0) \geq \alpha + g(\lambda), & \text{iff} \quad \beta_i = 0, \\
f_i(\phi_i) > \alpha + g(\lambda) > f_i(0), & \text{iff} \quad 0 < \beta_i < \phi_i, \\
\alpha \leq f_i(\phi_i) \leq \alpha + g(\lambda), & \text{iff} \quad \beta_i = \phi_i, \\
\alpha > f_i(\phi_i), & \text{iff} \quad \beta_i > \phi_i.
\end{array} \tag{1.7}$$

PROOF. From the original assumption, note that $f_i(\beta_i)$ and $g(\lambda)$ are increasing and nondecreasing, respectively, and both are positive. It is clear from the relations noted at (1.3) that $\alpha = \min_i f_i(\beta_i)$, which is the same as eq.(1.5). Thus, both α's are the same. Note that λ defined by eq.(1.6) is the network traffic in the optimal solution.

The necessity in (1.7) : It can be deriven easily from the stated set of relations 1.3,

$$\begin{array}{ll}
f_i(0) \geq \alpha + g(\lambda), & \text{if} \quad \beta_i = 0, \\
f_i(\phi_i) > \alpha + g(\lambda) > f_i(0), & \text{if} \quad 0 < \beta_i < \phi_i, \\
\alpha \leq f_i(\phi_i) \leq \alpha + g(\lambda), & \text{if} \quad \beta_i = \phi_i, \\
\alpha > f_i(\phi_i), & \text{if} \quad \beta_i > \phi_i.
\end{array}$$

The sufficiency in (1.7) : The reverse can be shown by contradiction. (For example, assume that, for some i, $\beta_i > 0$ when $f_i(0) \geq \alpha + g(\lambda)$. Then from the above necessity we see that for i, $\alpha + g(\lambda) > f_i(0)$, which would contradict the assumption.) $\square$

Note the following definitions in the optimal solution.

$R_d = \{i | \beta_i = 0\}$ (idle sources),

$R_a = \{i | 0 < \beta_i < \phi_i\}$ (active sources),

$N = \{i | \beta_i = \phi_i\}$ (neutrals),

$S = \{i | \beta_i > \phi_i\}$ (sinks).

Property 1.2 *If β is an optimal solution to the problem (1.1) then we have*

$$\begin{array}{ll}
\beta_i = 0, & i \in R_d, \\
\beta_i = f_i^{-1}(\alpha + g(\lambda)), & i \in R_a, \\
\beta_i = \phi_i, & i \in N, \\
\beta_i = f_i^{-1}(\alpha), & i \in S.
\end{array}$$

PROOF. This is clear from the Tantawi-Towsley theorem. $\square$

Property 1.3 *If β is an optimal solution to the problem (1.1) then we have*

$$\lambda = \lambda_S = \lambda_R,$$

where

$$\lambda_S = \sum_{i \in S} [f_i^{-1}(\alpha) - \phi_i], \tag{1.8}$$

$$\lambda_R = \sum_{i \in R_d} \phi_i + \sum_{i \in R_a} [\phi_i - f_i^{-1}(\alpha + g(\lambda_S))]. \tag{1.9}$$

PROOF. We see that eq.(1.4) is equivalent to the equality $\lambda_S = \lambda_R$, if we use the definitions given by eqs.(1.8) and (1.9) and note that $\sum_{i \in R_d} \phi_i + \sum_{i \in R_a} \phi_i + \sum_{i \in N} \phi_i + \sum_{i \in S} \phi_i = \Phi$.

Recall the above definitions on node partition. By noting that $\sum_{i=1}^{n} \beta_i = \sum_{i=1}^{n} \phi_i = \Phi$, we have

$$\lambda = \frac{1}{2} \sum_{i=1}^{n} |\phi_i - \beta_i| = \sum_{i \in S} (\beta_i - \phi_i).$$

Therefore, by noting Property 1.2 we see that $\lambda = \lambda_S$. $\square$

Let us consider the following definitions in the order shown below for an arbitrary $\alpha \ (\geq 0)$.

$$S(\alpha) = \{i | \alpha > f_i(\phi_i)\}, \tag{1.10}$$

$$\lambda_S(\alpha) = \sum_{i \in S(\alpha)} [f_i^{-1}(\alpha) - \phi_i], \tag{1.11}$$

$$R_d(\alpha) = \{i | f_i(0) \geq \alpha + g(\lambda_S(\alpha))\}, \tag{1.12}$$

$$R_a(\alpha) = \{i | f_i(\phi_i) > \alpha + g(\lambda_S(\alpha)) > f_i(0)\}, \tag{1.13}$$

$$\lambda_R(\alpha) = \sum_{i \in R_d(\alpha)} \phi_i + \sum_{i \in R_a(\alpha)} [\phi_i - f_i^{-1}(\alpha + g(\lambda_S(\alpha)))], \tag{1.14}$$

$$N(\alpha) = \{i | \alpha \leq f_i(\phi_i) \leq \alpha + g(\lambda_S(\alpha))\}. \tag{1.15}$$

From properties 1.1, 1.2 and 1.3 above, we see that, if an optimal α is given, the node partitions in the optimal solution are characterized as follows.

$$R_d = R_d(\alpha), \ R_a = R_a(\alpha), \ N = N(\alpha), \ \text{and} \ S = S(\alpha).$$

Thus we see that the node partitions are determined only by the conditions on $f_i(0)$, $f_i(\phi_i)$, and α. Furthermore, for an optimal α, we see that

$$\lambda = \lambda_S = \lambda_R = \lambda_S(\alpha) = \lambda_R(\alpha).$$

On the basis of properties 1.1, 1.2 and 1.3 above, the following load balancing algorithm is derived. The computational requirements of this algorithm are also given.

[ALGORITHM 1] :

1. Order nodes. $O(n \log n)$

   Order nodes such that $f_1(\phi_1) \leq f_2(\phi_2) \leq \dots \leq f_n(\phi_n)$.

   If $f_1(\phi_1) + g(0) \geq f_n(\phi_n)$, then no load balancing is required.

2. Determine α. $O(n)$ (See the following remarks.)

Find α such that $\lambda_S(\alpha) = \lambda_R(\alpha)$

(by using, for example, a binary search),

where, given α, each value is calculated in the following order.

$S(\alpha) = \{i | \alpha > f_i(\phi_i)\}$,

$\lambda_S(\alpha) = \sum_{i \in S(\alpha)} [f_i^{-1}(\alpha) - \phi_i]$,

$R_d(\alpha) = \{i | f_i(0) \geq \alpha + g(\lambda_S(\alpha))\}$,

$R_a(\alpha) = \{i | f_i(\phi_i) > \alpha + g(\lambda_S(\alpha)) > f_i(0)\}$,

$\lambda_R(\alpha) = \sum_{i \in R_d(\alpha)} \phi_i + \sum_{i \in R_a(\alpha)} [\phi_i - f_i^{-1}(\alpha + g(\lambda_S(\alpha)))]$.

3. Determine the optimal load. $O(n)$

$\beta_i = 0$, for $i \in R_d(\alpha)$,

$\beta_i = f_i^{-1}(\alpha + g(\lambda))$, for $i \in R_a(\alpha)$,

$\beta_i = f_i^{-1}(\alpha)$, for $i \in S(\alpha)$,

$\beta_i = \phi_i$, for $i \in N(\alpha)$,

where $N(\alpha) = \{i | \alpha \leq f_i(\phi_i) \leq \alpha + g(\lambda_S(\alpha))\}$.

Remarks: The main process of the algorithm is determining α (in step 2). The single-point algorithm of Tantawi and Towsley [TT85] determines the node partitions before calculating α exactly. Our algorithm does not need to provide the node partition process separately and can, instead, determine the node partitions during the process of obtaining α. The computational requirements given above are those with respect to the number of nodes n. In step 2, however, the computational requirements must increase as the acceptable tolerance for the relative error of α decreases. For example, if a binary search is used, the computational requirement then is $O(n \log 1/\varepsilon)$ where ε denotes the acceptable tolerance.

Continuing, consider the following additional properties.

Property 1.4 $\lambda_S(\alpha)$ *increases (from zero) and* $\lambda_R(\alpha)$ *decreases (from Φ) both monotonically as α increases.*

PROOF. It is simple to prove this from definitions (1.11) and (1.14) and the assumptions on $f_i(\alpha)$ and $g(\lambda)$. $\square$

Remark: This property assures that the algorithm can determine an α that satisfies $\lambda_S(\alpha) = \lambda_R(\alpha)$.

Property 1.5 *For an arbitrary α such that $\alpha_1 \leq \alpha \leq \alpha_2$,*

$$\begin{aligned} R_d(\alpha) &= R_d(\alpha_1), &&\text{if } R_d(\alpha_1) = R_d(\alpha_2), \\ R_a(\alpha) &= R_a(\alpha_1), &&\text{if } R_a(\alpha_1) = R_a(\alpha_2), \\ N(\alpha) &= N(\alpha_1), &&\text{if } N(\alpha_1) = N(\alpha_2), \\ S(\alpha) &= S(\alpha_1), &&\text{if } S(\alpha_1) = S(\alpha_2). \end{aligned}$$

PROOF. We first show $R_d(\alpha) = R_d(\alpha_1)$ for an arbitrary α ($\alpha_1 \leq \alpha \leq \alpha_2$) if $R_d(\alpha_1) = R_d(\alpha_2)$. Order nodes such that

$$f_1(0) \geq f_2(0) \geq f_3(0) \geq \ldots \geq f_n(0).$$

$R_d(\alpha_1) = R_d(\alpha_2)$ implies that $f_i(0) \geq \alpha_1 + g(\lambda_S(\alpha_1)) > f_{i+1}(0)$ and $f_i(0) \geq \alpha_2 + g(\lambda_S(\alpha_2)) > f_{i+1}(0)$ for the same i. From Property 1.4 and the assumption on $g(\lambda)$ we see that

$$\alpha_1 + g(\lambda_S(\alpha_1)) \leq \alpha + g(\lambda_S(\alpha)) \leq \alpha_2 + g(\lambda_S(\alpha_2)).$$

Thus, we see that

$$f_i(0) \geq \alpha + g(\lambda_S(\alpha)) > f_{i+1}(0),$$

which implies

$$R_d(\alpha) = R_d(\alpha_1) = R_d(\alpha_2).$$

The rest is shown similarly. □

Remarks: As an example, through a binary search, we can obtain a sequence $\alpha_1, \alpha_2, \ldots$ as candidates for optimal α. In the case where $(\lambda_S(\alpha_i) - \lambda_R(\alpha_i))(\lambda_S(\alpha_{i+1}) - \lambda_R(\alpha_{i+1})) < 0$ then α_{i+2} is selected between α_i and α_{i+1}. In this case we derive

$$\begin{aligned} R_d &= R_d(\alpha_{i+j}) = R_d(\alpha_{i+1}), && j = 2, 3, \ldots && \text{if } R_d(\alpha_i) = R_d(\alpha_{i+1}), \\ R_a &= R_a(\alpha_{i+j}) = R_a(\alpha_{i+1}), && j = 2, 3, \ldots && \text{if } R_a(\alpha_i) = R_a(\alpha_{i+1}), \\ N &= N(\alpha_{i+j}) = N(\alpha_{i+1}), && j = 2, 3, \ldots && \text{if } N(\alpha_i) = N(\alpha_{i+1}), \\ S &= S(\alpha_{i+j}) = S(\alpha_{i+1}), && j = 2, 3, \ldots && \text{if } S(\alpha_i) = S(\alpha_{i+1}). \end{aligned}$$

On the basis of the above, we derive another version of the above algorithm that may be a bit more complicated but may require less computational time.

1.2.4 Comparison of load balancing algorithm performance

FORTRAN programs are written for implementing Tantawi and Towsley's algorithm (T & T), Algorithm 1 (K & K 1) as above, and Algorithm 1' (K & K 1') remade from Algorithm 1 by incorporating Property 1.5. The numbers of program steps for Algorithms 1 and 1' are about a third and about two fifths of that of Tantawi and Towsley's algorithm, respectively.

Table 1.1: Algorithm compilation and computation times

Algorithm	Compilation time (sec)	Computation time (msec)				
		$\varepsilon = 10^{-2}$	10^{-3}	10^{-4}	10^{-5}	10^{-6}
T & T	2.11	3.07	3.24	3.66	3.80	4.17
K & K 1	0.78	1.88	2.10	2.65	2.86	3.30
K & K 1'	0.92	1.82	1.98	2.39	2.53	2.87

Table 1.1 compares the performance of the algorithms for a distributed computer system model that consists of ten nodes connected via a single channel communications network. The channel is modeled as an M/M/1 queuing system. Each node is modeled as a central server model. Optimal node partition is such that one sink (S), two neutrals (N), five active sources (R_a) and two idle sources (R_d) exist. The compilation and computation times of each algorithm are shown. ε indicates the acceptable tolerance for the relative error of α.

As we see from Table 1.1, for a large acceptable tolerance ε, the computational times of K & K 1 and 1' are nearly equal to each other and much less than that of T & T. The difference between those of T & T and K & K 1, however, becomes less as ε decreases. On the other hand, the difference between those of T & T and K & K 1' is rather independent of ε. Such a tendency can be observed for different parameter values. As such, if we desire a simple algorithm and accuracy (i.e. the size of ε) is not important, K & K 1 is recommended. If we require very high accuracy levels, K & K 1' is recommended.

1.2.5 An algorithm for star network configurations

The above idea can also be applied to the case of the star network configurations studied by Tantawi and Towsley [TT84]. The system consists of a set of n heterogeneous satellites connected to a central node (Figure 1.2). For notation and assumptions, see [TT84].

The problem of minimizing the mean job response time of a job is expressed in the following formulations, as stated by Tantawi and Towsley [TT84].

$$\text{minimize} \quad D(\boldsymbol{\beta}) = \frac{1}{\Phi}[\sum_{i=0}^{n} \beta_i F_i(\beta_i) + \sum_{i=1}^{n}(\phi_i - \beta_i)G_i(\phi_i - \beta_i)], \qquad (1.16)$$

subject to

$$\sum_{i=0}^{n} \beta_i = \Phi,$$

$$\beta_i \leq \phi_i, \quad i = 1, 2, ..., n, \text{ and}$$

$$\beta_i \geq 0, \quad i = 0, 1, 2, ..., n.$$

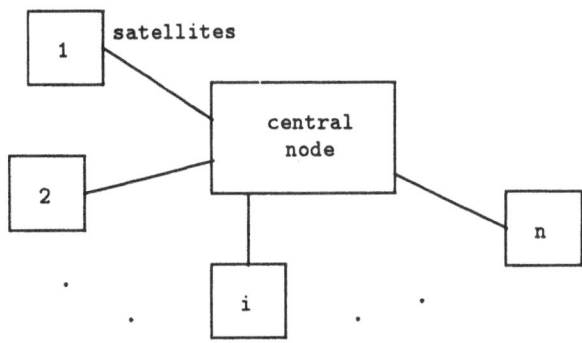

Figure 1.2: A star network configuration system.

The difference between the marginal node dealy and the marginal communication delay is denoted by $h_i(\beta_i)$,

$$h_i(\beta_i) = f_i(\beta_i) - g_i(\phi_i - \beta_i), \quad i = 1, 2, .., n. \tag{1.17}$$

Tantawi and Towsley [TT84] derived the following theorem by using the Kuhn-Tucker theorem (see Theorem 1 of [TT84]):

[Tantawi-Towsley Theorem for star network configurations] *The optimal solution to problem (1.16) satisfies the relations,*

$$
\begin{aligned}
h_i(\beta_i) &\geq \alpha, \quad \beta_i = 0 & (i \in R_d), \\
h_i(\beta_i) &= \alpha, \quad 0 < \beta_i < \phi_i & (i \in R_a), \\
h_i(\beta_i) &\leq \alpha, \quad \beta_i = \phi_i & (i \in N), \quad i = 1, 2, .., n,
\end{aligned}
$$

and

$$f_0(\beta_0) = \alpha, \tag{1.18}$$

subject to the total flow constraint,

$$\sum_{i \in R_a} h_i^{-1}(\alpha) + \sum_{i \in N} \phi_i + f_0^{-1}(\alpha) = \Phi, \tag{1.19}$$

where α is the Lagrange multiplier.

Let β be an optimal solution for the star network configurations. Then we have the following three properties.

Property 1.6

$$
\begin{aligned}
h_i(0) &\geq \alpha, & \textit{iff} \quad \beta_i = 0 & \quad (i \in R_d), \\
h_i(\phi_i) &> \alpha > h_i(0), & \textit{iff} \quad 0 < \beta_i < \phi_i & \quad (i \in R_a), \\
h_i(\phi_i) &\leq \alpha, & \textit{iff} \quad \beta_i = \phi_i & \quad (i \in N), \\
& & i = 1, 2, .., n. &
\end{aligned}
\tag{1.20}
$$

PROOF. From the original assumption, note that $f_i(\beta_i)$ is an increasing function and $g(\phi_i - \beta_i)$ is a nondecreasing function which implies that $h_i(\beta_i)$ is an increasing function.

The necessity in (1.20) : It can be derived easily from the stated set of relations (1.18),

$$
\begin{array}{llll}
h_i(0) \geq \alpha, & if \ \beta_i = 0 & (i \in R_d), \\
h_i(\phi_i) > \alpha > h_i(0), & if \ 0 < \beta_i < \phi_i & (i \in R_a), \\
h_i(\phi_i) \leq \alpha, & if \ \beta_i = \phi_i & (i \in N), \\
& i = 1, 2, .., n. &
\end{array}
$$

The sufficiency in (1.20) : The reverse can be shown by contradiction. (For example, assume that, for some i, $\beta_i > 0$ when $h_i(0) \geq \alpha$. Then from the above necessity we see that for $i = 1, 2, .., n$, $\alpha > h_i(0)$, which would contradict the assumption.) □

Property 1.7 *If β is an optimal solution to the problem (1.16) then we have*

$$
\begin{array}{ll}
\beta_i = 0, & i \in R_d, \\
\beta_i = h_i^{-1}(\alpha), & i \in R_a, \\
\beta_i = \phi_i, & i \in N.
\end{array}
$$

PROOF. This is clear from the Tantawi-Towsley theorem. □

Property 1.8 *If β is an optimal solution to the problem (1.16) then we have*

$$
\lambda_0 = \lambda_R,
$$

where

$$
\lambda_0 = [f_0^{-1}(\alpha) - \phi_0], \tag{1.21}
$$

$$
\lambda_R = \sum_{i \in R_d} \phi_i + \sum_{i \in R_a} [\phi_i - h_i^{-1}(\alpha)]. \tag{1.22}
$$

PROOF. We see that eq.(1.19) is equivalent to the equality $\lambda_0 = \lambda_R$, if we use the definitions given by eqs.(1.21) and (1.22) and note that $\sum_{i \in R_d} \phi_i + \sum_{i \in R_a} \phi_i + \sum_{i \in N} \phi_i + \phi_0 = \Phi$.

Recall the above definitions on node partition. By noting that $\sum_{i=0}^{n} \beta_i = \sum_{i=0}^{n} \phi_i = \Phi$, we have

$$
\lambda_R = \sum_{i=1}^{n} (\phi_i - \beta_i) = \beta_0 - \phi_0.
$$

Therefore, by noting Property 1.7 we see that $\lambda_R = \lambda_0$. □

On the basis of these three properties, the following load balancing algorithm for star network configurations is derived.

[ALGORITHM 2] :

1. Order nodes.

Order nodes such that $h_1(\phi_1) \geq h_2(\phi_2) \geq ... \geq h_n(\phi_n)$.

If $h_1(\phi_1) \leq f_0(\phi_0)$, then no load balancing is required.

2. Determine α.

Find α such that $\lambda_0(\alpha) = \lambda_R(\alpha)$
   (by using, for example, a binary search),

where, given α, each value is calculated in the following order.

$\lambda_0(\alpha) = f_0^{-1}(\alpha) - \phi_0$,

$R_d(\alpha) = \{i | h_i(0) \geq \alpha\}$,

$R_a(\alpha) = \{i | h_i(\phi_i) > \alpha > h_i(0)\}$,

$\lambda_R(\alpha) = \sum_{i \in R_d(\alpha)} \phi_i + \sum_{i \in R_a(\alpha)} [\phi_i - h_i^{-1}(\alpha)]$.

3. Determine the optimal load.

$\beta_i = 0$, for $i \in R_d(\alpha)$,

$\beta_i = h_i^{-1}(\alpha)$, for $i \in R_a(\alpha)$,

$\beta_i = \phi_i$, for $i \in N(\alpha)$, where, $N(\alpha) = \{i | h_i(\phi_i) \leq \alpha\}$.

1.2.6 Conclusion

This section has proposed simpler versions of single-point algorithms than those of Tantawi and Towsley [TT85], [TT84] for the models of distributed computer systems with a single communication channel and star network configurations. The idea underlying our algorithms may also be used for some other related models that have a single job class. It would seem difficult to develop algorithms that yield the optimal solution of multiple job class models without iterative calculation of loads. It would be easy, however, to develop iterative algorithms for multiple job class models at each stage in the iteration utilizing these same ideas.

1.3 Load Balancing in the Multi-class Job Environment

1.3.1 Introduction

A distributed computer system (Figure 1.3) is considered. Each user of the system expects that the performance of the distributed computer system is superior to that of the collection of the separated stand-alone computers. The optimal static load balancing strategy which minimizes the mean job response time of the system is critical in meeting these expectations [dSeSG84] [KH88]. For such load balancing, we can consider job scheduling policies which determine whether a job that arrives at a local node should be processed at the node or should be forwarded to a different node for remote processing. We assume again transferring a job does not depend on current information about the system state (e.g. the queue length at each instant). This is *static* in nature.

This section extends the model of Tantawi and Towsley [TT85] to the model of multiple job classes and consider a load balancing strategy which minimizes the mean job response time of the system. We call this strategy the *optimal load balancing policy*. Some properties of the solution of the optimal load balancing policy for multi-class jobs are then derived. On the basis of these properties, a straightforward and efficient algorithm is proposed. Note that there already exist well-known algorithms which can be easily applied to load balancing of multi-class jobs: the flow deviation (FD) algorithm [dSeSG84] [FGK73] and the Dafermos algorithm [Daf72] [Mag84].

These algorithms are compared in their performance. One performance metric is the amount of the storage required and the other metric is the amount of the computation time required. The results of comparison with respect to storage requirements show that the proposed algorithm and the FD algorithm require almost the same amount of storage. The Dafermos algorithm, however, requires much more storage space than the other two algorithms. For the comparison of computation times, we performed numerical experiments. In the experiment, our proposed algorithm and the Dafermos algorithm required nearly the same computation times for obtaining the optimal solution, but the FD algorithm required much more computation time than the other two algorithms.

1.3.2 Model description

A distributed computer system as shown in Figure 1.3 is considered. The system consists of n nodes, which represent host computers, connected by a single channel communications network. Each node contains one or more resources (CPU, I/O, ...) contended for by the jobs processed at the node. It is assumed that the expected communication delay from node i to node j for class k jobs is independent of the source-destination pair (i, j).

The notation and the assumptions are shown in the following.

- n Number of nodes

- m Number of job classes

- $\phi_i^{(k)}$ External class k job arrival rate to node i

- $x_{ij}^{(k)}$ Class k job flow rate from node i to node j

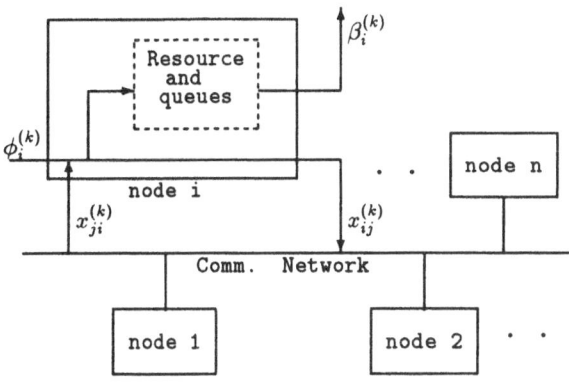

Figure 1.3: A distributed computer system with multiple classes of jobs

- $\beta_i^{(k)}$ Class k job processing rate (load) at node i $\left(\beta_i^{(k)} = \sum_{l=1}^{n} x_{li}^{(k)}\right)$

- $\lambda^{(k)}$ Class k job traffic through network

- $\phi^{(k)}$ Total class k job arrival rate $\left(\phi^{(k)} = \sum_{i=1}^{n} \phi_i^{(k)}\right)$

- Φ Total external job arrival rate $\left(\Phi = \sum_{k=1}^{m} \phi^{(k)}\right)$

- ϕ $[\phi_1, \phi_2, \ldots, \phi_n]$

- ϕ_i $[\phi_i^{(1)}, \phi_i^{(2)}, \ldots, \phi_i^{(m)}]$

- $\phi^{(k)}$ $[\phi_1^{(k)}, \phi_2^{(k)}, \ldots, \phi_n^{(k)}]$

- x $[x^{(1)}, x^{(2)}, \ldots, x^{(m)}]$

- $x^{(k)}$ $[x_1^{(k)}, x_2^{(k)}, \ldots, x_n^{(k)}]$

- $x_i^{(k)}$ $[x_{i1}^{(k)}, x_{i2}^{(k)}, \ldots, x_{in}^{(k)}]$

- β $[\beta_1, \beta_2, \ldots, \beta_n]$

- β_i $[\beta_i^{(1)}, \beta_i^{(2)}, \ldots, \beta_i^{(m)}]$

- $\beta^{(k)}$ $[\beta_1^{(k)}, \beta_2^{(k)}, \ldots, \beta_n^{(k)}]$

- $\boldsymbol{\lambda}$ $[\lambda^{(1)}, \lambda^{(2)}, \ldots, \lambda^{(m)}]$

- $F_i^{(k)}(\boldsymbol{\beta}_i)$ Expected node delay of class k job processed at node i (We assume that it is differentiable, increasing, and convex and that it is strictly convex as a function of a single variable $\beta_i^{(k)}$ for each k).

- $G^{(k)}(\boldsymbol{\lambda})$ Expected communication delay of class k job (We assume that it is source-destination-independent, differentiable, nondecreasing, and convex).

- $F_j^{(k)}(\boldsymbol{\beta}_j) + G^{(k)}(\boldsymbol{\lambda})$ Expected response time of a class k job which is sent from node i to node j and processed at node j.

Jobs arrive at each node according to a time-invariant Poisson process. A job arriving at node i (origin node) may either be processed at node i or transferred through the communications network to another node j (processing node). Also we assume that a transferred job from node i to node j receives its service at node j and is not transferred to other nodes.

We define $x_{ii}^{(k)} = \phi_i^{(k)} - \sum_{l \neq i} x_{il}^{(k)}$, and then we have $\beta_i^{(k)} = \sum_{l=1}^{n} x_{li}^{(k)}$. We refer to $\boldsymbol{x}$ as the job flow rate matrix of the system. Note that $\boldsymbol{x}$ determines $\boldsymbol{\beta}$, according to our definition.

We call nodes in the following way for each class k.

(1) class k idle source $(R_d^{(k)})$:   The node does not process any class k jobs, that is , $\beta_i^{(k)} = 0$.

(2) class k active source $(R_a^{(k)})$:   The node sends class k jobs and does not receive any class k jobs. But the node processes a part of class k jobs that arrive at the node.

(3) class k neutral $(N^{(k)})$:   The node processes class k jobs locally without sending or receiving class k jobs.

(4) class k sink $(S^{(k)})$:   The node receives class k jobs from other nodes but does not send out any class k jobs.

The problem of minimizing the system-wide mean job response time is expressed as follows:

$$\text{minimize}\quad D(\boldsymbol{\beta}) = \frac{1}{\Phi} \sum_{k=1}^{m} \sum_{i=1}^{n} [\beta_i^{(k)} F_i^{(k)}(\boldsymbol{\beta}_i) + \lambda^{(k)} G^{(k)}(\boldsymbol{\lambda})], \qquad (1.23)$$

subject to

$$\sum_{i=1}^{n} \beta_i^{(k)} = \phi^{(k)}, \quad k = 1, 2, \ldots, m,$$

$$\beta_i^{(k)} \geq 0, \quad i = 1, 2, \ldots, n, k = 1, 2, \ldots, m,$$

where the class k network traffic $\lambda^{(k)}$ may be expressed in terms of variable $\beta_i^{(k)}$ as

$$\lambda^{(k)} = \frac{1}{2} \sum_{i=1}^{n} |\phi_i^{(k)} - \beta_i^{(k)}|.$$

We note that the class k job flow rates $x_{ij}^{(k)}$ may be chosen arbitrarily as long as the job flow balance constraint is satisfied. In particular, we may assign the class k flow rate from node i to node j proportionate to the contribution of node j to the class k network traffic; that is,

$$x_{ij}^{(k)} = \begin{cases} \frac{(\beta_j^{(k)} - \phi_j^{(k)})}{\lambda^{(k)}} (\phi_i^{(k)} - \beta_i^{(k)}), & i \in R_d^{(k)}, R_a^{(k)} \ j \in S^{(k)}, \\ 0, & \text{otherwise.} \end{cases}$$

Since node j is a class k sink, the quantity $\beta_j^{(k)} - \phi_j^{(k)} (> 0)$ represents the class k traffic sent to node j. Node i, on the other hand, is a class k source and therefore $\phi_i^{(k)} - \beta_i^{(k)} (> 0)$ represents the class k traffic sent from node i.

1.3.3 Optimal solution

From the assumptions, $D(\boldsymbol{\beta})$ is shown to be a differentiable, increasing, and convex function of $\boldsymbol{\beta}$. Since the feasible region of $\boldsymbol{\beta}$ is a convex set, any local solution point of the problem is a global solution point.

Now we define two functions, the class k *marginal node delay* $f_i^{(k)}(\boldsymbol{\beta}_i)$ and the class k *marginal communication delay* $g^{(k)}(\boldsymbol{\lambda})$, as follows.

$$f_i^{(k)}(\boldsymbol{\beta}_i) = \frac{\partial}{\partial \beta_i^{(k)}} \sum_{l=1}^{m} \beta_i^{(l)} F_i^{(l)}(\boldsymbol{\beta}_i), \qquad (1.24)$$

$$g^{(k)}(\boldsymbol{\lambda}) = \frac{\partial}{\partial \lambda^{(k)}} \sum_{l=1}^{m} \lambda^{(l)} G^{(l)}(\boldsymbol{\lambda}). \qquad (1.25)$$

Following our assumptions about $F_i^{(k)}(\boldsymbol{\beta}_i)$ and $G^{(k)}(\boldsymbol{\lambda})$, we note from eqs. (1.24) and (1.25) that $f_i^{(k)}(\boldsymbol{\beta}_i)$ is an increasing function and $g^{(k)}(\boldsymbol{\lambda})$ is a nondecreasing function. The inverse of the marginal node delay is defined as

$$f_i^{(k)^{-1}}(\boldsymbol{\beta}_i|_{\beta_i^{(k)}=x}) = \begin{cases} a, & f_i^{(k)}(\boldsymbol{\beta}_i|_{\beta_i^{(k)}=a}) = x, \\ 0, & f_i^{(k)}(\boldsymbol{\beta}_i|_{\beta_i^{(k)}=0}) \geq x, \end{cases}$$

where $\boldsymbol{\beta}_i|_{\beta_i^{(k)}=x}$ denotes the vector whose elements are the same as those of $\boldsymbol{\beta}_i$ except that the element $\beta_i^{(k)}$ is replaced by x.

The optimal solution of the Tantawi and Towsley single job class model [TT85] can now easily be extended to the optimal solution of the multiple job class model. We can derive the following theorem by using the Kuhn-Tucker conditions.

Theorem 1.9 *The optimal solution to problem (1.23) satisfies the following relations,*
$k = 1, 2, .., m,$

$$
\begin{aligned}
f_i^{(k)}(\beta_i) \geq \alpha^{(k)} + g^{(k)}(\lambda), &\qquad \beta_i^{(k)} = 0, &\qquad (i \in R_d^{(k)}), \\
f_i^{(k)}(\beta_i) = \alpha^{(k)} + g^{(k)}(\lambda), &\qquad 0 < \beta_i^{(k)} < \phi_i^{(k)}, &\qquad (i \in R_a^{(k)}), \\
\alpha^{(k)} \leq f_i^{(k)}(\beta_i) \leq \alpha^{(k)} + g^{(k)}(\lambda), &\qquad \beta_i^{(k)} = \phi_i^{(k)}, &\qquad (i \in N^{(k)}), \\
\alpha^{(k)} = f_i^{(k)}(\beta_i), &\qquad \beta_i^{(k)} > \phi_i^{(k)}, &\qquad (i \in S^{(k)}),
\end{aligned}
\tag{1.26}
$$

subject to the total flow constraint,

$$
\sum_{i \in S^{(k)}} f_i^{(k)^{-1}}(\beta_i|_{\beta_i^{(k)} = \alpha^{(k)}}) + \sum_{i \in N^{(k)}} \phi_i^{(k)} + \sum_{i \in R_a^{(k)}} f_i^{(k)^{-1}}(\beta_i|_{\beta_i^{(k)} = \alpha^{(k)} + g^{(k)}(\lambda)}) = \phi^{(k)}, \tag{1.27}
$$

where $\alpha^{(k)}$ is the Lagrange multiplier

PROOF. We can easily extend the proof of Tantawi and Towsley [TT85] for the single job class model to a proof for the multiple job class model as follows. We denote the class k network traffic into node i and the network traffic out of node i by $u_i^{(k)}$ and $v_i^{(k)}$, respectively. They may be written in terms of the class k traffic $x_{ij}^{(k)}$ from node i to node j as

$$
u_i^{(k)} = \sum_{j=1, j \neq i}^{n} x_{ji}^{(k)}, \quad i = 1, 2, .., n,
$$

$$
v_i^{(k)} = \sum_{j=1, j \neq i}^{n} x_{ij}^{(k)}, \quad i = 1, 2, .., n.
$$

Clearly, from the balance of the total class k traffic in the network, we have that

$$
\lambda^{(k)} = \sum_{i=1}^{n} u_i^{(k)} = \sum_{i=1}^{n} v_i^{(k)}, \quad k = 1, 2, .., m. \tag{1.28}
$$

The class k load, $\beta_i^{(k)}$, on node i may be written as

$$
\beta_i^{(k)} = u_i^{(k)} - v_i^{(k)} + \phi_i^{(k)}, \quad k = 1, 2, .., m. \tag{1.29}
$$

Substituting eqs. (1.28) and (1.29) into problem (1.23) yields

$$
\text{minimize } \; D(\boldsymbol{u}, \boldsymbol{v}) = \frac{1}{\Phi} \sum_{k=1}^{m} \sum_{i=1}^{n} [(u_i^{(k)} - v_i^{(k)} + \phi_i^{(k)}) F_i^{(k)}(u_i - v_i + \psi_i) + \sum_{i=1}^{n} v_i^{(k)} G^{(k)}(\sum_{i=1}^{n} v_i)],
\tag{1.30}
$$

subject to

$$
-\sum_{i=1}^{n} u_i^{(k)} + \sum_{i=1}^{n} v_i^{(k)} = 0, \quad k = 1, 2, ..., m,
$$

$$
u_i^{(k)} - v_i^{(k)} + \phi_i^{(k)} \geq 0, \quad i = 1, 2, ..., n, \; k = 1, 2, ..., m,
$$

$$
u_i^{(k)} \geq 0, v_i^{(k)} \geq 0, \quad i = 1, 2, ..., n, \; k = 1, 2, ..., m.
$$

Since the objective function is convex by assumption and the feasible region is a convex set, any local solution point of the problem is a global solution point. To obtain the optimal solution, we form the Lagrangian function (note that each class k total arrival rate $\phi^{(k)}$ is constant).

$$L(\boldsymbol{u}, \boldsymbol{v}, \boldsymbol{\alpha}, \boldsymbol{\gamma}, \boldsymbol{\psi}, \boldsymbol{\eta}) = \Phi D(\boldsymbol{u}, \boldsymbol{v}) + \sum_{k=1}^{m} [\alpha^{(k)}(-\sum_{i=1}^{n} u_i^{(k)} + \sum_{i=1}^{n} v_i^{(k)}) +$$

$$\sum_{i=1}^{n} \gamma_i^{(k)}(u_i^{(k)} - v_i^{(k)} + \phi_i^{(k)}) + \sum_{i=1}^{n} \psi_i^{(k)} u_i^{(k)} + \sum_{i=1}^{n} \eta_i^{(k)} v_i^{(k)}]. \tag{1.31}$$

The optimal solution satisfies the Kuhn-Tucker condition,

$$\frac{\partial L}{\partial u_i^{(k)}} = f_i^{(k)}(u_i - v_i + \phi_i) - \alpha^{(k)} + \gamma_i^{(k)} + \psi_i^{(k)} = 0, \tag{1.32}$$

$$\frac{\partial L}{\partial v_i^{(k)}} = -f_i^{(k)}(u_i - v_i + \phi_i) + g^{(k)}(\sum_{i=1}^{n} v_i) + \alpha^{(k)} - \gamma_i^{(k)} + \eta_i^{(k)} = 0, \tag{1.33}$$

$$\frac{\partial L}{\partial \alpha^{(k)}} = -\sum_{i=1}^{n} u_i^{(k)} + \sum_{i=1}^{n} v_i^{(k)} = 0, \tag{1.34}$$

$$u_i^{(k)} - v_i^{(k)} + \phi_i^{(k)} \geq 0, \ \gamma_i^{(k)}(u_i^{(k)} - v_i^{(k)} + \phi_i^{(k)}) = 0, \ \gamma_i^{(k)} \leq 0, \tag{1.35}$$

$$u_i^{(k)} \geq 0, \ \psi_i^{(k)} u_i^{(k)} = 0, \ \psi_i^{(k)} \leq 0, \tag{1.36}$$

$$v_i^{(k)} \geq 0, \ \eta_i^{(k)} v_i^{(k)} = 0, \ \eta_i^{(k)} \leq 0, \tag{1.37}$$

Similarly as Tantawi and Towsley [TT85], we consider each case separately.
Case 1. $v_i^{(k)} = u_i^{(k)} + \phi_i^{(k)}$

$$f_i^{(k)}(\boldsymbol{\beta}_i) \geq \alpha^{(k)} + g^{(k)}(\boldsymbol{\lambda}), \ \beta_i^{(k)} = 0, \ k = 1, 2, .., m. \tag{1.38}$$

Case 2. $v_i^{(k)} < u_i^{(k)} + \phi_i^{(k)}$
  Case 2.1. $v_i^{(k)} > 0$

$$f_i^{(k)}(\boldsymbol{\beta}_i) = \alpha^{(k)} + g^{(k)}(\boldsymbol{\lambda}), \ 0 < \beta_i^{(k)} < \phi_i^{(k)}, \ k = 1, 2, .., m. \tag{1.39}$$

Case 2.2. $v_i^{(k)} = 0$
  Case 2.2.1. $u_i^{(k)} = 0$

$$\alpha^{(k)} \leq f_i^{(k)}(\boldsymbol{\beta}_i) \leq \alpha^{(k)} + g^{(k)}(\boldsymbol{\lambda}), \ \beta_i^{(k)} = \phi_i^{(k)}, \ k = 1, 2, .., m. \tag{1.40}$$

Case 2.2.2. $u_i^{(k)} > 0$

$$\alpha^{(k)} = f_i^{(k)}(\boldsymbol{\beta}_i), \ \beta_i^{(k)} > \phi_i^{(k)}, \ k = 1, 2, .., m. \tag{1.41}$$

Eq. (1.34) may be written in terms of $\beta_i^{(k)}$ as

$$\sum_{i=1}^{n} \beta_i^{(k)} = \phi^{(k)}, \ k = 1, 2, .., m.$$

Using eqs. (1.39) and (1.41), the above equation becomes

$$\sum_{i \in S^{(k)}} f_i^{(k)^{-1}}(\beta_i|_{\beta_i^{(k)}=\alpha^{(k)}}) + \sum_{i \in N^{(k)}} \phi_i^{(k)} + \sum_{i \in R_a^{(k)}} f_i^{(k)^{-1}}(\beta_i|_{\beta_i^{(k)}=\alpha^{(k)}+g^{(k)}(\lambda)}) = \phi^{(k)},$$

$$k = 1, 2, .., m. \tag{1.42}$$

Once the node partition is known, class k traffic can be obtained as

$$\lambda^{(k)} = \sum_{i \in S^{(k)}} [f_i^{(k)^{-1}}(\beta_i|_{\beta_i^{(k)}=\alpha^{(k)}}) - \phi_i^{(k)}], \ k = 1, 2, .., m.$$

Substituting $\lambda^{(k)}$ into eq. (1.42) yields a single equation in the unknown $\alpha^{(k)}$. The class k rates $\beta_i^{(k)}$ $(i = 1, 2, .., n, \ k = 1, 2, .., m)$, may then be obtained by solving eqs. (1.39) and (1.41). □

Remarks. Theorem 1.9 implies the following: Assume that the load on each node is fixed for all classes except class k. Then, in the optimal solution of the system, nodes are classified into either class k sources, class k neutrals, or class k sinks for each k. Furthermore, from the assumption, the class k load on each node is uniquely determined by the set of relations (1.26), since we have assumed $F_i^{(k)}(\beta_i)$ is strictly convex with respect to each $\beta_i^{(k)}$.

Let β be an optimal solution to problem (1.23) and let

$$\alpha^{(k)} = \min_i f_i^{(k)}(\beta_i), \tag{1.43}$$

$$\lambda^{(k)} = \frac{1}{2} \sum_{i=1}^{n} |\phi_i^{(k)} - \beta_i^{(k)}|. \tag{1.44}$$

We can then show from the above theorem that the following three properties for multiple job classes hold true in the optimal solution as well as the three properties for single job class obtained by Kim and Kameda [KK92a].

Property 1.10

$$
\begin{aligned}
f_i^{(k)}(\beta_i|_{\beta_i^{(k)}=0}) &\geq \alpha^{(k)} + g^{(k)}(\lambda), && \textit{iff } \ \beta_i^{(k)} = 0, \\
f_i^{(k)}(\beta_i|_{\beta_i^{(k)}=\phi_i^{(k)}}) &> \alpha^{(k)} + g^{(k)}(\lambda) > f_i^{(k)}(\beta_i|_{\beta_i^{(k)}=0}), && \textit{iff } \ 0 < \beta_i^{(k)} < \phi_i^{(k)}, \\
\alpha^{(k)} &\leq f_i^{(k)}(\beta_i|_{\beta_i^{(k)}=\phi_i^{(k)}}) \leq \alpha^{(k)} + g^{(k)}(\lambda), && \textit{iff } \ \beta_i^{(k)} = \phi_i^{(k)}, \\
f_i^{(k)}(\beta_i|_{\beta_i^{(k)}=\phi_i^{(k)}}) &< \alpha^{(k)}, && \textit{iff } \ \beta_i^{(k)} > \phi_i^{(k)}.
\end{aligned}
\tag{1.45}
$$

PROOF. From the assumption, note that $f_i^{(k)}(\beta_i)$ is (1.24) and $g^{(k)}(\lambda)$ is (1.25). It is clear from the relations (1.26) that $\alpha^{(k)} = \min_i f_i^{(k)}(\beta_i)$, which is the same as eq. (1.43). The necessity : We easily derive relations (1.45) from the set of relations (1.26).

The sufficiency : We can show the reverse by contradiction. For example, assume that, for some i, $\beta_i^{(k)} > 0$ when $f_i^{(k)}(\beta_i|_{\beta_i^{(k)}=0}) \geq \alpha^{(k)} + g^{(k)}(\lambda)$. Then from the above necessity, $\alpha^{(k)} + g^{(k)}(\lambda) > f_i^{(k)}(\beta_i|_{\beta_i^{(k)}=0})$, when $\beta_i^{(k)} > 0$, which would contradict the assumption. We can show all of the other cases like this. $\square$

Note the following definitions in the optimal solution:
$R_d^{(k)} = \{i | \beta_i^{(k)} = 0\}$ (class k idle sources),
$R_a^{(k)} = \{i | 0 < \beta_i^{(k)} < \phi_i^{(k)}\}$ (class k active sources),
$N^{(k)} = \{i | \beta_i^{(k)} = \phi_i^{(k)}\}$ (class k neutrals),
$S^{(k)} = \{i | \beta_i^{(k)} > \phi_i^{(k)}\}$ (class k sinks).

Property 1.11

$$
\begin{aligned}
\beta_i^{(k)} &= 0, & i &\in R_d^{(k)}, \\
\beta_i^{(k)} &= f_i^{(k)^{-1}}(\beta_i|_{\beta_i^{(k)}=\alpha^{(k)}+g^{(k)}(\lambda)}), & i &\in R_a^{(k)}, \\
\beta_i^{(k)} &= \phi_i^{(k)}, & i &\in N^{(k)}, \\
\beta_i^{(k)} &= f_i^{(k)^{-1}}(\beta_i|_{\beta_i^{(k)}=\alpha^{(k)}}), & i &\in S^{(k)}.
\end{aligned}
$$

PROOF. This is clear from Theorem 1.9. $\square$

Property 1.12

$$\lambda^{(k)} = \lambda_{sen}^{(k)} = \lambda_{rec}^{(k)},$$

where

$$\lambda_{sen}^{(k)} = \sum_{i \in R_d^{(k)}} \phi_i^{(k)} + \sum_{i \in R_a^{(k)}} (\phi_i^{(k)} - f_i^{(k)^{-1}}(\beta_i|_{\beta_i^{(k)}=\alpha^{(k)}+g^{(k)}(\lambda)})), \tag{1.46}$$

$$\lambda_{rec}^{(k)} = \sum_{i \in S^{(k)}} (f_i^{(k)^{-1}}(\beta_i|_{\beta_i^{(k)}=\alpha^{(k)}}) - \phi_i^{(k)}). \tag{1.47}$$

PROOF. We can see easily that eq. (1.27) is equivalent to the equation $\lambda_{rec}^{(k)} = \lambda_{sen}^{(k)}$, if we use the definitions given by eqs. (1.46) and (1.47), and $\sum_i \phi_i^{(k)} = \phi^{(k)}$. Recall the above definitions on node partition. By noting $\sum_{i=1}^{n} \beta_i^{(k)} = \sum_{i=1}^{n} \phi_i^{(k)} = \phi^{(k)}$, we have

$$\lambda^{(k)} = \frac{1}{2} \sum_{i=1}^{n} |\phi_i^{(k)} - \beta_i^{(k)}| = \sum_{i \in S^{(k)}} (\beta_i^{(k)} - \phi_i^{(k)}).$$

Therefore, we see that $\lambda^{(k)} = \lambda_{rec}^{(k)}$ by property 1.11 . $\square$

1.3.4 Optimal load balancing algorithm

The optimal load balancing algorithm is derived using the above three properties. This algorithm (called a single-point algorithm) can obtain the optimal solution for an arbitrary set of parameter values.

o Single-point algorithm

1. Initialize.

   $l := 0$ (l: iteration number)

   $\beta_{(l)} := \phi$

2. $l := l + 1$ and execute step 3.

3. Execute the following sub-algorithm for $k = 1, 2, .., m$ by using $\beta_{(l)}^{(1)}, \beta_{(l)}^{(2)}, .., \beta_{(l)}^{(k-1)}$, $\beta_{(l-1)}^{(k+1)}, .., \beta_{(l-1)}^{(m)}$ as the values of $\beta^{(1)}, .., \beta^{(k-1)}, \beta^{(k+1)}, .., \beta^{(m)}$.

   Denote by $\beta_{(l)}^{(k)}$, the value of $\beta^{(k)}$ obtained by the sub-algorithm. Thus the resulting value of β at this step becomes as follows:

   $(\beta_{(l)}^{(1)}, \beta_{(l)}^{(2)}, .., \beta_{(l)}^{(k-1)}, \beta_{(l)}^{(k)}, \beta_{(l-1)}^{(k+1)}, .., \beta_{(l-1)}^{(m)})$.

   The final resulting value (at step $k = m$) of β is denoted by $\beta_{(l)}$.

4. (Stopping rule) Compare $D(\beta_{(l-1)})$ and $D(\beta_{(l)})$.

   If $|D(\beta_{(l)}) - D(\beta_{(l-1)})| < \varepsilon$ where $\varepsilon > 0$ is a properly chosen acceptance tolerance, then STOP.

   Otherwise, go to Step 2.

o Sub-algorithm that computes the value of $\beta^{(k)}$ for class k jobs with $\beta^{(1)}, .., \beta^{(k-1)}$, $\beta^{(k+1)}, .., \beta^{(m)}$ being fixed.

1. Order nodes.

   Order nodes such that $f_1^{(k)}(\beta_1|_{\beta_1^{(k)}=\phi_1^{(k)}}) \leq f_2^{(k)}(\beta_2|_{\beta_2^{(k)}=\phi_2^{(k)}}) \leq \cdots \leq f_n^{(k)}(\beta_n|_{\beta_n^{(k)}=\phi_n^{(k)}})$.

   If $f_1^{(k)}(\beta_1|_{\beta_1^{(k)}=\phi_1^{(k)}}) + g^{(k)}(\lambda|_{\lambda^{(k)}=0}) \geq f_n^{(k)}(\beta_n|_{\beta_n^{(k)}=\phi_n^{(k)}})$,

   then stop the sub-algorithm.

2. Determine $\alpha^{(k)}$.

   Find $\alpha^{(k)}$ such that $\lambda_{rec}^{(k)}(\alpha^{(k)}) = \lambda_{sen}^{(k)}(\alpha^{(k)})$.

   (by using, for example, a binary search),

where, given $\alpha^{(k)}$, each value is calculated in the following order.

$$S^{(k)}(\alpha^{(k)}) = \{i | f_i^{(k)}(\beta_i|_{\beta_i^{(k)}=\phi_i^{(k)}}) < \alpha^{(k)}\},$$

$$\lambda_{rec}^{(k)}(\alpha^{(k)}) = \sum_{i \in S^{(k)}(\alpha^{(k)})} (f_i^{(k)-1}(\beta_i|_{\beta_i^{(k)}=\alpha^{(k)}}) - \phi_i^{(k)}),$$

$$R_d^{(k)}(\alpha^{(k)}) = \{i | f_i^{(k)}(\beta_i|_{\beta_i^{(k)}=0}) \geq \alpha^{(k)} + g^{(k)}(\lambda|_{\lambda^{(k)}=\lambda_{rec}^{(k)}(\alpha^{(k)})})\},$$

$$R_a^{(k)}(\alpha^{(k)}) = \{i | f_i^{(k)}(\beta_i|_{\beta_i^{(k)}=\phi_i^{(k)}}) > \alpha^{(k)} + g^{(k)}(\lambda|_{\lambda^{(k)}=\lambda_{rec}^{(k)}(\alpha^{(k)})}) > f_i^{(k)}(\beta_i|_{\beta_i^{(k)}=0})\},$$

$$\lambda_{sen}^{(k)}(\alpha^{(k)}) = \sum_{i \in R_d^{(k)}(\alpha^{(k)})} \phi_i^{(k)} +$$

$$\sum_{i \in R_a^{(k)}(\alpha^{(k)})} (\phi_i^{(k)} - f_i^{(k)-1}(\beta_i|_{\beta_i^{(k)}=\alpha^{(k)}+g^{(k)}(\lambda|\lambda^{(k)}=\lambda_{rec}^{(k)}(\alpha^{(k)})})).$$

3. Determine optimal load.

$$\beta_i^{(k)} = f_i^{(k)-1}(\beta_i|_{\beta_i^{(k)}=\alpha^{(k)}}), \text{ for } i \in S^{(k)},$$

$$\beta_i^{(k)} = 0, \text{ for } i \in R_d^{(k)},$$

$$\beta_i^{(k)} = f_i^{(k)-1}(\beta_i|_{\beta_i^{(k)}=\alpha^{(k)}+g^{(k)}(\lambda)}), \text{ for } i \in R_a^{(k)}$$

$$\beta_i^{(k)} = \phi_i^{(k)}, \text{ for } i \in N^{(k)},$$

where $N^{(k)}(\alpha^{(k)}) = \{i | \alpha^{(k)} \leq f_i^{(k)}(\beta_i|_{\beta_i^{(k)}=\phi_i^{(k)}}) \leq \alpha^{(k)} + g^{(k)}(\lambda|_{\lambda^{(k)}=\lambda_{rec}^{(k)}(\alpha^{(k)})})\}.$

Another sub-algorithm for multiple job classes that would be extended directly from the single-point algorithm for single class jobs given by Tantawi and Towsley [TT85] could be obtained. However, it is easy to see that our sub-algorithm is more straightforward and efficient and has a better performance than the sub-algorithm directly extended from Tantawi and Towsley's. Because the sub-algorithm directly extended from Tantawi and Towsley's has to determine the node partition first, and then it may obtain $\alpha^{(k)}$ using eqs. (1.26) and (1.27), but our proposed sub-algorithm does not need to provide the node partition process separately, and can, instead, determine the node partitions during the process of obtaining $\alpha^{(k)}$ (step 2 of sub-algorithm).

Continuing, consider the following additional property.

Property 1.13 $\lambda_{rec}^{(k)}(\alpha^{(k)})$ *increases (from zero) and* $\lambda_{sen}^{(k)}(\alpha^{(k)})$ *decreases (from* $\phi^{(k)}$*) both monotonically as* $\alpha^{(k)}$ *increases,* $k = 1, 2, ..., m.$

PROOF. It is simple to prove this from the following two definitions shown in the sub-algorithm

$$\lambda_{rec}^{(k)}(\alpha^{(k)}) = \sum_{i \in S^{(k)}(\alpha^{(k)})} (f_i^{(k)-1}(\beta_i|_{\beta_i^{(k)}=\alpha^{(k)}}) - \phi_i^{(k)}),$$

$$\lambda_{sen}^{(k)}(\alpha^{(k)}) = \sum_{i \in R_d^{(k)}(\alpha^{(k)})} \phi_i^{(k)} + \sum_{i \in R_a^{(k)}(\alpha^{(k)})} (\phi_i^{(k)} - f_i^{(k)^{-1}}(\beta_i|_{\beta_i^{(k)}=\alpha^{(k)}} + g^{(k)}(\lambda|\lambda^{(k)}=\lambda_{rec}^{(k)}(\alpha^{(k)})))),$$

and the assumptions on $f_i^{(k)}(\beta_i|_{\beta_i^{(k)}=\alpha^{(k)}})$ and $g^{(k)}(\lambda|\lambda^{(k)} = \lambda_{rec}^{(k)}(\alpha^{(k)}))$. $\square$

Remark: This property assures that the sub-algorithm can determine an $\alpha^{(k)}$ that satisfies $\lambda_{rec}^{(k)}(\alpha^{(k)}) = \lambda_{sen}^{(k)}(\alpha^{(k)})$.

The convergence of the proposed algorithm will be shown. Roughly, the method of solution can be described as follows: Starting from an initial feasible load pattern, we can show a sequence of feasible load patterns which converges to the optimal solution.

For the simple explanation, we call the proposed sub-algorithm "equilibration operator",

$$E^{(k)} : F \to F, \ k = 1, 2, ..., m,$$

where F is the set of all feasible load β.

Then we can have an equilibration operator

$$E = E^{(m)} \circ E^{(m-1)} \circ ... \circ E^{(2)} \circ E^{(1)}. \tag{1.48}$$

E has a map

$$E : F \to F.$$

We will define that $E_{(l)}$ means l times iterative application of E.

We can say that if E has the following property, then the proposed algorithm can obtain the optimal solution of problem (1.23) for any $\beta_{(0)} \in F$.

$$\beta_{(l)} \to \beta^*, \ l \to \infty \tag{1.49}$$

where

$$\beta_{(l)} \equiv E_{(l)} \beta_{(0)}, \ l = 1, 2, ... \tag{1.50}$$

and β^* is an optimal solution of problem (1.23).

The following theorem shows sufficient conditions for the equilibration operator which induce the proposed algorithm for the optimal solution of problem (1.23). Dafermos [DS69] proved the convergence of an algorithm induced by an equilibration operator for the traffic assignment problem of the transportation network. The operator defined by Dafermos for the traffic assignment problem have four properties (as shown in Theorem (2.1) of [DS69]) almost identical with those of the following Theorem 1.14. Therefore, the proof of convergence can be easily derived by modifying the proof of Theorem (2.1) of [DS69], slightly.

Theorem 1.14 *If E has the following properties, then the proposed algorithm can have the optimal solution of problem (1.23).*

(1) $E\beta = \beta$ *for some* $\beta \in F$ *implies that* β *satisfies (1.26) and (1.27), so that* $\beta = \beta^*$.

(2) E *is a continuous mapping from* F *to* F.

(3) $D(\beta) \geq D(E\beta)$ *for all* $\beta \in F$.

(4) $D(\beta) = D(E\beta)$ *for some* $\beta \in F$ *implies that* $E\beta = \beta$.

PROOF. Let $\beta_{(0)} \in F$ and $\beta_{(l)} \equiv E_{(l)}\beta_{(0)}$, $l = 1, 2, ...$ We have to prove that

$$\beta_{(l)} \to \beta^*, \ l \to \infty. \tag{1.51}$$

We first prove that every convergent subsequence $\{\beta_{(l_k)}\}$ of $\{\beta_{(l)}\}$ converges to a solution β^* of problem (1.23). In fact, let

$$\beta_{(l_k)} \to \beta, \ k \to \infty. \tag{1.52}$$

Since F is closed, $\beta \in F$. The sequence $\{D(\beta_{(l)})\}$ is decreasing and bounded from below 0 and the set of feasible solution F is closed. By Cauchy's theorem, given $\epsilon > 0$,

$$|D(E\beta_{(l_k)}) - D(\beta_{(l_k)})| = |D(\beta_{(l_k+1)}) - D(\beta_{(l_k)})| < \epsilon/3,$$

if $k \geq k_1(\epsilon)$. Since $D(\beta)$ is continuous, $\lim_{k \to \infty} D(\beta_{(l_k)}) = D(\beta)$. Hence

$$|D(\beta_{(l_k)}) - D(\beta)| < \epsilon/3,$$

if $k \geq k_2(\epsilon)$. Furthermore, the continuity of $D(\beta)$ implies the existence of δ_ϵ some such that

$$|D(\beta) - D(\beta')| < \epsilon/3, \text{ if } ||\beta - \beta'|| < \delta_\epsilon,$$

where $||\beta - \beta'|| \equiv \sum_{k=1}^{m} \sum_{i=1}^{n} |\beta_i^{(k)} - \beta_i^{(k)'}|$. On the other hand, since E is a continuous mapping, given $\delta > 0$ there exists $\eta(\delta)$ such that

$$||E\beta - E\beta'|| < \delta, \text{ if } ||\beta - \beta'|| < \eta(\delta).$$

Finally, from (1.52) it follows that given $\eta > 0$, there exists $k_3(\eta)$ such that

$$||\beta - \beta_{(l_k)}|| < \eta, \text{ if } k \geq k_3(\eta).$$

Suppose now that $k \geq \max\{k_1(\epsilon), k_2(\epsilon), k_3(\eta)\}$. Combining the above results we obtain

$$|D(E\beta) - D(\beta)| \leq |D(E\beta) - D(E\beta_{(l_k)})| + |D(E\beta_{(l_k)}) - D(\beta_{(l_k)})| + |D(\beta_{(l_k)}) - D(\beta)| < \epsilon.$$

But the left-hand side of the above inequality is independent of k and hence

$$D(E\beta) - D(\beta) = 0,$$

whence β is a solution β^* of problem (1.23) by properties (1) and (4).

We now proceed to the proof of (1.51). Suppose that it is false. Then there exists a positive number δ and a subsequence $\{\beta_{(l_k)}\}$ such that

$$||\beta_{(l_k)} - \beta^*|| \equiv \sum_{k=1}^{m} \sum_{i=1}^{n} |\beta_{i(l_k)}^{(k)} - \beta_i^{(k)*}| > \delta. \tag{1.53}$$

The sequence $\{\beta_{(l_k)}\}$ is bounded. By the theorem of Bolzano-Weierstrass there exists a converging subsequence $\{\beta_{(l_{k_r})}\}$. As proved above

$$\beta_{(l_{k_r})} \to \beta^*, \ r \to \infty,$$

where β^* is a solution of problem (1.23) and this is a contradiction to (1.53). $\square$

We now show the equilibration operator E has the following four properties.

Theorem 1.15 *E has the following properties:*

(1) $E\beta = \beta$ *for some* $\beta \in F$ *implies that* β *satisfies (1.26) and (1.27), so that* $\beta = \beta^*$.

(2) *E is a continuous mapping from F to F.*

(3) $D(\beta) \geq D(E\beta)$ *for all* $\beta \in F$.

(4) $D(\beta) = D(E\beta)$ *for some* $\beta \in F$ *implies that* $E\beta = \beta$.

PROOF.
(1) Note that β consists of $\beta^{(k)}$, $k = 1, 2, ..., m$, i.e. $\beta = [\beta^{(1)}, \beta^{(2)}, ..., \beta^{(m)}]$. We first show that $E\beta = \beta$ implies $E^{(k)}\beta = \beta$, $k = 1, 2, ..., m$. It is proven by contradiction.

Consider the smallest k such that $E^{(k)}\beta \neq \beta$ if it exists. By noting that the operator $E^{(k)}$ changes only the value of $\beta^{(k)}$ with other elements of β being fixed. Thus $\beta^{(k)}$ of $E^{(k)}\beta$ is not equal to the corresponding elements of $E\beta$, which would contradict the assumption.

Now, we will show β satisfies relations (1.26) and (1.27). By the definition of $E^{(k)}$ and the proposed sub-algorithm, we see that $E^{(k)}\beta = \beta$ implies that β should satisfy the relations (1.26) and (1.27), $k = 1, 2, ..., m$. Therefore, by noting the (1.48) we see that the value of such β that $E\beta = \beta$, satisfies the relations (1.26) and (1.27).

(2) By carefully examining the sub-algorithm and by noting that $f_i^{(k)}(\beta_i)$ is a continuous function of β_i and increasing with the increase of $\beta_i^{(k)}$, $i = 1, 2, ..., n$, $k = 1, 2, ..., m$, we see that there exists $\delta'(\epsilon') > 0$ for $\epsilon' > 0$ such that

$$|\alpha^{(k)}(\beta') - \alpha^{(k)}(\beta)| < \epsilon', \ \text{if} \ ||\beta' - \beta|| < \delta'(\epsilon')$$

and also there exists $\delta''(\epsilon) > 0$ for $\epsilon > 0$ such that

$$||E^{(k)}\beta\,' - E^{(k)}\beta|| < \epsilon, \text{ if } |\alpha^{(k)}(\beta\,') - \alpha^{(k)}(\beta)| < \delta''(\epsilon).$$

Therefore we see that there exists $\delta(\epsilon) > 0$ for $\epsilon > 0$ such that

$$||E^{(k)}\beta\,' - E^{(k)}\beta|| < \epsilon, \text{ if } ||\beta\,' - \beta|| < \delta(\epsilon)$$

where $\delta = \delta' \circ \delta''$.

That is, $E^{(k)}$ is a continuous mapping from F to F. Therefore, by noting (1.48) we see that E is a continuous mapping from F to F.

(3) Consider an arbitrary $\beta \in F$. Let us define

$$\beta^k = E^{(k)} \circ E^{(k-1)} \circ ... \circ E^{(1)}\beta, \quad k = 1, 2, ..., m.$$

Then we have

$$\beta^k = E^{(k)}\beta^{k-1}, \quad k = 1, 2, ..., m.$$

By carefully examining the sub-algorithm, we easily see

$$D(\beta^{k-1}) \geq D(E^{(k)}\beta^{k-1}), \quad k = 1, 2, ..., m,$$

or

$$D(\beta^{k-1}) \geq D(\beta^k), \quad k = 1, 2, ..., m.$$

Therefore, we have

$$D(\beta^0) \geq D(\beta^1) \geq ... \geq D(\beta^k) \geq ... \geq D(\beta^m),$$

where $\beta^0 = \beta$ and $E\beta = \beta^m$.

Thus, we have the following relation:

$$D(\beta) \geq D(E\beta).$$

(4) If $D(E\beta) = D(\beta)$, from property (3) above

$$D(\beta^0) = D(\beta^1) = D(\beta^2) = ... = D(\beta^k) = ... = D(\beta^m).$$

β consists of $\beta^{(k)}$, $k = 1, 2, ..., m$, i.e. $\beta = [\beta^{(1)}, \beta^{(2)}, ..., \beta^{(m)}]$, and $E\beta = \beta$ implies $E^{(k)}\beta = \beta$, $k = 1, 2, ..., m$, by the proof of the property (1) above. Therefore, if β satisfies (1.26) and (1.27), then each $\beta^{(k)}$, $k = 1, 2, ..., m$, also satisfies (1.26) and (1.27).

The value of $\beta^{(k)}$ obtained by the $E^{(k)}\beta$, when all the elements of β except $\beta^{(k)}$ are fixed, is unique. By noting property (1), $\beta^{(k)}$ satisfies (1.26) and (1.27). Therefore, we can have

$$\beta^0 = \beta^1 = \beta^2 = ... = \beta^{m-1} = \beta^m.$$

That is $E\beta = \beta$. □

From Theorems 1.14 and 1.15 we can conclude that by using the proposed algorithm, we can reach the minimum of $D(\beta)$.

Table 1.2: Storage requirements of algorithms

Algorithm	Storage requirement
FD	$O(n \times m)$
DAFERMOS	$O(n^2 \times m)$
K & K	$O(n \times m)$

1.3.5 Comparison of algorithm performance

The performance of our proposed algorithm and the other two well-known algorithms, the FD algorithm and the Dafermos algorithm, are compared. The FD algorithm and the Dafermos algorithm for load balancing of multi-class jobs are provided in Appendix A. Two metrics of performance, the amount of storage required and the computational time required, are examined.

Comparison of storage requirements

The proposed algorithm (K & K) and FD algorithm (FD) need not calculate x, but only β for each step of iteration. Note that the numbers of elements of x and β are $n^2 \times m$ and $n \times m$, respectively. Therefore, these two algorithms require the amount of storage of $O(n \times m)$. In the case of the Dafermos algorithm (DAFERMOS), x must be obtained at each iteration because the origin-destination job flow rate, $x_{ij}^{(k)}$, must be used in the process of calculation. Therefore, the Dafermos algorithm requires the amount of storage of $O(n^2 \times m)$ which is much larger than those of other two algorithms. Table 1.2 summarizes the storage requirements of these three algorithms.

Comparison of computational time requirements

The structures of these three algorithms differ greatly from one another. Therefore, it seems difficult to compare their computation time requirements in terms of complexity theory. Therefore, we compare computation time requirements of the three algorithms numerically. We examine some system models such as the following example for numerical examination.

The models used in the numerical examination   The model of a distributed computer system that consists of an arbitrary number of host computers (nodes) connected via a single channel communications network is considered. A multi-class central server model is used as each node model (Figure 1.4). Server 0 is a CPU that processes jobs according to the processor sharing discipline. Servers 1, 2, ..., d are I/O devices which

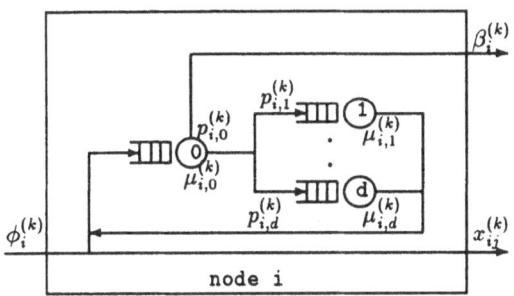

Figure 1.4: Multi-class central server model

process jobs according to FCFS. Let $p_{i,0}^{(k)}$ and $p_{i,j}^{(k)}$, $j = 1, 2, ..., d$, denote the transition probabilities that, after departing from the CPU, a class k job leaves the node i or requests an I/O service at device j, $j = 1, 2, ..., d$, respectively.

The expected class k node delay of the central server model is given as follows.

$$F_i^{(k)}(\beta_i) = \sum_{j=0}^{d} \frac{q_{i,j}^{(k)} \frac{1}{\mu_{i,j}^{(k)}}}{1 - (q_{i,j}^{(1)} \frac{\beta^{(1)}}{\mu_{i,j}^{(1)}} + ... + q_{i,j}^{(m)} \frac{\beta^{(m)}}{\mu_{i,j}^{(m)}})}, \tag{1.54}$$

where $q_{i,0}^{(k)} = \frac{1}{p_{i,0}^{(k)}}$, and $q_{i,j}^{(k)} = \frac{p_{i,j}^{(k)}}{p_{i,0}^{(k)}}, (j = 1, .., d)$, and $\mu_{i,j}^{(k)}, (j = 0, 1, .., d)$ denotes the class k processing rate for server j at node i. We assume that the scheduling discipline on the servers $(j = 1, 2, .., d)$ is FCFS. Therefore, all of the $\mu_{i,j}^{(k)}$ $(j = 1, .., d, k = 1, .., m)$ should be the same so that the BCMP theorem [BCMP75] can be applied. We consider processor sharing M/G/1 model for the single channel communications network. Expected class k communications delay is given as follows.

$$G^{(k)}(\lambda) = \frac{t^{(k)}}{1 - (t^{(1)}\lambda^{(1)} + ... + t^{(m)}\lambda^{(m)})}, \tag{1.55}$$

where $t^{(k)}$, $k = 1, 2, .., m$, denotes the mean communication time (excluding the queueing time) for class k jobs.

The functions of node and communications delays are convex but not strictly so. Therefore, there are some possibilities that the optimal load may not be unique. But, since we assume that the functions are convex, we can obtain the minimum system-wide response time on the basis of Theorem 1.9.

As one example, we have the following parameter setting. In this example, only two job classes are treated. Table 1.3 shows the category (due to the size of the capacity) and

Table 1.3: An example of the set of parameter values of a system model

Node	Category of node (due to processing capacity) class 1	class 2	Job arrival rate (jobs/sec) class 1 (I/O bound)	class 2 (CPU bound)
1	large	extra-large	11.0	9.0
2			10.0	0.5
3	middle	middle	5.0	1.0
4			1.0	2.0
5			5.0	0.02
6	small	small	2.0	0.05
7			0.5	0.10

Table 1.4: Parameters of node models of the example model

Node	Processing rates of servers(jobs/sec) $\mu_{i,0}^{(1)}$ $\mu_{i,0}^{(2)}$	$\mu_{i,1}^{(1)}$ $\mu_{i,1}^{(2)}$	$\mu_{i,2}^{(1)}$ $\mu_{i,2}^{(2)}$	$\mu_{i,3}^{(1)}$ $\mu_{i,3}^{(2)}$	Probabilities of a job leaving CPU $p_{i,0}^{(1)}$ $p_{i,0}^{(2)}$	$p_{i,1}^{(1)}$ $p_{i,1}^{(2)}$	$p_{i,2}^{(1)}$ $p_{i,2}^{(2)}$	$p_{i,3}^{(1)}$ $p_{i,3}^{(2)}$
1	1000	200	200	200	0.1	0.3	0.3	0.3
	100	200	200	200	0.2	0.266	0.266	0.266
2	300	100	100	·	0.1	0.45	0.45	·
	15	100	100	·	0.2	0.4	0.4	·
3-4	300	100	·	·	0.1	0.9	·	·
	15	100	·	·	0.2	0.8	·	·
5-7	150	100	·	·	0.1	0.9	·	·
	3	100	·	·	0.2	0.8	·	·

the external job arrival rate of each node. Table 1.4 shows an example of the set of the values of processing rates $\mu_{ij}^{(k)}$ and the transition probabilities $p_{ij}^{(k)}$.

The results of the experiments Table 1.5 shows computation times and the number of iterations for obtaining the optimal solutions by using the three algorithms in the above example. Here $t^{(1)} = t^{(2)}$, and ε indicates the acceptance tolerance that gives the convergence condition of algorithms.

Table 1.6 shows computation times and the number of iterations for obtaining the optimal solutions in some other two-job class models that consist of other numbers of nodes. In the case of the 5 node model, the parameter values of nodes 1, 2, .., 5 are the same as those of nodes 1, 2, .., 5 in the above example, respectively. In the 14, 21, and 28

Table 1.5: Computation times of algorithms (1)

Algo.	Computation time in sec (number of iterations)			
	$t^{(1)} = t^{(2)} = 0.01$	0.03	0.1	0.2
FD	25.66(647)	25.10(636)	36.57(997)	99.74(4233)
DAF.	0.95(10)	0.76(9)	0.33(5)	0.31(4)
K & K	0.90(7)	0.65(6)	0.48(5)	0.22(3)

$$[\varepsilon = 0.00001]$$

Table 1.6: Computation times of algorithms (2)

Num.	Algo.	Computation time in sec (number of iterations)			
		$t^{(1)} = t^{(2)} = 0.01$	0.03	0.1	0.2
	FD	94.17(8350)	85.98(7136)	174.64(30995)	186.91(34970)
5	DAF.	1.09(11)	0.71(10)	0.31(4)	0.22(4)
	K & K	0.62(7)	0.62(6)	0.30(4)	0.16(3)
	FD	186.89(3911)	158.31(3297)	188.20(5515)	218.37(5565)
14	DAF.	4.93(22)	1.25(8)	0.76(6)	0.83(6)
	K & K	1.97(7)	1.55(6)	1.17(6)	0.45(3)
	FD	203.33(2267)	173.70(2144)	205.12(3106)	254.87(3233)
21	DAF.	3.80(12)	2.21(8)	1.33(6)	1.73(8)
	K & K	3.03(7)	2.33(6)	1.34(5)	0.65(3)
	FD	248.20(2058)	206.31(1850)	267.15(2316)	283.47(2462)
28	DAF.	5.24(11)	2.37(7)	2.27(7)	1.63(5)
	K & K	4.07(7)	3.88(7)	1.47(4)	0.81(3)

$$[\varepsilon = 0.00001]$$

node models, the parameter values of node i are the same as those of node $(i \bmod 7)+1$ in the above example, where $m \bmod n$ means the arithmetic that the number m is replaced by the remainder after division by number n. That is, in the 14, 21, and 28 node models, the values of parameters of the above 7 node example are used twice, three times, and four times, respectively.

Figure 1.5 shows the value of the system-wide mean job response time for each step of the iterative calculation in the above example ($t^{(1)} = t^{(2)} = 0.03$).

In Figure 1.5, it is observed that the proposed algorithm and the Dafermos algorithm converge to the optimal solution rapidly, but that the FD algorithm converges to the optimal solution very slowly.

When the processor sharing M/G/1 model was used as a node model in a distributed computer system model, the proposed algorithm and the Dafermos algorithm also required

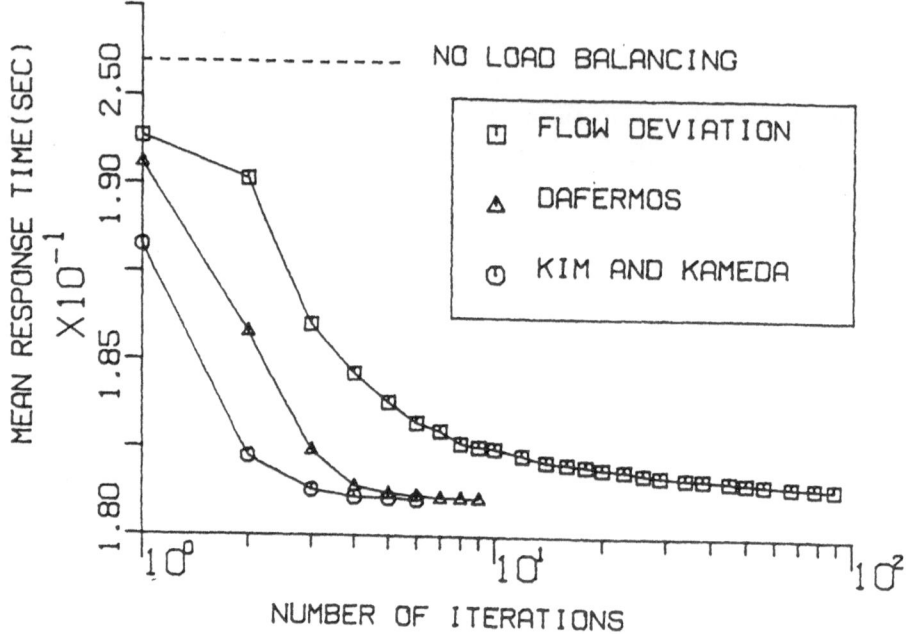

Figure 1.5: The mean response times obtained by the three algorithms for each step of the iterative calculation

mutually comparable and much less computation times than the FD algorithm, but the difference was much greater than those shown in Tables 1.5 and 1.6.

Discussion

The Dafermos algorithm requires much more storage than the other two algorithms, that is, it may occupy a larger working set. And as a result of the experiment, the proposed algorithm and Dafermos algorithm are mutually comparable and superior to the FD algorithm with respect to computation time requirements.

Some system models that differ from the above example were also used during numerical experiments. In those experiments, almost the same trends as above were observed that showed that the proposed algorithm and the Dafermos algorithm require nearly the same computation times, but that the FD algorithm requires much more computation time than the other two algorithms. Therefore, in this particular distributed computer system model, the proposed algorithm is recommended for obtaining the optimal solution.

1.3.6 Conclusion

The problem of statically balancing the load in a distributed computer system with multiple job classes has been studied. The multiple job class model considered is an extension of the single job class model developed by Tantawi and Towsley. Some properties of the optimal solution have been shown and an algorithm for obtaining the optimal load that minimizes the system-wide mean job response time has been proposed. The newly proposed algorithm is very straightforward and efficient comparing with multiple job class algorithm which can be obtained directly from Tantawi and Towsley's single job class algorithm.

The proposed algorithm and two other well-known multi-class algorithms with respect to storage and computational time requirements have been compared. The result of this comparison with respect to the storage requirements shows that the proposed algorithm and the FD algorithm need almost the same amount of storage, but the Dafermos algorithm needs much more storage than the other two algorithms. In comparing computation times, our proposed algorithm and the Dafermos algorithm require nearly the same computation times for obtaining the optimal solution, but the FD algorithm requires much more computation times than the other two algorithms.

Chapter 2

Overall Optimal Load Balancing vs. Individually Optimal Load Balancing

2.1   Introduction

We can think of two load balancing policies each of which has a distinct performance objective. One is the *overall optimal policy* and the other is the *individually optimal policy* . The overall optimal load balancing policy is the policy whereby job scheduling is determined so as to minimize the system-wide mean job response time. We refer to the solution as the *optimum* . This policy has been studied by Tantawi and Towsley [TT85]. We also studied the policy in sections 1.1 and 1.2 this chapter. Of particular note, the conditions that the overall optimal solution satisfies and an efficient load balancing algorithm for multiple job classes are discussed. On the other hand, the individually optimal load balancing policy is the policy whereby each individual job is scheduled so as to minimize its own expected job response time, given the expected node and communication delays. We refer to the solution as the *equilibrium* . Wardrop considered this policy in traffic assignment problems [War52]. In the context of distributed computer systems, the policy has been studied by Kameda and Hazeyama [KH88].

We can consider three important parameters in a distributed computer system: the communication time of the network, the processing capacity of each node, and the job arrival rate of each node. The effects of the three system parameters on the behavior of the system can be studied by parametric analysis.

This chapter provides comparative study on the overall optimal load balancing and individually optimal load balancing policies in single channel and star network configurations by parametric analysis.

The individually optimal load balancing policies in the same models for the single channel and star network configurations in chapter 1 are newly considered in sections 2.2, 2.3 and 2.4. The conditions that the optimal solution of the individually optimal policy satisfies are derived.

A striking parallel between the conditions on the optimum and those on the equilibrium can be observed. That is, the individually optimal policy would be realized by an overall optimal policy, if the values of the marginal node and communication delays were given as the expected node and communication delay, and vice versa.

Furthermore, the parametric analysis of the overall and individually optimal policies, that is, the effects of varying the system parameters on the performance variables of these policies, are studied in detail.

It is very difficult, however, to show the effects of the parameters by mathematical parametric analysis in the multi-class job environment. In section 2.4, we study the effects of changing these system parameters on the two load balancing policies by numerical experiments. For the numerical experiments, the optimum load balancing algorithm for multi-class jobs proposed at section 1.2 of chapter 1 is used for the overall optimal policy and an equilibrium load balancing algorithm that is very similar to that of the overall optimal policy is proposed here for the individually optimal policy.

We observe that decreasing the arrival rate, for particular class k and node i, has an effect similar to that of increasing the processing capacity for the same class and node on the system behavior. Increasing the arrival rate, for particular class k and node i, also has an effect similar to that of decreasing the processing capacity for the same class and node on the system behavior. The above tendencies may be observed both in the overall optimal policy and in the individually optimal policy.

For all the models studied in this chapter we see the following. In the experimental result, it is observed that each class jobs are processed at remote nodes that have high processing capacity for the class when the communication time is short. As the result, it may happen that jobs of a class are mainly processed a node and jobs of another class are mainly processed on another node. This is a kind of specialization. It is observed also that in the overall optimal policy more jobs are processed remotely than in the individually optimal policy when the communication time is long

In the two load balancing policies, two groups of jobs (of the same class) that arrive at the same node and that are processed by different nodes according to the policies are considered. In the optimum, the two groups of jobs have shown mutually different expected response times, which seems *unfair* . On the other hand, in the equilibrium, the two groups of jobs have shown mutually identical expected response times and each job could expect no improvement in its expected response time even if it were processed by a node different from the processing node, which seems *fair* .

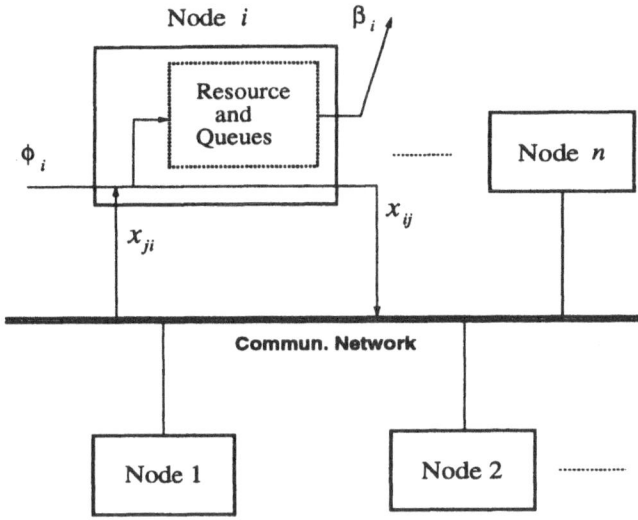

Figure 2.1: A model of a distributed computer system

Furthermore, for all the models studied in this chapter we observe an *anomalous* or *counter-intuitive* phenomenon in the equilibrium that the system-wide mean job response time, when there exist some jobs processed remotely, at some values of communication time is longer than the system-wide mean job response time when there exists no job processed remotely.

2.2 Static Load Balancing in Single Channel Communications Networks

2.2.1 Model Description

A distributed computer system model as shown in Fig. 2.1 is considered. The system consists of n nodes connected by a single channel communications network. The key assumptions of the model are the same as those of Tantawi and Towsley [TT85]. Nodes may be heterogeneous; that is, they may have different configurations, number of resources, and processing capacities. We assume that the expected communication delay from node i to node j is independent of source-destination pair (i, j). Let us have the following notation.

- n Number of nodes.

- ϕ_i External arrival rate to node i.

- Φ Total external job arrival rate, that is, $\Phi = \sum_{i=1}^{n} \phi_i$.

- x_{ij} Job flow rate from node i to node j.

- β_i Job processing rate (load) at node i.

- $\boldsymbol{\beta}$ $[\beta_1, \beta_2, \ldots, \beta_n]$.

- λ Total traffic through the network, that is, $\lambda = \sum_{i} \sum_{j, j \neq i} x_{ij}$.

- $F_i(\beta_i)$ Expected node delay of jobs processed at node i — a differentiable, increasing, and convex function with respect to β_i.

- $G(\lambda)$ Expected communication delay of jobs (including waiting time) — a differentiable, nondecreasing, and convex function with respect to λ.

- $T(\boldsymbol{\beta})$ Overall mean job response time, that is, the mean length of the time period that starts when a job arrives in the system and ends when it leaves the system.

Jobs arrive at each node according to a time-invariant Poisson process. A job that arrives at node i (origin node) may either be processed at node i or be transferred to another node j (processing node). After the job is processed at node j, a response is sent back to the origin node. Also we assume that a transferred job from node i to node j receives its service at node j and is not transferred to other nodes. The overall mean job response time can be written as the sum of the mean node and communication delays; that is

$$T(\boldsymbol{\beta}) = \frac{1}{\Phi} \Big[\sum_{i=1}^{n} \beta_i F_i(\beta_i) + \lambda G(\lambda) \Big], \tag{2.1}$$

subject to

$$\sum_{i=1}^{n} \beta_i = \Phi, \tag{2.2}$$

$$\beta_i \geq 0, \quad i = 1, 2, \ldots, n, \tag{2.3}$$

where the network traffic λ may be expressed in terms of variable β_i as

$$\lambda = \frac{1}{2} \sum_{i=1}^{n} \big| \phi_i - \beta_i \big|. \tag{2.4}$$

Tantawi and Towsley [TT85] classified the nodes in the following way. Node i is said to be an *idle source* if it does not process any jobs; that is, $\beta_i = 0$. Node i is said to be an *active source* if it sends jobs and does not receive any jobs; that is, $\phi_i > \beta_i > 0$. Node i is said to be a *neutral* if it processes jobs locally without sending or receiving jobs; that is, $\beta_i = \phi_i$. Node i is said to be a *sink* if it receives jobs from other nodes but does not send any jobs out; that is, $\beta_i > \phi_i$. We denote the sets of sinks, idle sources, active sources, and neutrals by S, R_d, R_a, and N, respectively.

We note that the job flow rates x_{ij} may be chosen arbitrarily as long as the job flow balance constraint is satisfied. In particular, we may assign the flow rate from node i to node j proportionate to the contribution of node j to the network traffic; that is,

$$
x_{ij} = \begin{cases} \dfrac{(\beta_j - \phi_j)}{\lambda}(\phi_i - \beta_i), & i \in R_d,\, R_a \ j \in S, \\ 0, & \text{otherwise.} \end{cases}
$$

Since node j is a sink, the quantity $\beta_j - \phi_j (> 0)$ represents the traffic sent to node j. Node i, on the other hand, is a source and therefore $\phi_i - \beta_i (> 0)$ represents the traffic sent from node i.

2.2.2 Optimal Solutions

The Overall Optimal Policy

By the overall optimal policy we mean the policy whereby jobs are scheduled so as to minimize the overall mean job response time; that is, to solve problem (2.1) characterized by load vector β with constraints (2) and (3).

Now two functions, the *incremental node delay* $f_i(\beta_i)$ and the *incremental communication delay* $g(\lambda)$, are introduced as Tantawi and Towsley [TT85] as follows.

$$
f_i(\beta_i) = \frac{d}{d\beta_i}\beta_i F_i(\beta_i), \tag{2.5}
$$

$$
g(\lambda) = \frac{d}{d\lambda}\lambda G(\lambda). \tag{2.6}
$$

The inverse of the incremental node delay f_i^{-1} is defined by

$$
f_i^{-1}(x) = \begin{cases} a, & f_i(a) = x, \\ 0, & f_i(0) \geq x. \end{cases}
$$

According to the results of Tantawi and Towsley [TT85], we have the following theorem by which we can determine β that implements the overall optimal policy.

Theorem 2.1 *The optimal solution, β, to problem (2.1) satisfies the relations*

$$
\begin{aligned}
f_i(\beta_i) &\geq \alpha + g(\lambda), & \beta_i = 0 & & (i \in R_d), & (2.7)\\
f_i(\beta_i) &= \alpha + g(\lambda), & 0 < \beta_i < \phi_i & & (i \in R_a), & (2.8)\\
\alpha &\leq f_i(\beta_i) \leq \alpha + g(\lambda), & \beta_i = \phi_i & & (i \in N), & (2.9)\\
\alpha &= f_i(\beta_i), & \beta_i \geq \phi_i & & (i \in S), & (2.10)
\end{aligned}
$$

subject to the total flow constraint

$$\sum_{i \in S} f_i^{-1}(\alpha) + \sum_{i \in R_a} f_i^{-1}(\alpha + g(\lambda)) + \sum_{i \in N} \phi_i = \Phi, \tag{2.11}$$

where α is the Lagrange multiplier.

Remark. By this theorem, β that implements the overall optimal policy can be determined. The above relations can be interpreted in the same way as done by Tantawi and Towsley as follows. Eq. (2.10) states that the incremental node delays of all sinks are the same. Similarly, eq. (2.8) states that the incremental node delays for active sources, are all equal; they consist of the incremental node delay at a sink and the incremental communication delay due to sending a job through the network to a sink. The incremental node delay for idles, as given by eq. (2.7), is greater than the incremental node delay of active sources; this makes idle sources send all their jobs to sinks. Finally, eq. (2.9) states that the incremental node delay for neutrals, is greater than the incremental node delay for sinks but less than that for active sources; this makes neutrals neither receive any jobs from other nodes nor send any jobs out.

The Individually Optimal Policy

According to the individually optimal policy, jobs are scheduled so that every job may feel that its own expected response time is minimum if it knows the expected node delay at each node and the expected communication delay. In other words, when the individually optimal policy is realized, the expected response time of a job cannot be improved further when the scheduling decisions for other jobs are fixed, and the system reaches an equilibrium. The equilibrium conditions are defined by the following definition.

Definition: β is said to satisfy the equilibrium conditions for the individually optimal policy, if the following relations hold:

$$F_i(\beta_i) \geq R + G(\lambda), \quad \beta_i = 0 \qquad (i \in R_d), \tag{2.12}$$

$$F_i(\beta_i) = R + G(\lambda), \quad 0 < \beta_i < \phi_i \qquad (i \in R_a), \tag{2.13}$$

$$R \leq F_i(\beta_i) \leq R + G(\lambda), \; \beta_i = \phi_i \quad (i \in N), \tag{2.14}$$

$$R = F_i(\beta_i), \qquad \beta_i \geq \phi_i \qquad (i \in S), \tag{2.15}$$

subject to the total flow constraint

$$\sum_{i \in S} F_i^{-1}(R) + \sum_{i \in R_a} F_i^{-1}(R + G(\lambda)) + \sum_{i \in N} \phi_i = \Phi. \tag{2.16}$$

β is called the optimal solution of the individually optimal policy if it satisfies the equilibrium conditions (2.12) - (2.15). Such a β is also referred to as the *equilibrium*.

Remark. In the above definition, R and $R + G(\lambda)$ represent the expected response time of jobs arriving at sinks and the expected response time of jobs sent to sinks, respectively. $F_i(\beta_i)$ denotes the expected node delay of node i. For example, let us examine a job that arrives at an idle source. According to eq. (2.12), we see that the job should be sent to a sink. If the job decides to receive its service locally, its expected response time cannot be improved because the expected node delay of the idle source is greater than the expected response time of jobs sent to sinks. Furthermore, if the job decides to be sent to another sink, its expected response time can not be improved because the expected node delays of all sinks are equal according to eq. (2.15), either. For a job arriving at an active source or a neutral, we see that the expected response time of the job cannot be improved according to eqs. (2.13) and (2.14).

For the individually optimal policy, we have the following theorem.

Theorem 2.2 *The individually optimal policy has a solution. That is, there exists one and only one β that satisfies the equilibrium conditions (2.12) - (2.15).*

PROOF. A proof of Theorem 2.2.

As pointed out by Magnanti [Mag84], the individually optimization problem can be expressed by using an equivalent overall optimization problem. Denote the expected node and communication delays of the equivalent overall optimization problem by $F_i^*(\beta_i)$ and $G^*(\lambda)$, respectively, and define $F_i^*(\beta_i)$ and $G^*(\lambda)$ as follows.

$$F_i^*(\beta_i) = \frac{1}{\beta_i} \int_0^{\beta_i} F_i(\beta_i) \, d\beta_i, \quad F_i^*(0) = F_i(0),$$
$$G^*(\lambda) = \frac{1}{\lambda} \int_0^{\lambda} G(\lambda) \, d\lambda, \quad G^*(0) = G(0).$$

$F_i^*(\beta_i)$ and $G^*(\lambda)$ are different from $F_i(\beta_i)$ and $G(\lambda)$, respectively, but they have relationship with $F_i(\beta_i)$ and $G(\lambda)$ as defined above. By using a similar way as that in problem (2.1), the equivalent overall optimization problem is formulated as follows.

$$\min \ T^*(\beta) = \frac{1}{\Phi} \Big[\sum_{i=1}^{n} \beta_i F_i^*(\beta_i) + \lambda G^*(\lambda) \Big]. \tag{2.17}$$

subject to

$$\sum_{i=1}^{n} \beta_i = \Phi, \tag{2.18}$$
$$\beta_i \geq 0, \quad i = 1, 2, \ldots, n. \tag{2.19}$$

where the network traffic λ is expressed in terms of variables β_i as $\lambda = \frac{1}{2}\sum_{i=1}^{n}|\phi_i - \beta_i|$.

Noting problems (2.1) and (2.17), we see that they are similar problems except for the differences between $F_i^*(\beta_i), G^*(\lambda)$ and $F_i(\beta_i), G(\lambda)$. Here, we need to check the convexity of problem (2.17). Since $dF_i^*/d\beta_i > 0$ and $d^2F_i^*/d\beta_i^2 > 0$ for all i, and $dG^*/d\lambda \geq 0$ and $d^2G^*/d\lambda^2 \geq 0$ for λ, we may immediately conclude that $T^*(\boldsymbol{\beta})$ is a strictly convex function of variables β_i. Furthermore, the variables β_i belong to a convex polyhedron. Thus we may conclude that if the problem is feasible at all, then any local minimum is a global minimum for $T^*(\boldsymbol{\beta})$.

We can use the same techniques as these of Tantawi and Towsley [TT85] to solve problem (2.17). We denote the network traffic into node i and the network traffic out of node i by u_i and v_i, respectively. They may be written in terms of the traffic x_{ij} from node i to node j as

$$u_i = \sum_{j=1,j\neq i}^{n} x_{ji}, \quad i = 1, 2, .., n,$$

$$v_i = \sum_{j=1,j\neq i}^{n} x_{ij}, \quad i = 1, 2, .., n.$$

Clearly, from the balance of the total traffic in the network, we have that

$$\lambda = \sum_{i=1}^{n} u_i = \sum_{i=1}^{n} v_i. \tag{2.20}$$

The load, β_i, on node i may be written as

$$\beta_i = u_i - v_i + \phi_i. \tag{2.21}$$

Substituting eqs. (2.20) and (2.21) into problem (2.17) yields

$$\min \quad T^*(\boldsymbol{u}, \boldsymbol{v}) = \frac{1}{\Phi}[\sum_{i=1}^{n}(u_i - v_i + \phi_i)F_i^*(u_i - v_i + \phi_i) + \sum_{i=1}^{n} v_i G^*(\sum_{i=1}^{n} v_i)], \tag{2.22}$$

subject to

$$-\sum_{i=1}^{n} u_i + \sum_{i=1}^{n} v_i = 0,$$

$$u_i - v_i + \phi_i \geq 0, \quad i = 1, 2, \ldots, n,$$

$$u_i \geq 0, v_i \geq 0, \quad i = 1, 2, ..., n.$$

To obtain the optimal solution, we form the Lagrangian function as follows.

$$L^*(\boldsymbol{u}, \boldsymbol{v}, R, \boldsymbol{\gamma}, \boldsymbol{\psi}, \boldsymbol{\eta}) = \Phi T^*(\boldsymbol{u}, \boldsymbol{v}) + R(-\sum_{i=1}^{n} u_i + \sum_{i=1}^{n} v_i)$$

$$+ \sum_{i=1}^{n} \gamma_i(u_i - v_i + \phi_i) + \sum_{i=1}^{n} \psi_i u_i + \sum_{i=1}^{n} \eta_i v_i. \tag{2.23}$$

The optimal solution satisfies the Kuhn-Tucker condition,

$$\frac{\partial L^*}{\partial u_i} = F_i(u_i - v_i + \phi_i) - R + \gamma_i + \psi_i = 0, \tag{2.24}$$

$$\frac{\partial L^*}{\partial v_i} = -F_i(u_i - v_i + \phi_i) + G(\sum_{i=1}^{n} v_i) + R - \gamma_i + \eta_i = 0, \tag{2.25}$$

$$\frac{\partial L^*}{\partial R} = -\sum_{i=1}^{n} u_i + \sum_{i=1}^{n} v_i = 0, \tag{2.26}$$

$$u_i - v_i + \phi_i \geq 0, \ \gamma_i(u_i - v_i + \phi_i) = 0, \gamma_i \leq 0, \tag{2.27}$$

$$u_i \geq 0, \psi_i u_i = 0, \psi_i \leq 0, \tag{2.28}$$

$$v_i \geq 0, \eta_i v_i = 0, \eta_i \leq 0. \tag{2.29}$$

Similarly as Tantawi and Towsley [TT85], we consider each case separately.

Case 1. $v_i = u_i + \phi_i$

$$F_i(\beta_i) \geq R + G(\lambda), \ \beta_i = 0. \tag{2.30}$$

Case 2. $v_i < u_i + \phi_i$

    Case 2.1. $v_i > 0$

$$F_i(\beta_i) = R + G(\lambda), \ 0 < \beta_i < \phi_i. \tag{2.31}$$

    Case 2.2. $v_i = 0$

        Case 2.2.1. $u_i = 0$

$$R \leq F_i(\beta_i) \leq R + G(\lambda), \ \beta_i = \phi_i. \tag{2.32}$$

        Case 2.2.2. $u_i > 0$

$$R = F_i(\beta_i), \ \beta_i > \phi_i. \tag{2.33}$$

Eq. (2.26) may be written in terms of β_i as

$$\sum_{i=1}^{n} \beta_i = \Phi.$$

Using eqs. (2.31) and (2.33), the above equation becomes

$$\sum_{i \in S} F_i^{-1}(R) + \sum_{i \in N} \phi_i + \sum_{i \in R_a} F_i^{-1}(R + G(\lambda)) = \Phi. \tag{2.34}$$

Once the node partition is known, the traffic can be obtained as

$$\lambda = \sum_{i \in S} [F_i^{-1}(R) - \phi_i].$$

Substituting λ into eq. (2.34) yields a single equation in the unknown R. The rates $\beta_i (i = 1, 2, .., n)$, may then be obtained by solving eqs. (2.31) and (2.33).

The above relations can be written in the following way.

$$F_i(\beta_i) \;\geq\; R + G(\lambda) \qquad\qquad \beta_i = 0 \quad (i \in R_d). \tag{2.35}$$

$$F_i(\beta_i) \;=\; R + G(\lambda), \qquad\qquad 0 < \beta_i < \phi_i \quad (i \in R_a), \tag{2.36}$$

$$R \;\leq\; F_i(\beta_i) \leq R + G(\lambda), \quad \beta_i = \phi_i \quad (i \in N), \tag{2.37}$$

$$R \;=\; F_i(\beta_i), \qquad\qquad \beta_i \geq \phi_i \quad (i \in S), \tag{2.38}$$

subject to the total flow constraint

$$\sum_{i \in S} F_i^{-1}(R) + \sum_{i \in R_a} F_i^{-1}(R + G(\lambda)) + \sum_{i \in N} \phi_i = \Phi. \tag{2.39}$$

By noting the equilibrium conditions (2.12) - (2.15) for the individually optimal policy in subsection 2.2.2, we see that the conditions that the optimal solution of the problem (2.17) satisfies are equivalent to those equilibrium conditions. Therefore we conclude that the individually optimal policy has one and only one solution and that its solution satisfies the conditions (2.12) - (2.15). □

Remark. It can be observed that the optimal solutions of the overall and individually optimal policies have a striking parallelism in the forms of the optimal solutions of the two policies. The parallelism between Theorems 2.2 and 2.1 gives us an intuitive explanation of one in terms of the other. That is, the overall optimal policy would be realized by an individually optimal policy, if the values of the incremental node and communication delays were given as the expected node and communication delays, and vice versa. Therefore, the individually optimal policy can be implemented by using the algorithm which is extended from that of Kim and Kamada [KK92a] by replacing $f_i(\beta_i)$ and $g(\lambda)$ with $F_i(\beta_i)$ and $G(\lambda)$, respectively.

2.2.3 Parametric Analysis

In this subsection, the effects of the system parameters on the behavior of the system in both the optimum and the equilibrium are studied while node partition remains the same. We consider three main parameters: the communication time t (excluding the waiting time), the node i processing time $u_i (i = 1, 2, \ldots, n)$, and the node i job arrival rate $\phi_i (i = 1, 2, \ldots, n)$ as the system parameters. A vector p is used to denote $[t, u_1, u_2, \ldots, u_n, \phi_1, \phi_2, \ldots, \phi_n]$. $F_i(\beta_i, u_i)$ and $G(\lambda, t)$ denote the expected node and communication delays in order to make the parameters t and u_i explicit, respectively. Similarly, $f_i(\beta_i, u_i)$ and $g(\lambda, t)$ denote the incremental node and communication delays, respectively. We assume

$$\frac{\partial G(\lambda, t)}{\partial t} > 0, \qquad \frac{\partial g(\lambda, t)}{\partial t} > 0,$$

$$\frac{\partial F_i(\beta_i, u_i)}{\partial u_i} > 0, \quad \frac{\partial f_i(\beta_i, u_i)}{\partial u_i} > 0.$$

Parameters λ, β_i, and α in the optimum are determined when p is given and we write them as $\lambda(p), \beta_i(p)$, and $\alpha(p)$. Similarly we may write parameters λ, β_i, and R in the equilibrium as $\lambda(p), \beta_i(p)$, and $R(p)$.

Given u_i, $F_i(\beta_i, u_i)$ is monotonically increasing with β_i. An inverse function E_i of the expected node delay F_i is defined as follows.

$$E_i(R, u_i) = \beta_i, \tag{2.40}$$

if and only if

$$F_i(\beta_i, u_i) = R. \tag{2.41}$$

That is $E_i(F_i(\beta_i, u_i), u_i) = \beta_i$. $E_i(R, u_i)$ is monotonically increasing with in R. From eq. (2.41), we can say that, given R, β_i is monotonically decreasing with the increase in u_i. Therefore we have for all i

$$\frac{\partial E_i(R, u_i)}{\partial R} > 0, \tag{2.42}$$

$$\frac{\partial E_i(R, u_i)}{\partial u_i} < 0. \tag{2.43}$$

Similarly, an inverse function e_i of incremental node delay f_i is defined as follows.

$$e_i(\alpha, u_i) = \beta_i, \tag{2.44}$$

if and only if

$$f_i(\beta_i, u_i) = \alpha. \tag{2.45}$$

We have for all i

$$\frac{\partial e_i(\alpha, u_i)}{\partial \alpha} > 0, \tag{2.46}$$

$$\frac{\partial e_i(\alpha, u_i)}{\partial u_i} < 0. \tag{2.47}$$

Let us define

$$\delta_i(X) = \begin{cases} 1, & i \in X, \\ 0, & i \notin X. \end{cases}$$

The Individually Optimal Policy

For the individually optimal policy, we have the following theorems.

Lemma 2.3 *For a given set of sinks S and for all i,*

$$\frac{\partial \lambda(\boldsymbol{p})}{\partial t} = A(\boldsymbol{p}) \frac{\partial R(\boldsymbol{p})}{\partial t}, \tag{2.48}$$

$$\frac{\partial \lambda(\boldsymbol{p})}{\partial u_i} = A(\boldsymbol{p}) \frac{\partial R(\boldsymbol{p})}{\partial u_i} + \delta_i(S) \frac{\partial E_i}{\partial u_i}, \tag{2.49}$$

$$\frac{\partial \lambda(\boldsymbol{p})}{\partial \phi_i} = A(\boldsymbol{p}) \frac{\partial R(\boldsymbol{p})}{\partial \phi_i} - \delta_i(S), \tag{2.50}$$

where

$$A(\boldsymbol{p}) = \sum_{i \in S} \frac{\partial E_i(R, u_i)}{\partial R}\Big|_{R=R(\boldsymbol{p})}. \tag{2.51}$$

PROOF. Eq. (2.48) is derived in the same way as eq. (17) of [TT85]. To derive eq. (2.49) we have from relation (2.15) and definition (2.40)

$$E_i(R(\boldsymbol{p}), u_i) = \beta_i(\boldsymbol{p}), \qquad i \in S. \tag{2.52}$$

Therefore we have

$$\frac{\partial \beta_i(\boldsymbol{p})}{\partial u_j} = \frac{\partial E_i(R, u_i)}{\partial R}\Big|_{R=R(\boldsymbol{p})} \frac{\partial R(\boldsymbol{p})}{\partial u_j} + \frac{\partial E_i(R, u_i)}{\partial u_j}\Big|_{R=R(\boldsymbol{p})}. \tag{2.53}$$

By noting that $\lambda(\boldsymbol{p}) = \sum_{i \in S}(\beta_i(\boldsymbol{p}) - \phi_i)$, we easily have eq. (2.49) from eq. (2.53).
We have (2.50) in the similar way as above. $\square$

Lemma 2.4 *For a given set of active sources R_a and for all i,*

$$\frac{\partial \lambda(\boldsymbol{p})}{\partial t} = -\frac{(\partial R(\boldsymbol{p})/\partial t) + (\partial G(\lambda, t)/\partial t)}{(1/B(\boldsymbol{p})) + (\partial G(\lambda, t)/\partial \lambda)}, \tag{2.54}$$

$$\frac{\partial \lambda(\boldsymbol{p})}{\partial u_i} = -\frac{(\partial R(\boldsymbol{p})/\partial u_i) + \delta_i(R_a)(\partial E_i(\hat{R}, u_i)/\partial u_i)}{(1/B(\boldsymbol{p})) + (\partial G(\lambda, t)/\partial \lambda)}, \tag{2.55}$$

$$\frac{\partial \lambda(\boldsymbol{p})}{\partial \phi_i} = -\frac{(\partial R(\boldsymbol{p})/\partial \phi_i) - \delta_i(R_a)}{(1/B(\boldsymbol{p})) + (\partial G(\lambda, t)/\partial \lambda)}, \tag{2.56}$$

where

$$\hat{R}(\boldsymbol{p}) = R(\boldsymbol{p}) + G(\lambda(\boldsymbol{p}), t), \tag{2.57}$$

$$B(\boldsymbol{p}) = \sum_{i \in R_a} \frac{\partial E_i(\hat{R}, u_i)}{\partial \hat{R}}\Big|_{\hat{R}=\hat{R}(\boldsymbol{p})}. \tag{2.58}$$

PROOF. Eq. (2.54) is derived in the same way as eq. (20) of [TT85]. To derive eq. (2.55) we have from relation (2.13) by using definition (2.40)

$$\beta_i(\boldsymbol{p}) = E_i(R(\boldsymbol{p}) + G(\lambda, t), u_i), \qquad i \in R_a.$$

Then we have

$$\frac{\partial \beta_i(\boldsymbol{p})}{\partial u_j} = \frac{\partial E_i(\hat{R}, u_i)}{\partial \hat{R}}\left(\frac{\partial R}{\partial u_j} + \frac{\partial G}{\partial \lambda}\frac{\partial \lambda}{\partial u_j}\right) + \frac{\partial E_i(\hat{R}, u_i)}{\partial u_j}. \tag{2.59}$$

Note that $\lambda(\boldsymbol{p}) = \sum_{i \in R_a}(\phi_i - \beta_i(\boldsymbol{p})) + \sum_{i \in R_d}\phi_i$. From these, we have eq. (2.55).
We have (2.56) in the similar way as above. $\square$

Theorem 2.5 *The following relations hold for the expected node delay $R(\boldsymbol{p})$ at sinks.*

$$\frac{\partial R(\boldsymbol{p})}{\partial t} < 0. \tag{2.60}$$

$$\frac{\partial R(\boldsymbol{p})}{\partial u_i} > 0, \qquad i \in S \cup R_a,$$

$$= 0, \qquad i \in N \cup R_d. \tag{2.61}$$

$$\frac{\partial R(\boldsymbol{p})}{\partial \phi_i} > 0, \qquad i \in S \cup R_a,$$

$$= 0, \qquad i \in N \cup R_d. \tag{2.62}$$

PROOF. By combining each equation in Lemma 2.3 with the corresponding equation in Lemma 2.4, we can derive equations for $\partial R(\boldsymbol{p})/\partial t$, $\partial R(\boldsymbol{p})/\partial u_i$, and $\partial R(\boldsymbol{p})/\partial \phi_i$. Then we can derive the relations (2.60) - (2.62). $\square$

Remark. This theorem implies that the expected node delay at sinks will decrease as the communication time increases, and that it will increase with the increase in the processing time or in the arrival rate, at a sink or at an active source. These agree with our intuition. The corresponding results on the optimum can be derived simply by replacing R with α in eqs. (2.60), (2.61), and (2.62).

Corollary 2.6 *The following relations hold for the network traffic $\lambda(\boldsymbol{p})$.*

$$\frac{\partial \lambda(\boldsymbol{p})}{\partial t} < 0. \tag{2.63}$$

$$\frac{\partial \lambda(\boldsymbol{p})}{\partial u_i} < 0, \qquad i \in S,$$

$$= 0, \qquad i \in N \cup R_d,$$

$$> 0, \qquad i \in R_a. \tag{2.64}$$

$$\frac{\partial \lambda(\boldsymbol{p})}{\partial \phi_i} < 0, \qquad i \in S,$$

$$= 0, \qquad i \in N \cup R_d,$$

$$> 0, \qquad i \in R_a. \tag{2.65}$$

PROOF. We can derive these from Lemmas 2.3 and 2.4 and Theorem 2.5. □

Remark. This corollary implies that the network traffic decreases with the increase in the communication time, or with the increase in the processing time or the arrival rate at sinks, and that it increases with the increase in the processing time or in the arrival rate at an active source. These agree with our intuition. The corresponding results on the optimum have the same form as relations (2.63), (2.64), and (2.65).

Theorem 2.7 *The following relations hold for the expected response time for jobs arriving at active sources* $\hat{R}(p) = R(p) + G(\lambda(p), t)$.

$$\frac{\partial \hat{R}(p)}{\partial t} > 0. \tag{2.66}$$

$$\frac{\partial \hat{R}(p)}{\partial u_i} > 0, \qquad i \in S \cup R_a,$$
$$= 0, \qquad i \in N \cup R_d. \tag{2.67}$$

$$\frac{\partial \hat{R}(p)}{\partial \phi_i} > 0, \qquad i \in S \cup R_a,$$
$$= 0, \qquad i \in N \cup R_d. \tag{2.68}$$

PROOF. Relation (2.66) can be derived in the same way as Theorem 7 of [TT85].
   Note that

$$\frac{\partial \hat{R}(p)}{\partial u_i} = \frac{\partial R(p)}{\partial u_i} + \frac{\partial G(\lambda, t)}{\partial \lambda} \frac{\partial \lambda(p)}{\partial u_i}. \tag{2.69}$$

From Theorem 2.5, Corollary 2.6, and eq. (2.69) we easily see that

$$\frac{\partial \hat{R}(p)}{\partial u_i} > 0, i \in R_a,$$
$$= 0, i \in N \cup R_d.$$

For $i \in S$, we have from eqs. (2.55) and (2.69)

$$\frac{\partial \hat{R}(p)}{\partial u_i} = \left(\frac{\partial R(p)}{\partial u_i}\right) \frac{1}{1 + B(p)(\partial G(\lambda, t)/\partial \lambda)} > 0. \tag{2.70}$$

Therefore we have relation (2.67).
   Relation (2.68) is derived similarly as above. □

Remark. This theorem implies that the expected response time for jobs arriving at active sources will increase with the increase in the communication time, and that it will increase with the increase in the processing time or in the job arrival rate at a sink or an active source. These agree with our intuition. The corresponding results on the optimum can be derived simply by replacing $\hat{R}(\boldsymbol{p}) = R(\boldsymbol{p}) + G(\lambda(\boldsymbol{p}), t)$ with $\hat{\alpha}(\boldsymbol{p}) = \alpha(\boldsymbol{p}) + g(\lambda(\boldsymbol{p}), t)$.

Denote the overall mean job response time in the equilibrium under the individually optimal policy by $T(\boldsymbol{p})$. Then we have

$$\Phi T(\boldsymbol{p}) = \sum_{i \in S} \phi_i R(\boldsymbol{p}) + \sum_{R_a \cup R_d} \phi_i (R(\boldsymbol{p}) + G(\lambda(\boldsymbol{p}), t)) + \sum_{i \in N} \phi_i F_i(\phi_i). \qquad (2.71)$$

Theorem 2.8 *The following relations hold for the overall mean job response time in the equilibrium, $T(\boldsymbol{p})$.*

$$\frac{\partial T(\boldsymbol{p})}{\partial u_i} > 0, \quad i \in S \cup R_a \cup N,$$
$$= 0, \quad i \in R_d. \qquad (2.72)$$
$$\frac{\partial T(\boldsymbol{p})}{\partial \phi_i} > 0, \quad i \in R_a \cup R_d \cup N. \qquad (2.73)$$

PROOF. From eq. (2.71) we have

$$\frac{\partial T(\boldsymbol{p})}{\partial u_i} = \frac{1}{\Phi} \Big[\sum_{i \in S} \phi_i \frac{\partial R(\boldsymbol{p})}{\partial u_i} + \sum_{R_a \cup R_d} \phi_i \frac{\partial \hat{R}(\boldsymbol{p})}{\partial u_i} \Big], \quad i \in S \cup R_a,$$
$$\phi_i \frac{\partial F_i}{\partial u_i}, \qquad i \in N.$$

Therefore we have relation (2.72) by noting Theorems 2.5 and 2.7.

Relation (2.73) is derived similarly as above. □

Remark. This theorem implies that the overall mean job response time in the equilibrium will increase with the increase in the node processing times at sinks, sources, or neutrals, or with the increase in the job arrival rates at sources or neutrals. These agree with our intuition.

The Overall Optimal Policy

We analyze the behavior of the performance variables of the overall optimal policy as follows.

Theorem 2.9 *The following relations hold for the incremental node delay $\alpha(\boldsymbol{p})$ at sinks.*

$$\frac{\partial \alpha(\boldsymbol{p})}{\partial t} < 0. \qquad (2.74)$$
$$\frac{\partial \alpha(\boldsymbol{p})}{\partial u_i} > 0, \qquad i \in S \cup R_a,$$
$$= 0, \qquad i \in N \cup R_d. \qquad (2.75)$$
$$\frac{\partial \alpha(\boldsymbol{p})}{\partial \phi_i} > 0, \qquad i \in S \cup R_a,$$
$$= 0, \qquad i \in N \cup R_d. \qquad (2.76)$$

PROOF. Relation (2.74) is relation (23) of [TT85]. By using a similar way as that in Theorem 2.5, we can derive equations on $\partial \alpha(\boldsymbol{p})/\partial u_i$, and $\partial \alpha(\boldsymbol{p})/\partial \phi_i$. Then we can derive the above relations. $\square$

Remark. This theorem implies that the incremental node delay at sinks will decrease as the communication time increases, and that it will increase with the increase in the processing time or in the arrival rate, at a sink or at an active source.

Corollary 2.10 *The following relations hold for the network traffic $\lambda(\boldsymbol{p})$.*

$$\frac{\partial \lambda(\boldsymbol{p})}{\partial t} < 0. \tag{2.77}$$

$$\begin{aligned} \frac{\partial \lambda(\boldsymbol{p})}{\partial u_i} &< 0, i \in S, \\ &= 0, i \in N \cup R_d, \\ &> 0, i \in R_a. \end{aligned} \tag{2.78}$$

$$\begin{aligned} \frac{\partial \lambda(\boldsymbol{p})}{\partial \phi_i} &< 0, i \in S, \\ &= 0, i \in N \cup R_d, \\ &> 0, i \in R_a. \end{aligned} \tag{2.79}$$

PROOF. By using a similar way as that in Corollary 2.6, we can derive these from Theorem 2.9. $\square$

Remark. This corollary implies that the network traffic decreases with the increase in the communication time, or with the increase in the processing time or the arrival rate at a sink, and that it increases with the increase in the processing time or in the arrival rate at an active source.

Denote the overall mean job response time in the optimum under the overall optimal policy by $T(\boldsymbol{p})$. Then we have the following theorem.

Theorem 2.11 *The following relations hold for the overall mean job response time in the optimum, $T(\boldsymbol{p})$.*

$$\frac{\partial T(\boldsymbol{p})}{\partial t} > 0, \tag{2.80}$$

$$\begin{aligned} \frac{\partial T(\boldsymbol{p})}{\partial u_i} &> 0, \quad i \in S \cup R_a \cup N, \\ &= 0, \quad i \in R_d. \end{aligned} \tag{2.81}$$

PROOF. From eq. (2.1) and relations (2.8) and (2.10), we have

$$\frac{\partial T(\boldsymbol{p})}{\partial t} = \frac{1}{\Phi}\Big[\sum_{i=1}^{n}\frac{\partial(\beta_i F_i)}{\partial \beta_i}\frac{\partial \beta_i}{\partial t} + \frac{\partial(\lambda G(\lambda))}{\partial \lambda}\frac{\partial \lambda}{\partial t} + \lambda\frac{\partial G}{\partial t}\Big]$$
$$= \frac{\lambda}{\Phi}\frac{\partial G}{\partial t}.$$

Therefore we have relation (2.80).

We have (2.81) in the similar way as above. □

Remark. This theorem implies that the overall mean job response time in the optimum will increase with the increase in the communication time, or with the increase in the node processing times at sinks, sources, or neutrals.

2.2.4 Anomalous Behaviors of the Optimum and the Equilibrium

In the previous subsection, the effects of the system parameters on the performance variables of the overall and individually optimal policies have been analyzed. Note that $\partial T(\boldsymbol{p})/\partial \phi_i$ has not been presented in subsection 2.2.3 and that $\partial T(\boldsymbol{p})/\partial t$ and $\partial T(\boldsymbol{p})/\partial \phi_i$ for $i \in S$ have not been presented in subsection 2.2.3. Let us examine these in the following.

The Individually Optimal Policy

We have in deriving the theorem 2.5,

$$\frac{\partial R(\boldsymbol{p})}{\partial t} = \frac{-B(\boldsymbol{p})(\partial G(\lambda, t)/\partial t)}{A(\boldsymbol{p}) + B(\boldsymbol{p}) + A(\boldsymbol{p})B(\boldsymbol{p})(\partial G(\lambda, t)/\partial \lambda)},$$
$$\frac{\partial \hat{R}(\boldsymbol{p})}{\partial t} = \frac{A(\boldsymbol{p})(\partial G(\lambda, t)/\partial t)}{A(\boldsymbol{p}) + B(\boldsymbol{p}) + A(\boldsymbol{p})B(\boldsymbol{p})(\partial G(\lambda, t)/\partial \lambda)}.$$

From eq. (2.71) and the above relations, we have

$$\frac{\partial T(\boldsymbol{p})}{\partial t} = \frac{1}{\Phi}\Big\{\sum_{i\in S}\phi_i\frac{\partial R(\boldsymbol{p})}{\partial t} + \sum_{R_a \cup R_d}\phi_i\frac{\partial \hat{R}(\boldsymbol{p})}{\partial t}\Big\}$$
$$= \frac{1}{\Phi}\frac{\sum_{R_a \cup R_d}\phi_i A(\boldsymbol{p}) - \sum_{i\in S}\phi_i B(\boldsymbol{p})}{A(\boldsymbol{p}) + B(\boldsymbol{p}) + A(\boldsymbol{p})B(\boldsymbol{p})(\partial G(\lambda, t)/\partial \lambda)}\frac{\partial G(\lambda, t)}{\partial t}.$$

Note that if all sinks are congested, $A(\boldsymbol{p}) = \sum_{i\in S}(\partial E_i(R, u_i)/\partial R)$ will be small; that is $A(\boldsymbol{p}) \approx 0$. Thus, for example, if all sinks are nearly saturated or the arrival rates $\sum_{i\in S}\phi_i$ at sinks are high, an *anomalous* behavior of the equilibrium such that the overall mean job response time decreases even though the communication time increases can be observed.

Similarly from the theorems 2.5, and eq. (2.71), we have for $i \in S$

$$\frac{\partial T(\boldsymbol{p})}{\partial \phi_i} = \frac{1}{\Phi^2} \frac{1}{A(\boldsymbol{p}) + B(\boldsymbol{p}) + A(\boldsymbol{p})B(\boldsymbol{p})(\partial G/\partial \lambda)} \times$$

$$\left\{ \Phi \left(\sum_{i \in S} \phi_i + \sum_{R_a \cup R_d} \phi_i + \sum_{i \in S} \phi_i B(\boldsymbol{p}) \frac{\partial G}{\partial \lambda} \right) \right.$$

$$- \left[\sum_{R_a \cup R_d} \phi_i G + \sum_{i \in N} \phi_i (F_i(\phi_i) - R(\boldsymbol{p})) \right]$$

$$\times \left. [A(\boldsymbol{p}) + B(\boldsymbol{p}) + A(\boldsymbol{p})B(\boldsymbol{p})(\partial G/\partial \lambda)] \right\}.$$

Although the above relation is quite complex, we note that it can be negative. Thus, in such cases, the overall mean job response time of the equilibrium may have the chances to decrease as the job arrival rate at a sink increases.

The Overall Optimal Policy

From eq. (2.1) and relations (2.8) and (2.10), we have

$$\frac{\partial T(\boldsymbol{p})}{\partial \phi_i} = \frac{1}{\Phi^2} \left[\Phi \left(\sum_{j=1}^{n} \frac{\partial(\beta_j F_j)}{\partial \beta_j} \frac{\partial \beta_j}{\partial \phi_i} + \frac{\partial(\lambda G(\lambda))}{\partial \lambda} \frac{\partial \lambda}{\partial \phi_i} \right) - \Phi T(\boldsymbol{p}) \right]$$

$$= \frac{1}{\Phi^2} \left[\Phi \phi_i \frac{\partial F_i(\beta_i, u_i)}{\partial \beta_i} \bigg|_{\beta_i = \phi_i} + \sum_{j=1}^{n} \beta_j \left(F_i(\beta_i, u_i) - F_j(\beta_j, u_j) \right) \right.$$

$$\left. - \lambda G(\lambda, t) \right] \qquad\qquad i \in N$$

$$\frac{1}{\Phi^2} \left(\sum_{j=1}^{n} \beta_j (\alpha + g - F_j(\beta_j, u_j)) - \lambda G(\lambda, t) \right) \qquad i \in R_a \cup R_d$$

$$\frac{1}{\Phi^2} \left(\sum_{j=1}^{n} \beta_j (\alpha - F_j(\beta_j, u_j)) - \lambda G(\lambda, t) \right) \qquad i \in S$$

Let us see the case $i \in S$ as an example. When the communication time is long, and sinks are lightly loaded, that is, the communication delay is large and the incremental node delay at a sink is small, we may have $\alpha < F_j(\beta_j, u_j)$ and $G(\lambda) \gg 0$ and hence $\partial T/\partial \phi_i < 0$. In such cases, the overall mean job response time has the chances to decrease as the job arrival rate at a sink increases.

2.2.5   Numerical Examination

The effects of the system parameters in several examples of a distributed computer system that consists of four host computers (nodes) connected via a single channel communications network were examined numerically. The nodes are modeled as central server models. Let us present one typical example among them.

The expected node delays of the nodes are given as

$$F_1(\beta_1) = \frac{4}{150 - \beta_1}, \quad F_2(\beta_2) = \frac{3}{10 - \beta_2}, \quad F_3(\beta_3) = \frac{4}{10 - \beta_3}, \quad F_4(\beta_4) = \frac{3}{12 - \beta_4}.$$

The single channel communications network is considered as an M/M/1 queueing system. The expected communication delay is given by

$$G(\lambda) = \frac{t}{1 - \lambda t},$$

where t is the mean communication time for sending and receiving a job.

The algorithm used in our numerical calculation program is developed by Kim and Kameda [KK92a]. It is observed (results not presented here) that, in most cases, the results of the numerical examination agree with our intuition and that the overall mean job response time of the equilibrium is close to that of the optimum. Some results of the numerical examination are given in Figs. 2.2, 2.3, 2.4, and 2.5. The overall mean job response times $(T(\boldsymbol{\beta}))$ of the optimum and of the equilibrium are denoted by '*OOP*' with solid line and '*IOP*' with dotted line, respectively.

Figs. 2.2 and 2.4 show how the overall mean job response times $(T(\boldsymbol{\beta}))$ of the optimum and of the equilibrium vary as the communication time (t) changes. The values of ϕ_2, ϕ_3, and ϕ_4 are fixed to be 7, 7, and 7.5 (jobs/sec), respectively. In Fig. 2.2, ϕ_1 equals 80 (jobs/sec). From this figure, we can observe that the overall mean job response time of the equilibrium is close to that of the optimum. The rightmost end of each curve shows the case where all nodes are neutral; that is, the case of no load balancing. In this figure, we see how static load balancing improves the mean response time in an example.

In Fig. 2.4, ϕ_1 equals 140 (jobs/sec). We can observe such an *anomalous* behavior that the overall mean job response time of the equilibrium decreases even though the communication time increases up to a certain value. This is what we noted in subsection 2.2.4. Furthermore, we can even find such a seemingly extraordinary case that the overall mean job response time of the equilibrium is minimum only when all nodes are neutral (no load balancing).

In Fig. 2.3, the parameter setting is the same as that in Fig. 2.2. This figure shows the expected response time of two groups of jobs arriving at node 3 under the two policies. The two groups of jobs that arrive at node 3 and that processed at different nodes (nodes 1 and 3) have distinct expected job response times under the overall optimal policy. This seems unfair. On the other hand, under the individually optimal policy, the two groups of jobs that arrive at node 3 and that are processed at different nodes (nodes 1 and 3) have the same expected response times. This also seems fair.

Fig. 2.5 shows how the overall mean job response times $(T(\boldsymbol{\beta}))$ of the optimum and of the equilibrium vary as the job arrival rate of node 1 (ϕ_1) is changed from 0 to 146

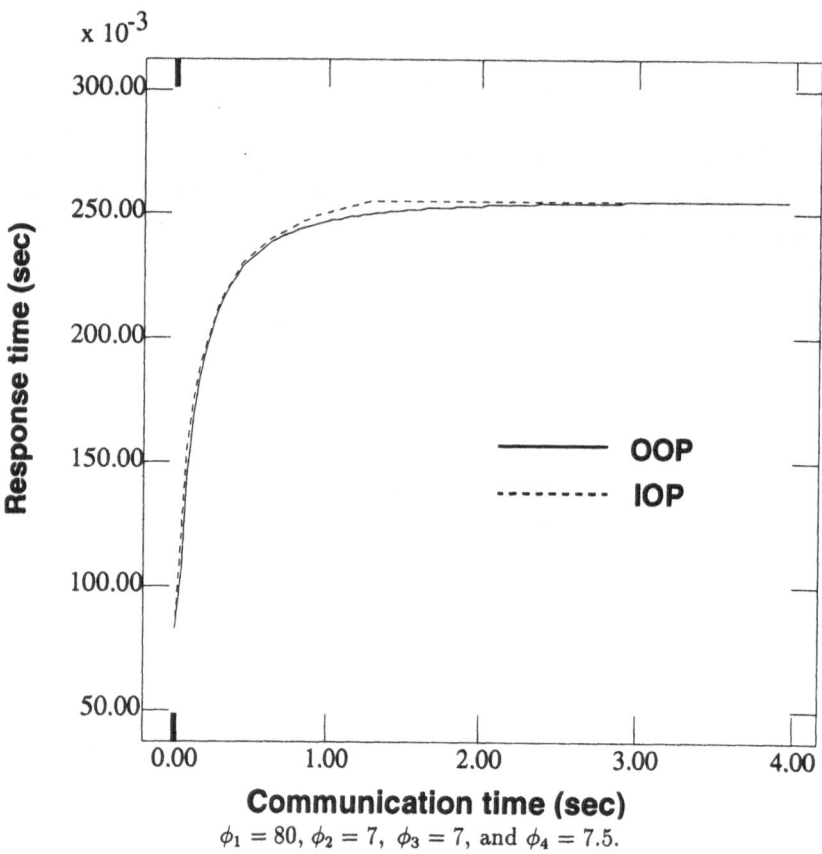

$\phi_1 = 80,\ \phi_2 = 7,\ \phi_3 = 7,\ \text{and}\ \phi_4 = 7.5.$

Figure 2.2: Effect of the communication time on the overall mean job response times of the optimum and of the equilibrium.

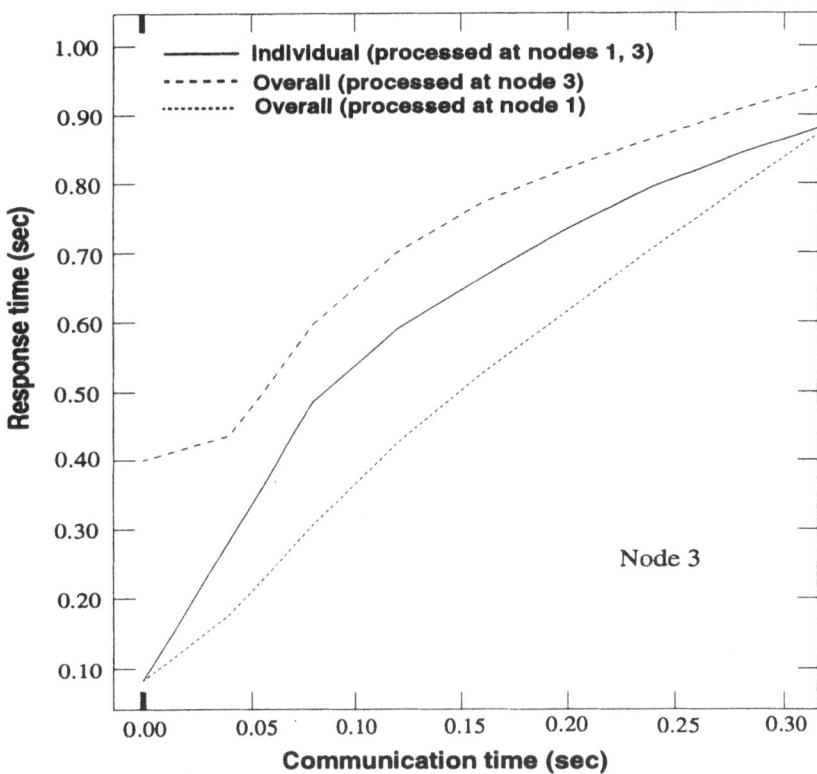

Figure 2.3: Unfairness in the optimum and fairness in the equilibrium for the jobs arriving at node 3.

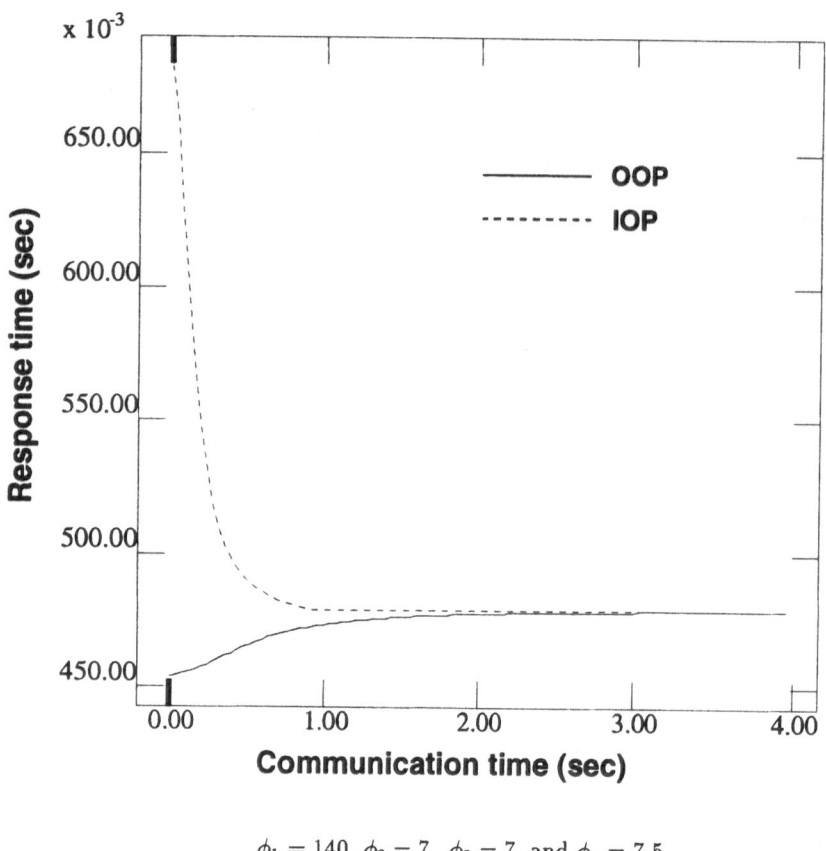

$\phi_1 = 140,\ \phi_2 = 7,\ \phi_3 = 7,$ and $\phi_4 = 7.5.$

Figure 2.4: Effect of the communication time on the overall mean job response times of the optimum and of the equilibrium.

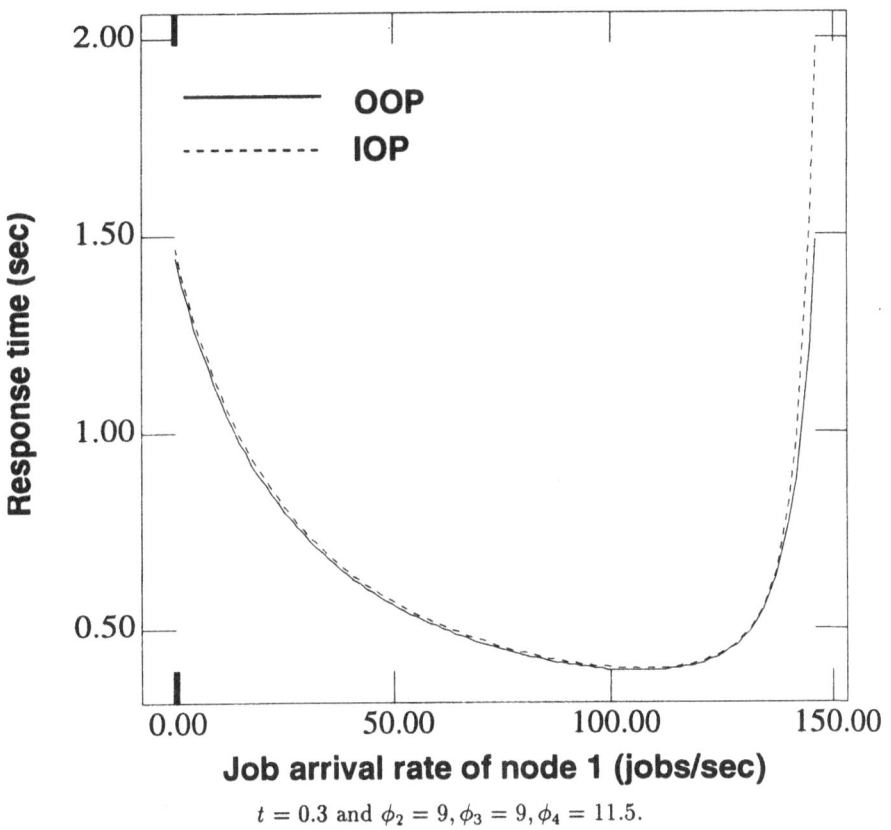

Figure 2.5: Effect of the job arrival rate of node 1 on the overall mean job response times of the optimum and of the equilibrium.

(jobs/sec). The values of ϕ_2, ϕ_3, ϕ_4 and t are fixed as 9, 9, 11.5 (jobs/sec) and 0.3 (sec), respectively. In Fig. 2.5, we can observe that the overall mean job response time of the equilibrium is close to that of the optimum. Under both policies, however, we can observe such anomalous behavior that overall mean job response times of the optimum and of the equilibrium decrease even though the job arrival rate of node 1 increases up to a certain large value. These are what we noted in subsection 2.2.4.

2.2.6   Conclusion

We studied two contrastive policies, the overall and individually optimal policies, for statistically balancing the load on a set of heterogeneous host computers connected by a single channel communications network. The conditions that the optimal solution of the individually optimal policy satisfies are newly derived. We found that the forms of the optimal solutions of the two policies are similar to each other.

We studied the effects of changing various system parameters including the communication time, the job arrival rate, and processing capacity on the behavior of the optimum and of the equilibrium. We found the behaviors of the optimum and of the equilibrium are similar to each other. We examined their behaviors numerically. We found the overall mean job response time of the equilibrium is close to that of the optimum. We examined two groups of jobs that arrive at the same node and that processed by different nodes under the overall and individually optimal policies. We found that, in the optimum, the two groups of jobs may show mutually different expected response times, and that, in the equilibrium, the two groups of jobs both show the same expected response times.

We observed, however, several anomalous phenomena that there are cases where in the equilibrium, the overall mean job response time decreases even though the communication time increases. We also observed that there are cases where in the optimum and in the equilibrium, the overall mean job response time decreases even though the job arrival rates increase.

2.3   Static Load Balancing in Star Network Configurations

2.3.1   Model Description

The system consists of n heterogeneous satellites connected to the central site through the communication links (as shown in Fig. 2.6). Satellites and the central site can contain a number of resources such as processing devices and input/output devices.

Jobs arrive at satellite i, $i = 1, 2, \cdots, n$, according to a time-invariant Poisson process.

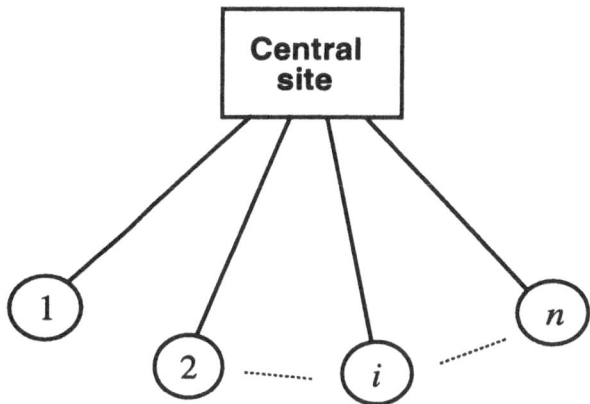

Figure 2.6: A star network model.

After arrival to satellite i, a job may be either processed locally or sent to the central site where it is processed remotely. We assume that the decision of processing a job locally or remotely does not depend on the system state, and hence is *static*. We also assume that once a job is sent to the central site, it will be processed at the central site and it is not transferable to other satellites.

The notation and the assumptions of the model are shown as follows.

n Number of satellite nodes.

ϕ_i External job arrival rate of node i.

Φ Total job arrival rate of all nodes, that is, $\Phi = \sum_{i=1}^{n} \phi_i$.

β_i Job processing rate (load) at node i.

β $[\beta_0, \beta_1, \ldots, \beta_n]$.

λ_i Traffic rate over communication link i, that is, $\lambda_i = \phi_i - \beta_i$.

$F_i(\beta_i)$ Mean (expected) node delay of jobs processed at node i, that is, the mean length of the time period that starts when a job joins the queue of node i and ends when it goes out of node i (we assume that it is differentiable, increasing, and convex with respect to β_i).

$G_i(\lambda_i)$ Mean (expected) communication delay (waiting time for transmission is included) between satellite i and the central site (we assume that it is differentiable, nondecreasing, and convex with respect to λ_i).

$T_i(\beta_i)$ Mean response time of jobs arriving at node i, that is, the mean length of the
time period that starts when a job comes to node i and ends when it goes out of
the system.

$T(\beta)$ Overall mean job response time, that is, the mean length of the time period that
starts when a job arrives in the system and ends when it leaves the system.

The mean response time of jobs arriving at node, $T_i(\beta_i)$, is given as follows.

$$T_i(\beta_i) = \frac{1}{\phi_i}[\beta_i F_i(\beta_i) + \lambda_i(F_0(\beta_0) + G_i(\lambda_i))], i = 1, 2, \ldots, n. \qquad (2.82)$$

For the central site, we have $T_0(\beta_0) = F_0(\beta_0)$.

Therefore, the overall mean job response time, $T(\beta)$, can be written as

$$T(\beta) = \frac{1}{\Phi} \sum_{i=0}^{n} \phi_i T_i(\beta_i) = \frac{1}{\Phi}[\sum_{i=0}^{n} \beta_i F_i(\beta_i) + \sum_{i=1}^{n} \lambda_i G_i(\lambda_i)], \qquad (2.83)$$

subject to

$$\sum_{i=0}^{n} \beta_i = \Phi, \qquad (2.84)$$

$$\beta_i + \lambda_i = \phi_i, i = 1, 2, \ldots, n, \qquad (2.85)$$

$$\beta_i \leq \phi_i, i = 1, 2, \ldots, n, \qquad (2.86)$$

$$\beta_0 \geq \phi_0, \qquad (2.87)$$

$$\beta_i \geq 0, i = 1, \ldots, n. \qquad (2.88)$$

Tantawi and Towsley [TT84] classified satellites in the following ways: *idle sources*,
active sources, and *neutrals* which are denoted by R_d, R_a and N, respectively. An idle
source does not process any jobs; that is, $\beta_i = 0$. An active source processes some jobs
and sends the remaining jobs to the central site; that is, $0 < \beta_i < \phi_i$. A neutral processes
jobs locally without sending any jobs to the central site; that is, $\beta_i = \phi_i$.

2.3.2   Optimal Solutions

The Overall Optimal Policy

By the overall optimal policy, we mean the policy whereby job scheduling is determined
so as to minimize the overall mean job response time. From eq. (2.83) and relations (2.84)
- (2.88), we see that the problem of minimizing the overall mean job response time is to
solve the eq. (2.83) characterized by the load vector β.

First we define the *incremental node delay* and the *incremental communication delay* as follows.

$$f_i(\beta_i) = \frac{d}{d\beta_i} \beta_i F_i(\beta_i), \tag{2.89}$$

$$g_i(\lambda_i) = \frac{d}{d\lambda_i} \lambda_i G_i(\lambda_i). \tag{2.90}$$

The difference between the incremental node delay and the incremental communication delay is denoted by $h_i(\beta_i)$,

$$h_i(\beta_i) = f_i(\beta_i) - g_i(\lambda_i), \quad i = 1, 2, \ldots, n. \tag{2.91}$$

The inverse of the incremental node delay is defined by

$$f_i^{-1}(x) = \begin{cases} a, & f_i(a) = x, \\ 0, & f_i(0) \geq x. \end{cases}$$

Tantawi and Towsley [TT84] solved problem (2.83) and derived the following theorem by using the Kuhn-Tucker theorem (See Theorem 1 of [TT84]).

Theorem 2.12 *The optimal solution, β, to problem (2.83) satisfies the following relations,*

$$h_i(\beta_i) \geq \alpha, \ \beta_i = 0 \qquad (i \in R_d), \tag{2.92}$$

$$h_i(\beta_i) = \alpha, \ 0 < \beta_i < \phi_i \ (i \in R_a), \tag{2.93}$$

$$h_i(\beta_i) \leq \alpha, \ \beta_i = \phi_i \qquad (i \in N), \tag{2.94}$$

$$\beta_0 = f_0^{-1}(\alpha), \tag{2.95}$$

subject to the total flow constraint,

$$f_0^{-1}(\alpha) + \sum_{i \in R_a} h_i^{-1}(\alpha) + \sum_{i \in N} \phi_i = \Phi. \tag{2.96}$$

where α is the Lagrange multiplier.

The Individually Optimal Policy

The job scheduling decision by the individually optimal policy is such that each individual job is scheduled so as to minimize its expected response time, given the mean node delay at each node and the expected communication delay of each link. When the individually optimal policy is realized, each job could expect no improvement in its expected response time even if it were processed by a node different from the processing node that had been determined by the policy, and the system reaches an equilibrium. The equilibrium conditions can be defined by the following definition.

Definition: β *is said to satisfy the equilibrium conditions for the individually optimal policy if the following relations hold:*

$$H_i(\beta_i) \geq F_0(\beta_0), \ \beta_i = 0 \qquad (i \in R_d), \tag{2.97}$$

$$H_i(\beta_i) = F_0(\beta_0), \ 0 < \beta_i < \phi_i \ (i \in R_a), \tag{2.98}$$

$$H_i(\beta_i) \leq F_0(\beta_0), \ \beta_i = \phi_i \qquad (i \in N), \tag{2.99}$$

subject to the flow constraint,

$$\sum_{i=0}^{n} \beta_i = \Phi. \tag{2.100}$$

H_i, F_i^{-1} are defined as follows.

$$H_i(\beta_i) = F_i(\beta_i) - G_i(\lambda_i), \ i = 1, 2, \ldots, n. \tag{2.101}$$

$$F_i^{-1}(X) = \begin{cases} A, & F_i(A) = X, \\ 0, & F_i(0) \geq X. \end{cases} \tag{2.102}$$

Lemma 2.13 *The following relations are equivalent to the equilibrium conditions (2.97) - (2.99).*

$$H_i(\beta_i) \geq R, \ \beta_i = 0 \qquad (i \in R_d), \tag{2.103}$$

$$H_i(\beta_i) = R, \ 0 < \beta_i < \phi_i \ (i \in R_a), \tag{2.104}$$

$$H_i(\beta_i) \leq R, \ \beta_i = \phi_i \qquad (i \in N), \tag{2.105}$$

$$\beta_0 = F_0^{-1}(R), \tag{2.106}$$

subject to the flow constraint,

$$F_0^{-1}(R) + \sum_{i \in R_a} H_i^{-1}(R) + \sum_{i \in N} \phi_i = \Phi. \tag{2.107}$$

PROOF. Eq. (2.106) can be written as

$$R = F_0(\beta_0) \qquad \beta_0 > 0, \tag{2.108}$$

$$R \leq F_0(\beta_0) \qquad \beta_0 = 0. \tag{2.109}$$

The sufficiency. For $\beta_0 > 0$, we easily derive relations (2.97) - (2.99) from relations (2.103) - (2.106). For $\beta_0 = 0$, since $\beta_0 = \sum_{i=1}^{n}(\phi_i - \beta_i)$, we have $\beta_i = \phi_i$ for all i. From relations (2.105) and (2.109), we have

$$H_i(\beta_i) \leq R \leq F_0(\beta_0).$$

The necessity. For $\beta_0 > 0$, by letting $R = F_0(\beta_0)$, we easily derive relations (2.103) - (2.105), and (2.108) from relations (2.97) - (2.99). For $\beta_0 = 0$, since $\beta_0 = \sum_{i=1}^{n}(\phi_i - \beta_i)$, we have $\beta_i = \phi_i$ for all i. For relation (2.99), we can find such an R that

$$H_i(\beta_i) \le R \le F_0(\beta_0).$$

Thus we have relations (2.105) and (2.109).

It is easy to see that (2.107) is equivalent to eq. (2.100). □

A load pattern $\boldsymbol{\beta}$, which satisfies the above equilibrium conditions, is referred to as the optimal solution of the individually optimal policy or the *equilibrium*. For the individually optimal policy, the following theorem is derived.

Theorem 2.14 *The individually optimal policy has a solution. That is, there exists one and only one $\boldsymbol{\beta}$ that satisfy the equilibrium conditions (2.103) - (2.106).*

PROOF. The case of $\phi_0 = 0$ is shown. It is easy to show the case of $\phi_0 > 0$ by using $F_0(\beta_0 + \phi_0)$ instead of $F_0(\beta_0)$. Let us consider such an equivalent overall optimization problem with respect to the individually optimization problem as follows.

$$T^*(\boldsymbol{\beta}) = \frac{1}{\Phi}\left[\sum_{i=0}^{n} \beta_i F_i^*(\beta_i) + \sum_{i=1}^{n} \lambda_i G_i^*(\lambda_i)\right]. \qquad (2.110)$$

subject to

$$\sum_{i=0}^{n} \beta_i = \Phi, \qquad (2.111)$$

$$\beta_i + \lambda_i = \phi_i, \quad i = 1, 2, \ldots, n, \qquad (2.112)$$

$$\beta_i \le \phi_i, \quad i = 1, 2, \ldots, n, \qquad (2.113)$$

$$\beta_i \ge 0, \quad i = 0, 1, \ldots, n. \qquad (2.114)$$

From constraints (2.111) - (2.114), we see that they form a convex set and are nonempty. $F_i^*(\beta_i)$ and $G_i^*(\lambda_i)$ are the expected node and communication delays of the equivalent overall optimization problem and are defined as follows.

$$F_i^*(\beta_i) = \frac{1}{\beta_i} \int_0^{\beta_i} F_i(\beta_i)\, d\beta_i, \quad F_i^*(0) = F_i(0),$$

$$G_i^*(\lambda_i) = \frac{1}{\lambda_i} \int_0^{\lambda_i} G_i(\lambda_i)\, d\lambda_i, \quad G_i^*(0) = G_i(0).$$

From the above definitions, we see that $\dfrac{dF_i^*}{d\beta_i}$, $\dfrac{d^2F_i^*}{d\beta_i^2} > 0$ and $\dfrac{dG_i^*}{d\lambda_i}$, $\dfrac{d^2G_i^*}{d\lambda_i^2} \ge 0$ for $\beta_i \ge 0$, $\lambda_i \ge 0$. Thus, $T^*(\boldsymbol{\beta})$ is a strictly convex function. Therefore, problem (2.110) has one and only one optimal solution.

By using nonlinear programming techniques, we have the following Lagrangian function.

$$L^*(\boldsymbol{\beta}, R, \boldsymbol{\gamma}, \boldsymbol{\psi}) = \Phi T^*(\boldsymbol{\beta}) + R(\Phi - \sum_{i=0}^{n} \beta_i) + \sum_{i=1}^{n} \gamma_i(\phi_i - \beta_i) + \sum_{i=0}^{n} \psi_i \beta_i. \qquad (2.115)$$

According to the Kuhn-Tucker conditions, the following relations are obtained.

$$\frac{\partial L^*}{\partial \beta_i} = H_i(\beta_i) - R - \gamma_i + \psi_i = 0, \; i = 1, 2, \ldots, n, \qquad (2.116)$$

$$\frac{\partial L^*}{\partial \beta_0} = F_0(\beta_0) - R + \psi_0 = 0, \qquad (2.117)$$

$$\frac{\partial L^*}{\partial R} = \Phi - \sum_{i=0}^{n} \beta_i, \qquad (2.118)$$

$$\phi_i - \beta_i \geq 0, \; \gamma_i(\phi_i - \beta_i) = 0, \; \gamma_i \leq 0, \; i = 1, 2, \ldots, n, \qquad (2.119)$$

$$\beta_i \geq 0, \; \psi_i \beta_i = 0, \; \psi_i \leq 0, \; i = 0, 1, \ldots, n. \qquad (2.120)$$

where $H_i(\beta_i)$ is defined by eq. (2.101).

From conditions (2.119), we note that $(\phi_i - \beta_i)$ may be either zero or positive. We consider each case seperately.

Case 1. $\phi_i = \beta_i$.

From conditions (2.119) and (2.120), we have $\gamma_i \leq 0$ and $\psi_i = 0$. Thus eq. (2.116) becomes

$$H_i(\beta_i) \leq R \qquad \beta_i = \phi_i. \qquad (2.121)$$

Case 2. $\phi_i > \beta_i$.

From conditions (2.119), we have $\gamma_i = 0$.

Case 2.1. $\beta_i > 0$.

From conditions (2.120), we have $\psi_i = 0$. Thus eq. (2.116) yields

$$H_i(\beta_i) = R \qquad 0 < \beta_i < \phi_i. \qquad (2.122)$$

Case 2.2. $\beta_i = 0$.

It follows from conditions (2.120) $\psi_i \leq 0$. Thus from eq. (2.116), we have

$$H_i(\beta_i) \geq R \qquad \beta_i = 0. \qquad (2.123)$$

From eq. (2.117) and conditions (2.120), we have

$$\begin{aligned} R &= F_0(\beta_0) & \beta_0 > 0, \\ R &\leq F_0(\beta_0) & \beta_0 = 0. \end{aligned} \qquad (2.124)$$

By substituting eqs. (2.122) and (2.124) into eq. (2.118), we have

$$F_0^{-1}(R) + \sum_{i \in R_a} H_i^{-1}(R) + \sum_{i \in N} \phi_i = \Phi. \tag{2.125}$$

The above relations can be written in the following way.

$$H_i(\beta_i) \geq R, \ \beta_i = 0 \qquad (i \in R_d), \tag{2.126}$$

$$H_i(\beta_i) = R, \ 0 < \beta_i < \phi_i \ (i \in R_a), \tag{2.127}$$

$$H_i(\beta_i) \leq R, \ \beta_i = \phi_i \qquad (i \in N), \tag{2.128}$$

$$\beta_0 = F_0^{-1}(R). \tag{2.129}$$

subject to the flow constraint,

$$F_0^{-1}(R) + \sum_{i \in R_a} H_i^{-1}(R) + \sum_{i \in N} \phi_i = \Phi. \tag{2.130}$$

By noting the equilibrium conditions (2.103) - (2.106), we see that relations (2.126) - (2.129) are equivalent to the equilibrium conditions of the individually optimal policy. That is, the equilibrium conditions of the individually optimal policy are equivalent to the solution of problem (2.110). Therefore, there exists one and only one β that satisfies the equilibrium conditions of the individually optimal policy. $\square$

Remark. By noting Theorems 2.12 and 2.14, we see that between the forms of the optimal solutions of the overall and individually optimal policies there exists a striking parallelism, even though their objectives are completely different from each other. The parallelism between the two policies gives us an intuitive explanation of one in terms of the other. That is, the overall optimal policy would be realized by an individually optimal policy, if the values of the incremental node and communication delays were given as the expected node and communication delays, and vice versa. Therefore, the individually optimal policy can be implemented by using the algorithm that is extended from that of Tantawi and Towsley [TT84] where $h_i(\beta_i)$ and α are replaced with $H_i(\beta_i)$ and R, respectively.

2.3.3   Parametric Analysis

In the previous subsection, the conditions that the optimal solutions of the overall and individually optimal policies satisfy are shown. In this subsection, the effects of varying three main system parameters on the performance variables are studied. We consider following system parameters: the node i processing time u_i $(i = 0, 1, \ldots, n)$, the node i job arrival rate ϕ_i $(i = 0, 1, \ldots, n)$, the link i communication time t_i $(i = 1, 2, \ldots, n)$. A parameter vector $\mathbf{p}, \mathbf{p} = [t_1, t_2, \ldots, t_n, u_0, u_1, \ldots, u_n, \phi_0, \phi_1, \ldots, \phi_n]$ is used to denote the system parameters. $F_i(\beta_i(\boldsymbol{p}), u_i)$ denotes the mean node delay of node i. We assume that $F_i(\beta_i(\boldsymbol{p}), u_i)$ is an increasing function of u_i. We assume $\beta_0(\boldsymbol{p}) > 0$. The mean node delay of the central site is denoted by $R(\boldsymbol{p})$. $G_i(\lambda_i(\boldsymbol{p}), t_i)$ denotes the mean communication delay of link i. We assume that $G_i(\lambda_i(\boldsymbol{p}), t_i)$ is an increasing funciton of t_i.

The Individually Optimal Policy

When the individually optimal policy is realized, two groups of jobs that arrive at the same active source and that processed by different nodes (the central site and the local node) have the same expected response time, and therefore the expected response times of the two groups of jobs can both be written as $R(\boldsymbol{p}) + G_j(\lambda_j(\boldsymbol{p}))$. Since all of the jobs arriving at idle sources are processed at the central site, the expected response time of jobs arriving at the idle sources can also be written as $R(\boldsymbol{p}) + G_j(\lambda_j(\boldsymbol{p}))$. Therefore, the overall mean job response time can be expressed as:

$$T(\boldsymbol{p}) = \frac{1}{\Phi}\Big[\sum_{j \in R_a \cup R_d} \phi_j(R(\boldsymbol{p}) + G_j(\lambda_j(\boldsymbol{p}), t_i)) + \sum_{j \in N} \phi_j F_j(\phi_i) + \phi_0 R(\boldsymbol{p}) \Big]. \qquad (2.131)$$

For the individually optimal policy, the following relations are derived.

Lemma 2.15   *The following relations hold for the expected node delay of the central site, $R(\boldsymbol{p})$:*

$$\frac{\partial R(\boldsymbol{p})}{\partial t_i} \quad < \quad 0, \quad i \in R_a,$$
$$= \quad 0, \quad i \in N \cup R_d. \qquad (2.132)$$
$$\frac{\partial R(\boldsymbol{p})}{\partial u_i} \quad > \quad 0, \quad i = 0,\ i \in R_a,$$
$$= \quad 0, \quad i \in N \cup R_d. \qquad (2.133)$$
$$\frac{\partial R(\boldsymbol{p})}{\partial \phi_i} \quad > \quad 0, \quad i = 0,\ i \in R_a \cup R_d,$$
$$= \quad 0, \quad i \in N. \qquad (2.134)$$

PROOF. A proof of Lemma 2.15.

By differentiating the both sides of relation $R(\boldsymbol{p}) = F_0'(\beta_0(\boldsymbol{p}), u_0)$ with respect to t_i, we have

$$\frac{\partial R(\boldsymbol{p})}{\partial t_i} \quad = \quad \frac{\partial F_0}{\partial \beta_0}\frac{\partial \beta_0(\boldsymbol{p})}{\partial t_i}. \qquad (2.135)$$

From relations (2.101) and (2.104), we have

$$\frac{\partial R(\boldsymbol{p})}{\partial t_i} = \frac{\partial F_j}{\partial \beta_j}\frac{\partial \beta_j(\boldsymbol{p})}{\partial t_i} - \frac{\partial G_j}{\partial \lambda_j}\frac{\partial \lambda_j(\boldsymbol{p})}{\partial t_i} - \frac{\partial G_j}{\partial t_i}\delta_{ij}. \qquad (2.136)$$

where $\delta_{ij} = 1$, if $i = j$, otherwise $\delta_{ij} = 0$.

From relations (2.84) and (2.85), we have $\beta_0(\boldsymbol{p}) = \sum_{j \in R_a} [\phi_j - \beta_j(\boldsymbol{p})] + \sum_{j \in R_d} \phi_j + \phi_0$ and
$\lambda_j(\boldsymbol{p}) = \phi_j - \beta_j(\boldsymbol{p})$. Therefore, we have

$$\frac{\partial \beta_0(\boldsymbol{p})}{\partial t_i} = -\sum_{j \in R_a} \frac{\partial \beta_j(\boldsymbol{p})}{\partial t_i}. \tag{2.137}$$

$$\frac{\partial \lambda_j(\boldsymbol{p})}{\partial t_i} = -\frac{\partial \beta_j(\boldsymbol{p})}{\partial t_i}. \tag{2.138}$$

By subtituting eq. (2.138) into eq. (2.136), we have

$$\frac{\partial \beta_j(\boldsymbol{p})}{\partial t_i} = \left(\frac{\partial R(\boldsymbol{p})}{\partial t_i} + \frac{\partial G_j}{\partial t_i} \delta_{ij} \right) \Big/ A_j. \tag{2.139}$$

where $A_j = \dfrac{\partial F_j}{\partial \beta_j} + \dfrac{\partial G_j}{\partial \lambda_j}$.

By substituting eq. (2.139) into eq. (2.137), and by substituting eq. (2.137) into eq. (2.135), we have

$$\frac{\partial R(\boldsymbol{p})}{\partial t_i} = -\frac{\frac{\partial F_0}{\partial \beta_0} \frac{\partial G_i}{\partial t_i}}{A_i \left(1 + \frac{\partial F_0}{\partial \beta_0} \sum_{j \in R_a} \frac{1}{A_j} \right)}, \tag{2.140}$$

Because $F_i(\beta_i(\boldsymbol{p}), u_i)$ is an increasing and convex function of $\beta_i(\boldsymbol{p})$, $G_i(\lambda_i(\boldsymbol{p}), t_i)$ is a nondecreasing and convex function of $\lambda_i(\boldsymbol{p})$ and is an increasing function of t_i, it follows that $\partial R(\boldsymbol{p})/\partial t_i$ is negative.

It is easy to see that $\partial R(\boldsymbol{p})/\partial t_i = 0$ for $i \in R_d \cup N$.

Relations (2.133) and (2.134) can be derived similarly as above. $\square$

Theorem 2.16 *The following relations hold for the optimal load of active source j, $\beta_j(\boldsymbol{p})$:*

$$\frac{\partial \beta_j(\boldsymbol{p})}{\partial t_i} \begin{array}{l} > \; 0, \; i \in R_a, \; j = i, \\ < \; 0, \; i \in R_a, \; j \neq i, \\ = \; 0, \; i \in N \cup R_d. \end{array} \tag{2.141}$$

$$\frac{\partial \beta_j(\boldsymbol{p})}{\partial u_i} \begin{array}{l} < \; 0, \; i \in R_a, \; j = i, \\ > \; 0, \; i = 0, \; i \in R_a, \; j \neq i, \\ = \; 0, \; i \in N \cup R_d. \end{array} \tag{2.142}$$

$$\frac{\partial \beta_j(\boldsymbol{p})}{\partial \phi_i} \begin{array}{l} > \; 0, \; i = 0, \; i \in R_a \cup R_d, \\ = \; 0, \; i \in N. \end{array} \tag{2.143}$$

PROOF. A proof of Theorem 2.16.

It is easy to see that the optimal loads of both neutrals and idle sources do not depend on the changes of the system parameters t_i, u_i and ϕ_i. Here we only show the behavior of the optimal load at active sources.

By substituting relation (2.140) into relation (2.139), we have

$$\frac{\partial \beta_j(\boldsymbol{p})}{\partial t_i} = \begin{cases} \dfrac{\frac{\partial G_i}{\partial t_i}\left[1+\frac{\partial F_0}{\partial \beta_0}\sum_{\substack{j\in R_a \\ j\neq i}}\frac{1}{A_j}\right]}{\left[1+\frac{\partial F_0}{\partial \beta_0}\sum_{j\in R_a}\frac{1}{A_j}\right]A_i} & i \in R_a,\ j = i, \\[4ex] -\dfrac{\frac{\partial F_0}{\partial \beta_0}\frac{\partial G_i}{\partial t_i}}{A_i A_j\left(1+\frac{\partial F_0}{\partial \beta_0}\sum_{j\in R_a}\frac{1}{A_j}\right)} & i \in R_a,\ j \neq i. \end{cases}$$

where A_j is defined by (2.139).

Because $F_0(\beta_0(\boldsymbol{p}), u_0)$ is an increasing and convex function of $\beta_0(\boldsymbol{p})$, and because $G_i(\lambda_i(\boldsymbol{p}), t_i)$ is a nondecreasing and convex function of $\lambda_i(\boldsymbol{p})$ and is an increasing function of t_i, we have

$$\frac{\partial \beta_j(\boldsymbol{p})}{\partial t_i} \ > \ 0 \quad i \in R_a,\ j = i,$$
$$< \ 0 \quad i \in R_a,\ j \neq i.$$

Relations (2.142) and (2.143) can be derived similarly as above. □

Remark. This theorem can be interpreted as follows. When the communication time of link i increases, the optimal load of active source i increases whereas the optimal load of active source j ($j \neq i$) decreases. When the node processing time of active source i increases, the optimal load of node i decreases whereas the optimal load of active source $j(j \neq i)$ increases. When the node processing time of the central site increases the optimal load of all the active sources increases. The optimal load of an active source increases when the job arrival rate of the central site, an active source or an idle source increases. These agree with our intuition.

Theorem 2.17   *The following relations hold for the overall mean job response time in the equilibrium, $T(\boldsymbol{p})$.*

$$\frac{\partial T(\boldsymbol{p})}{\partial t_i} \ > \ 0, \quad i \in R_d,$$
$$= \ 0, \quad i \in N, \qquad\qquad (2.144)$$
$$\frac{\partial T(\boldsymbol{p})}{\partial u_i} \ > \ 0, \quad i = 0,\ i \in R_a \cup N,$$
$$= \ 0, \quad i \in R_d. \qquad\qquad (2.145)$$

PROOF. A proof of Theorem 2.17.

By differentiating both sides of relation (2.131), it follows that

$$\frac{\partial T(p)}{\partial t_i} = \frac{1}{\Phi}\Big[\sum_{j \in R_a \cup R_d} \phi_j\Big(\frac{\partial R}{\partial t_i} + \frac{\partial G_j}{\partial \lambda_j}\frac{\partial \lambda_j}{\partial t_i} + \frac{\partial G_j}{\partial t_i}\delta_{ij}\Big) + \sum_{j \in N}\phi_j\frac{\partial F_j}{\partial t_i} + \phi_0\frac{\partial R}{\partial t_i}\Big]. \quad (2.146)$$

From Lemma 2.15 and by noting that $T(p)$ is independent of t_i when $i \in N$, we have

$$\frac{\partial T(p)}{\partial t_i} = \begin{cases} \dfrac{\phi_i}{\Phi}\dfrac{\partial G_i}{\partial t_i} > 0 & i \in R_d, \\ 0 & i \in N. \end{cases} \quad (2.147)$$

From Lemma 2.15, we have

$$\begin{aligned}\frac{\partial T(p)}{\partial u_i} &= \frac{1}{\Phi}\Big[\sum_{j \in R_a \cup R_d}\phi_j\frac{\partial(R+G_j)}{\partial u_i} + \phi_0\frac{\partial R}{\partial u_i}\Big] \\ &> 0, \quad i = 0, i \in R_a \cup N, \\ &= 0, \quad i \in R_d.\,\square \end{aligned}$$

Remark. This theorem implies that the overall mean job response time in the equilibrium will increase with the increase of the communication time of link i which connect to an idle source or with the increase of the node processing time of the central site, active sources or neutrals. These agree with our intuition.

The Overall Optimal Policy

Let $f_i(\beta_i(p), u_i)$ and $g_i(\lambda_i(p), t_i)$ denote the incremental node delay of node i and the incremental communication delay of link i, respectively. We assume that $f_i(\beta_i(p), u_i)$ is an increasing function of u_i, and $g_i(\lambda_i(p), t_i)$ is an increasing function of t_i. We have $\alpha(p) = f_0(\beta_0(p), u_0)$ by noting relation (2.95). For the overall optimal policy, the following relations are derived.

Lemma 2.18 *The following relations hold for the incremental node delay of the central site, $\alpha(p)$.*

$$\begin{aligned}\frac{\partial \alpha(p)}{\partial t_i} &< 0, \quad i \in R_a, \\ &= 0, \quad i \in N \cup R_d. \quad (2.148)\end{aligned}$$

$$\begin{aligned}\frac{\partial \alpha(p)}{\partial u_i} &> 0, \quad i = 0, i \in R_a, \\ &= 0, \quad i \in N \cup R_d. \quad (2.149)\end{aligned}$$

$$\begin{aligned}\frac{\partial \alpha(p)}{\partial \phi_i} &> 0, \quad i = 0, i \in R_a \cup R_d, \\ &= 0, \quad i \in N. \quad (2.150)\end{aligned}$$

PROOF. By using a similar way as Lemma 2.15, we can derive equations on $\partial\alpha(\boldsymbol{p})/\partial t_i$, $\partial\alpha(\boldsymbol{p})/\partial u_i$, and $\partial\alpha(\boldsymbol{p})/\partial\phi_i$. Then the above relations can be derived. $\square$

Theorem 2.19   *The following relations hold for the optimal load of active source j, $\beta_j(\boldsymbol{p})$.*

$$
\begin{aligned}
\frac{\partial\beta_j(\boldsymbol{p})}{\partial t_i} \quad &> \quad 0, \ i \in R_a, \ j = i, \\
&< \quad 0, \ i \in R_a, \ j \neq i, \\
&= \quad 0, \ i \in N \cup R_d. \quad\quad (2.151)
\end{aligned}
$$

$$
\begin{aligned}
\frac{\partial\beta_j(\boldsymbol{p})}{\partial u_i} \quad &< \quad 0, \ \ i \in R_a, \ j = i, \\
&> \quad 0, \ i = 0, \ i \in R_a, \ j \neq i, \\
&= \quad 0, \ \ i \in N \cup R_d. \quad\quad (2.152)
\end{aligned}
$$

$$
\begin{aligned}
\frac{\partial\beta_j(\boldsymbol{p})}{\partial\phi_i} \quad &> \quad 0, \ \ i = 0, \ i \in R_a \cup R_d, \\
&= \quad 0, \ \ i \in N. \quad\quad (2.153)
\end{aligned}
$$

PROOF. These relations can be derived by using a similar way as that in the proof of Theorem 2.16. $\square$

Remark. By comparing Theorem 2.19 to Theorem 2.16, we see that the optimal loads in the optimum and in the equilibrium behave similarly.

Theorem 2.20   *The following relations hold for the overall mean job response time in the optimum, $T(\boldsymbol{p})$*

$$
\begin{aligned}
\frac{\partial T(\boldsymbol{p})}{\partial t_i} \quad &> \quad 0, \ \ i \in R_a \cup R_d, \\
&= \quad 0, \ \ i \in N, \quad\quad (2.154)
\end{aligned}
$$

$$
\begin{aligned}
\frac{\partial T(\boldsymbol{p})}{\partial u_i} \quad &> \quad 0, \ \ i = 0, \ i \in R_a \cup N, \\
&= \quad 0, \ \ i \in R_d. \quad\quad (2.155)
\end{aligned}
$$

PROOF. A proof of Theorem 2.20.

By differentiating the both sides of relation (2.83) and by noting relations (2.91), (2.95) and $\lambda_i = \phi_i - \beta_i$, we have

$$
\begin{aligned}
\frac{\partial T(\boldsymbol{p})}{\partial t_i} &= \frac{1}{\Phi}\Big[\sum_{j=0}^{n}\frac{\partial(\beta_j F_j)}{\partial\beta_j}\frac{\partial\beta_j}{\partial t_i} + \sum_{j=1}^{n}\frac{\partial(\lambda_j G_j)}{\partial\lambda_j}\frac{\partial\lambda_j}{\partial t_i} + \lambda_i\frac{\partial G_i}{\partial t_i}\Big] \\
&= \frac{1}{\Phi}\Big[\alpha\frac{\partial\beta_0}{\partial t_i} + \sum_{j\in R_a}h_j\frac{\partial\beta_j}{\partial t_i} + \sum_{j\in R_d}h_j\frac{\partial\beta_j}{\partial t_i} + \sum_{j\in N}h_j\frac{\partial\beta_j}{\partial t_i} + \lambda_i\frac{\partial G_i}{\partial t_i}\Big]
\end{aligned}
$$

Because $h_j = \alpha$ for $j \in R_a$, $\dfrac{\partial \beta_0(p)}{\partial t_i} = -\sum_{j \in R_a} \dfrac{\partial \beta_j(p)}{\partial t_i}$, and $\dfrac{\partial \beta_j(p)}{\partial t_i} = 0$ for $j \in R_d \cup N$, we have

$$\frac{\partial T(p)}{\partial t_i} = \frac{\lambda_i}{\Phi} \frac{\partial G_i}{\partial t_i}$$
$$> 0 \quad i \in R_a \cup R_d,$$
$$= 0 \quad i \in N.$$

We can derive relation (2.155) similarly as above. □

Remark. This theorem implies that the overall mean job response time of the optimum increases with the increase in the communication time of link i which connects to an active source or an idle source, or with the increase in the node processing time of the central site, active sources or neutrals.

2.3.4 Anomalous Behaviors of the Performance Variables

In subsection 2.3.3, the parametric analysis of the individually optimal policy has been studied and we see that the results agree with our intuition. In subsection 2.3.3, the parametric analysis of the overall optimal policy has been studied. These results show that there exist striking parallel characteristics for the two policies. In the parametric analysis, such counter-intuitive phenomena as follows can be observed.

There are cases where the overall mean job response time in the equilibrium may decrease even when the communication time of communication links increases.

From Lemma 2.15 and Theorem 2.17, we have the following relations.

$$\frac{\partial T(p)}{\partial t_i} = \frac{\partial G_i/\partial t_i}{\Phi A_i\left(1 + \frac{\partial F_0}{\partial \beta_0} \sum_{j \in R_a} 1/A_j\right)} \left[\left(\phi_i \frac{\partial F_i}{\partial \beta_i} - \phi_0 \frac{\partial F_0}{\partial \beta_0}\right)\right.$$
$$\left. + \frac{\partial F_0}{\partial \beta_0} \sum_{j \in R_a} \left(\phi_i \frac{\partial F_i}{\partial \beta_i} - \phi_j \frac{\partial F_j}{\partial \beta_j}\right)/A_j\right].$$

where A_j is defined by equation (2.139).

For example, when the job arrival rate at the central site is very big and the central site is very congested, we may have $\phi_0 \gg \phi_i$ and $\partial F_0/\partial \beta_0 \gg 0$. If the job arrival rate of active source j $(j \neq i)$ is much bigger than that of active source i, and active source j is also very congested, we may have $\phi_j > \phi_i$ and $\partial F_j/\partial \beta_j > \partial F_i/\partial \beta_i$. Under such situations, $\partial T(p)/\partial t_i$ may be negative, and hence the overall mean job response time in the equilibrium may decrease even when the communication time of communication link i increases.

There are cases where the overall mean job response time in the optimum and in the equilibrium may decrease even when the total external job arrival rate increases.

For the overall optimal policy, by differentiating the both sides of relation (2.83) with respect to ϕ_i, we have

$$\frac{\partial T(\boldsymbol{p})}{\partial \phi_i} = \frac{1}{\Phi^2}\Big[\Phi\Big(\sum_{j=0}^{n}\frac{\partial(\beta_j F_j)}{\partial \beta_j}\frac{\partial \beta_j}{\partial \phi_i} + \sum_{j=1}^{n}\frac{\partial(\lambda_j G_j)}{\partial \lambda_i}\frac{\partial \lambda_j}{\partial \phi_i}\Big) - \Phi T(\boldsymbol{p})\Big]$$

$$= \begin{cases} \dfrac{1}{\Phi^2}\Big[\Phi\phi_i\dfrac{\partial F_i}{\partial \beta_i}\Big|_{\beta_i=\phi_i} + \displaystyle\sum_{j=0}^{n}\beta_j(F_i - F_j) - \sum_{j=1}^{n}\lambda_j G_j\Big] & i \in N, \\[3mm] \dfrac{1}{\Phi^2}\Big[\displaystyle\sum_{j=0}^{n}\beta_j(\alpha + g_i - F_j) - \sum_{j=1}^{n}\lambda_j G_j\Big] & i \in R_a \cup R_d, \\[3mm] \dfrac{1}{\Phi^2}\Big[\displaystyle\sum_{j=0}^{n}\beta_j(\alpha - F_j) - \sum_{j=1}^{n}\lambda_j G_j\Big] & i = 0. \end{cases}$$

For $i \in N$, for example, when node j ($j \neq i$) and communication link j are congested, we may have $F_j > F_i$, and hence $\partial T(\boldsymbol{p})/\partial \phi_i$ may be negative. For $i \in R_a \cup R_d$, for example, when the central site and communication link i are lightly loaded; that is, when the expected node delay of the central site and the expected communication delay of link i are short, we may have $\alpha + g_i < F_j$, and hence $\partial T(\boldsymbol{p})/\partial \phi_i$ may be negative. For $i = 0$, for example, when the central site is lightly loaded; that is, the incremental node delay of the central site is short, we may have $\alpha < F_j$, and hence $\partial T(\boldsymbol{p})/\partial \phi_i$ may be nagative. Under these situations mentioned above, the overall mean job response time in the optimum may have the chances to decrease even when the total external job arrival rate increases.

For the individually optimal policy, by differentiating the both sides of relation (2.131) with respect to ϕ_i, we have

$$\frac{\partial T(\boldsymbol{p})}{\partial \phi_i} = \frac{1}{\Phi^2}\Big[\Phi\Big(\sum_{j\in R_a\cup R_d}\frac{\partial(\phi_j(R+G_j))}{\partial \phi_i} + \sum_{j\in N}\frac{\partial(\phi_j F_j)}{\partial \phi_i} + \frac{\partial(\phi_0 R)}{\partial \phi_i}\Big) - \Phi T(\boldsymbol{p})\Big]$$

$$= \begin{cases} \dfrac{1}{\Phi^2}\Big[\Phi\phi_i\dfrac{\partial F_i}{\partial \beta_i}\Big|_{\beta_i=\phi_i} + \displaystyle\sum_{j\in R_a\cup R_d}\phi_j\big(F_i - (R+G_j)\big) + \sum_{j\in N}\phi_j\big(F_i - F_j\big) \\ \qquad + \phi_0(F_i - R)\Big] & i \in N, \\[4mm] \dfrac{1}{\Phi^2}\Big[\Phi\Big(\displaystyle\sum_{j\in R_a\cup R_d}\phi_j\dfrac{\partial(R+G_j)}{\partial \phi_i} + \phi_0\dfrac{\partial R}{\partial \phi_i}\Big) + \sum_{j\in R_a\cup R_d}\phi_j(G_i - G_j) \\ \qquad + \displaystyle\sum_{j\in N}\phi_j\big(R+G_i - F_j\big) + \phi_0 G_i\Big] & i \in R_a \cup R_d, \\[4mm] \dfrac{1}{\Phi^2}\Big[\Phi\Big(\displaystyle\sum_{j\in R_a\cup R_d}\phi_j\dfrac{\partial(R+G_j)}{\partial \phi_i} + \phi_0\dfrac{\partial R}{\partial \phi_i}\Big) - \sum_{j\in R_a\cup R_d}\phi_j G_j \\ \qquad + \displaystyle\sum_{j\in N}\phi_j\big(R - F_j\big)\Big] & i = 0. \end{cases}$$

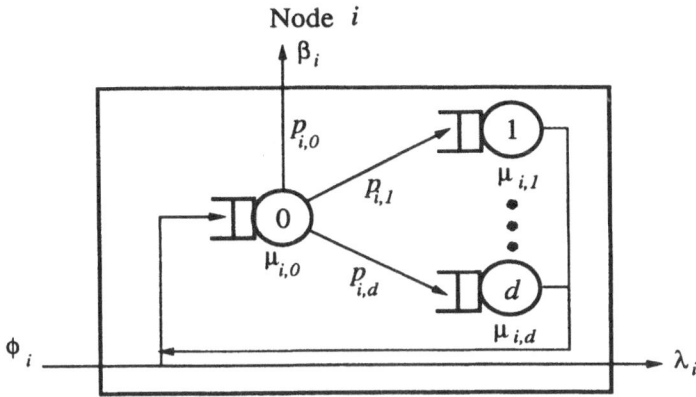

Figure 2.7: A node model.

For $i \in N$, for example, when node i is lightly loaded, we may have $F_i < R + G_j$, $F_i < F_j$ and $F_i < R$ ($j \neq i$), and hence we may have $\partial T(\boldsymbol{p})/\partial \phi_i < 0$. For $i \in R_a \cup R_d$, for example, when the central site and link i are lightly loaded, we may have $R + G_i < F_j$ and $G_i < G_j$, and hence we may have $\partial T(\boldsymbol{p})/\partial \phi_i < 0$. For $i = 0$, for example, when the expected communication delay of link i is long, the central site is lightly loaded, and the satellite nodes are congested, we may have $R < F_j$ and $G_j \gg 0$, and hence we may have $\partial T(\boldsymbol{p})/\partial \phi_i < 0$. Under these situations mentioned above, the overall mean job response time in the equilibrium may have the chances to decrease even when the total external job arrival rate increases.

2.3.5 Numerical Examination

The effects of varying the system parameters on the overall mean job response time have been examined numerically in several examples of a star network configuration. A model which consists of a central site (node 0) and 3 satellite nodes (nodes 1, 2, and 3) was considered. Each node is modeled as a central server model as shown in Fig. 2.7.

Server 0 is a CPU that processes jobs according to the processor sharing discipline. Servers $1, 2, \ldots, d$ are I/O devices which process jobs according to FCFS (first-come-first-served) discipline. Let $p_{i,j}(j = 0, 1, \cdots, d)$ denote the transition probabilities that after departing from the CPU, a job leaves node i and requests an I/O service at device $j, j = 1, 2, \ldots, d$, respectively. $\mu_{i,j}(j = 0, 1, \cdots, d)$ denotes the processing rate for server j

Table 2.1: Parameters of the node model

Node	Processing rate of CPU					Probabilities of a job CPU				
	$\mu_{i,0}$	$\mu_{i,1}$	$\mu_{i,2}$	$\mu_{i,3}$	$\mu_{i,4}$	$p_{i,0}$	$p_{i,1}$	$p_{i,2}$	$p_{i,3}$	$p_{i,4}$
0	500	100	100	100	100	0.2	0.2	0.2	0.2	0.2
1	40	16	16	-	-	0.2	0.4	0.4	-	-
2	50	20	20	-	-	0.2	0.4	0.4	-	-
3	50	20	20	-	-	0.2	0.4	0.4	-	-

at node i. The expected node delay of the central server model is given as

$$F_i(\beta_i) = \sum_{j=0}^{d} \frac{q_{i,j}}{\mu_{i,j} - q_{i,j}\beta_i}, \qquad (2.156)$$

where $q_{i,0} = 1/p_{i,0}$ and $q_{i,j} = p_{i,j}/p_{i,0}$.

The communication links are modeled as M/M/1 queueing systems; that is, the expected communication delay of link i is given as

$$G_i(\lambda_i) = \frac{t_i}{1 - \lambda_i t_i}, \qquad (2.157)$$

where t_i denotes the mean communication time (excluding the waiting time) of link i for sending and receiving a job.

The optimal solution of the overall optimal policy was calculated by using the algorithm of Tantawi and Towsley [TT84]. The optimal solution of the individually optimal policy was calculated by using the extension version of that of Tantawi and Towsley. That is, α and h_i in the algorithm of Tantawi and Towsley were replaced with R and H_i, respectively.

It is observed that the results of the numerical examination, in most cases, agree with our intuition and that the overall mean job response time of the equilibrium is close to that of the optimum. In this dissertation, we only show some examples using parameters given in Table 2.1. In Table 2.1, $\mu_{i,0}$ denotes the processing rate of CPU and $\mu_{i,j}$ denotes the processing rate of I/O devices (e.g. hard disks). $p_{i,j}$ denote the transition probabilities of a job to the CPU or the I/O devices after departing the CPU. Let us consider a job behavior in the central site as an example. The central site is constructed by a large scale computer system. A job spends an exponential distributed amount of time with mean 2 msec (=1/500sec) in CPU before issuing an I/O request. There are four identical I/O devices. The service time of each is exponential distributed with mean 10 msec (=1/100sec). On the average, a job visits CPU 5 times and an I/O device 4 times before leaving the central site.

The results of the numerical examination by using Table 2.1 are shown in Figs. 2.8 -
2.11. In these figures, 'AVERAGE' denotes the overall mean job response time, which is
shown by solid line. Dotted lines denote the mean response time of jobs arriving at the
nodes.

Figs. 2.8 and 2.9 show the effects of the communication time of each communication
link $t_i(t_1 = t_2 = t_3)$ on the overall and individually optimal polices. The communication
time of each link is changed from 0 to a certain large value until no jobs are processed
remotely under the two policies. The job arrival rates at the nodes are fixed as $\phi_0 =
75, \phi_1 = 6.9, \phi_2 = 8.0$ and $\phi_3 = 8.4$ (jobs/sec) where jobs/sec denotes the average number
of jobs arriving at a node in one second.

We can observe that, from Figs. 2.8 and 2.9, the behaviors of the mean response
time of jobs arriving at node i $(T_i(\beta_i))$ and the overall mean job response time $(T(\boldsymbol{\beta}))$
of the equilibrium is near to those of the optimum. These figures show that when the
communication time of each link increases, the overall mean job response time and the
mean response time of jobs arriving at satellite nodes increase, whereas the mean response
time of jobs processed at the central site decreases. We can also observe such a counter-
intuitive phenomenon that the overall mean job response time of the equilibrium decreases
and increases again when the communication times of the links increase.

Figs. 2.10 and 2.11 show the behaviors of the mean response time of jobs arriving at
node i $(T_i(\beta_i))$ and the overall mean job response time $(T(\boldsymbol{\beta}))$ of the equilibrium and of
the optimum when the job arrival rate of node 3 is changed from 0 to 19 (jobs/sec). The
other parameters are fixed as $\phi_0 = 0$, $\phi_1 = 7.4$, $\phi_2 = 9.1$ (jobs/sec) and $t_1 = 1.1$, $t_2 =
1$, $t_3 = 0.1$ (sec).

From Figs. 2.10 and 2.11, we see that the behaviors of the system in both the optimum
and the equilibrium are similar to each other. We can observe, however, that the overall
mean job response time of the optimum and of the equilibrium decrease when the job
arrival rate of node 3 increases from 0 to 15 (jobs/sec). When the job arrival rate of node
3 exceeds 15 (jobs/sec), node 3 becomes saturated, and we see, from Figs. 2.10 and 2.11,
that the mean response time of jobs arriving at node 3 increases rapidly. Consequently,
the overall mean job response time also increases.

2.3.6   Discussion

From Theorems 2.12 and 2.14, and the results of parametric analysis in subsection2.3.3,
we see that the characteristics of the individually optimal policy are close to those of the
overall optimal policy. The same results can also be observed in the numerical examination. It can be also observed that, in the numerical examination, the overall mean job
response time of the equilibrium is close to that of the optimum.

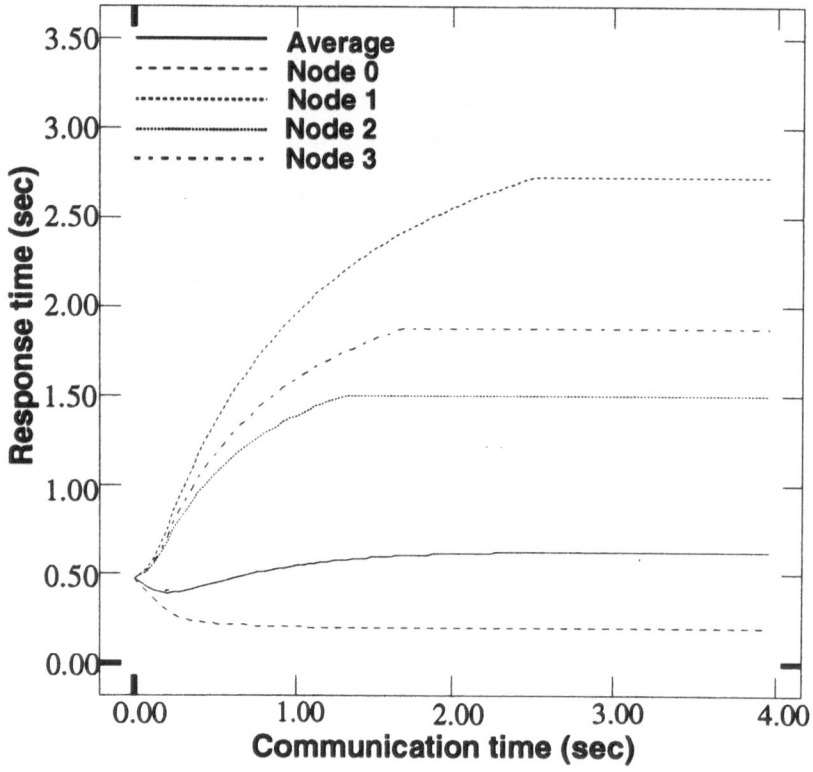

Figure 2.8: Effects of the communication times of the communication links on the equi-librium.

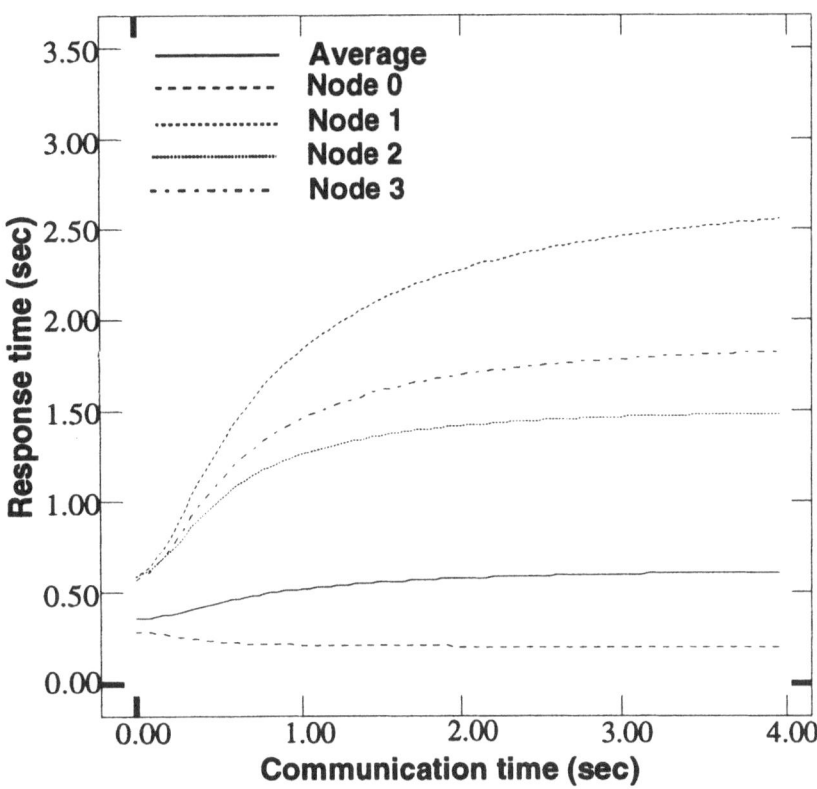

Figure 2.9: Effect of the communication times of the communication links on the optimum.

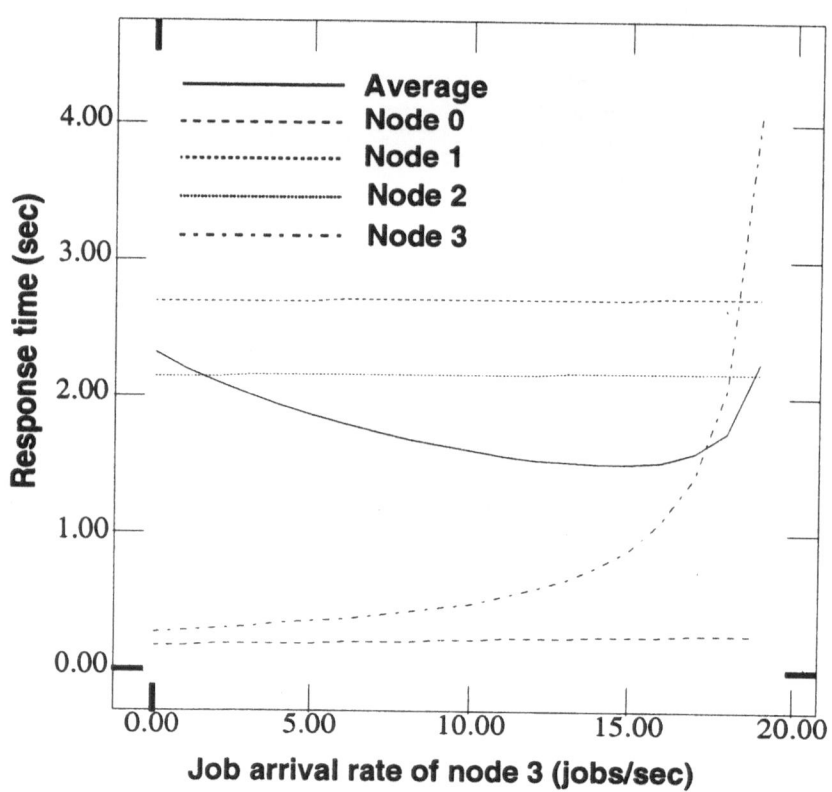

Figure 2.10: Effect of the job arrival rate of node 3 on the equilibrium.

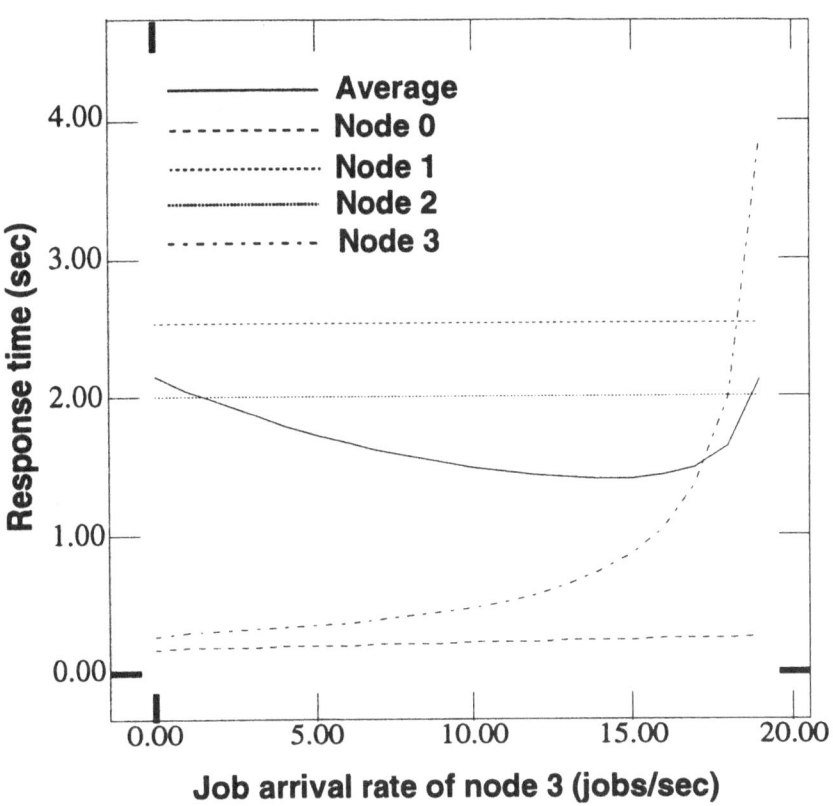

Figure 2.11: Effect of the job arrival rate of node 3 on the optimum.

The overall optimal policy requires the information such as the dependency of the expected node and communication delays on the loads of the nodes and of the communication links. It might not be easy, therefore, to implement the overall optimal policy in practice. On the other hand, the individually optimal policy is implemented on the basis of the information of the expected node and communication delays. Therefore, the implementation of this policy appears to be simple.

For the individually optimal policy, job scheduling is determined so that every job may feel that its own expected response time is minimized if it knows the expected node and communication delays. This appears to be closely related to the *competitive decentralized policy* whereby each arriving job decides for itself whether to be processed locally or remotely with the objective of minimizing its own expected response time. Unfortunately, this may lead to overutilization of the resources and may lead to poor overall performance. For example, when the communication times of the links are short; that is, almost all of the jobs arriving at satellite nodes are processed at the central site, the central site may be overutilized. As the value of the communication time increases, that is, the ratio of the jobs that arrive at satellite nodes and that are processed at the central site decreases, the situation may be alleviated. In this case, the overall mean job response time may have the chances to decrease (recall Fig. 2.8).

The overall mean job response time of load balancing policies is expressed as the sum of the mean node delay and the mean communication delay as shown in eq. (2.83). When the job arrival rate of a node with short expected node delay is increased; that is, the ratio of the jobs processed at a node with small expected node delay increases, the overall mean job response time may have the chances to decrease accordingly (recall Figs. 2.10 and 2.11).

These phenomena may show the lack of discriminatory power inherent in the overall mean job response time, as Kleinrock [Kle76] mentioned in his book: "One may therefore conclude that the average response time by itself is not a very good indicator of system performance."

2.3.7   Conclusion

In this section, we considered a model of star network configurations with the same assumptions as those in Tantawi and Towsley [TT84]. We first considered newly an individually optimal policy whereby jobs are scheduled so that every job may feel that its own expected response time is minimum. We found that the form of the optimal solution of the individually optimal policy is similar to that of the overall optimal policy.

Furthermore, we studied the parametric analysis, that is, the effects of changing various system parameters on the performance variables of the overall and individually optimal policies. We also examined the parametric analysis numerically. We observed that, in most cases, the overall mean job response time of the equilibrium is near to that of the optimum. We observed, however, several anomalous phenomena that there are cases where, in the equilibrium, the overall mean job response time happens to decrease even though the communication time increases, and that there are cases where, in the equilibrium and in the optimum, the overall mean job response time happens to decrease even though the job arrival rates of the nodes increase. These results are similar to those given in section 2.2.

2.4   Static Load Balancing in Multiclass Single Channel Networks

2.4.1   Policies and model

We consider a distributed computer system which consists of n nodes, which represent host computers, connected by a single channel communications network (Figure 2.12). Each node contains one or more resources (CPU, I/O, ...) contended for by the jobs processed at the node. We assume that the expected communication delay from node i to node j for class k jobs is independent of the source-destination pair but dependent on the class. This model is absolutely identical with the model which is considered throughout section 1.3 of chapter 1. Jobs arrive at each node according to a time-invariant Poisson process. A job arriving at node i (origin node) may either be processed at node i or transferred through the communications network to another node j (processing node). Also we assume that a transferred job from node i to node j receives its service at node j and is not transferred to other nodes. The notation and assumptions can be referenced from that of section 1.3 of chapter 1.

We classified nodes in the following way for each class k.

(1) class k idle source $(R_d^{(k)})$: The node does not process any class k jobs, that is , $\beta_i^{(k)} = 0$.

(2) class k active source $(R_a^{(k)})$: The node sends class k jobs and does not receive any class k jobs. But the node processes a part of class k jobs that arrive at the node.

(3) class k neutral $(N^{(k)})$: The node processes class k jobs locally without sending or receiving class k jobs.

(4) class k sink $(S^{(k)})$: The node receives class k jobs from other nodes but does not send out any class k jobs.

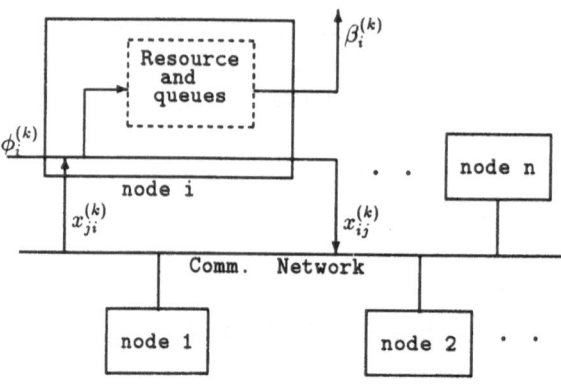

Figure 2.12: A model of a distributed computer system with multiple classes of jobs

Once again, there are two load balancing policies, each of which has a distinct performance objective. One is the *overall optimal policy* and the other is the *individually optimal*
policy. Tantawi and Towsley [TT85] [TT84] and Kim and Kameda (sections 1.2 and 1.3 of
chapter 1, [KK92a], [KK90a]) considered an optimal static load balancing strategy which
is the same policy that we refer to here as the overall optimal policy.

Overall optimal policy

For reference, we repeat here the problem and the optimal solution of the overall
optimal policy. The problem of minimizing the system-wide mean job response time is
expressed as follows:

$$\text{minimize} \quad D(\boldsymbol{\beta}) = \frac{1}{\Phi} \sum_{k=1}^{m} [\sum_{i=1}^{n} \beta_i^{(k)} F_i^{(k)}(\boldsymbol{\beta}_i) + \lambda^{(k)} G^{(k)}(\boldsymbol{\lambda})], \qquad (2.158)$$

subject to

$$\sum_{i=1}^{n} \beta_i^{(k)} = \phi^{(k)}, \quad k = 1, 2, ..., m,$$

$$\beta_i^{(k)} \geq 0, \quad i = 1, 2, ..., n, k = 1, 2, ..., m,$$

where the class k network traffic $\lambda^{(k)}$ may be expressed in terms of variable $\beta_i^{(k)}$ as

$$\lambda^{(k)} = \frac{1}{2} \sum_{i=1}^{n} |\phi_i^{(k)} - \beta_i^{(k)}|.$$

From the assumptions, $D(\boldsymbol{\beta})$ is shown to be a differentiable, increasing, and convex
function of $\boldsymbol{\beta}$. Since the feasible region of $\boldsymbol{\beta}$ is a convex set, any local solution point of the

problem is a global solution point. We defined two functions and showed the following conditions that the solution of overall optimal policy satisfies at section 1.3 of chapter 1

$$f_i^{(k)}(\boldsymbol{\beta}_i) = \frac{\partial}{\partial \beta_i^{(k)}} \sum_{l=1}^{m} \beta_i^{(l)} F_i^{(l)}(\boldsymbol{\beta}_i), \qquad (2.159)$$

$$g^{(k)}(\boldsymbol{\lambda}) = \frac{\partial}{\partial \lambda^{(k)}} \sum_{l=1}^{m} \lambda^{(l)} G^{(l)}(\boldsymbol{\lambda}). \qquad (2.160)$$

The optimal solution to problem (2.158) satisfies the following relations, $k = 1, 2, .., m$,

$$
\begin{array}{llll}
f_i^{(k)}(\boldsymbol{\beta}_i) \geq \alpha^{(k)} + g^{(k)}(\boldsymbol{\lambda}), & \beta_i^{(k)} = 0, & (i \in R_d^{(k)}), & \\
f_i^{(k)}(\boldsymbol{\beta}_i) = \alpha^{(k)} + g^{(k)}(\boldsymbol{\lambda}), & 0 < \beta_i^{(k)} < \phi_i^{(k)}, & (i \in R_a^{(k)}), & \\
\alpha^{(k)} \leq f_i^{(k)}(\boldsymbol{\beta}_i) \leq \alpha^{(k)} + g^{(k)}(\boldsymbol{\lambda}), & \beta_i^{(k)} = \phi_i^{(k)}, & (i \in N^{(k)}), & (2.161) \\
\alpha^{(k)} = f_i^{(k)}(\boldsymbol{\beta}_i), & \beta_i^{(k)} > \phi_i^{(k)}, & (i \in S^{(k)}), &
\end{array}
$$

subject to the total flow constraint,

$$\sum_{i \in S^{(k)}} f_i^{(k)^{-1}}(\boldsymbol{\beta}_i|_{\beta_i^{(k)}=\alpha^{(k)}}) + \sum_{i \in N^{(k)}} \phi_i^{(k)} + \sum_{i \in R_a^{(k)}} f_i^{(k)^{-1}}(\boldsymbol{\beta}_i|_{\beta_i^{(k)}=\alpha^{(k)}+g^{(k)}(\lambda)}) = \phi^{(k)}, \qquad (2.162)$$

where $\alpha^{(k)}$ is the Lagrange multiplier.

In section 1.3 of chapter 1, we derived from the above conditions the three properties for optimal solution of multiple job classes and an efficient load balancing algorithm using the three properties for the overall optimal policy.

Individually optimal policy

The job scheduling decision by the individually optimal policy is such that each individual job is scheduled so as to minimize its own expected job response time, given the expected node delay at each node, $F_i^{(k)}(\boldsymbol{\beta}_i)$, and the expected communication delay, $G^{(k)}(\boldsymbol{\lambda})$. As the result of the policy, every job may feel that its own expected job response time is minimized if it knows the expected node delay at each node and the expected communication delay. In a load pattern which corresponds to an equilibrium state, two jobs of the same class that arrive at the same node and are processed by different node have the same expected job response time.

The inverse of the node delay is defined as

$$F_i^{(k)^{-1}}(\boldsymbol{\beta}_i|_{\beta_i^{(k)}=x}) = \begin{cases} a, & F_i^{(k)}(\boldsymbol{\beta}_i|_{\beta_i^{(k)}=a}) = x, \\ 0, & F_i^{(k)}(\boldsymbol{\beta}_i|_{\beta_i^{(k)}=0}) \geq x, \end{cases}$$

where $\boldsymbol{\beta}_i|_{\beta_i^{(k)}=x}$ denotes the vector whose elements are the same as those of $\boldsymbol{\beta}_i$ except that the element $\beta_i^{(k)}$ is replaced by x.

The following are the conditions that the solution of the individually optimal policy satisfies [KH88].

Individually optimal load β satisfies the following relations, $k = 1, 2, .., m$,

$$
\begin{array}{llll}
F_i^{(k)}(\beta_i) \geq K^{(k)} + G^{(k)}(\lambda), & \beta_i^{(k)} = 0, & (i \in R_d^{(k)}), \\
F_i^{(k)}(\beta_i) = K^{(k)} + G^{(k)}(\lambda), & 0 < \beta_i^{(k)} < \phi_i^{(k)}, & (i \in R_a^{(k)}), \\
K^{(k)} \leq F_i^{(k)}(\beta_i) \leq K^{(k)} + G^{(k)}(\lambda), & \beta_i^{(k)} = \phi_i^{(k)}, & (i \in N^{(k)}), \\
K^{(k)} = F_i^{(k)}(\beta_i), & \beta_i^{(k)} > \phi_i^{(k)}, & (i \in S^{(k)}),
\end{array}
\qquad (2.163)
$$

subject to the total flow constraint,

$$
\sum_{i \in S^{(k)}} F_i^{(k)^{-1}}(\beta_i|_{\beta_i^{(k)} = K^{(k)}}) + \sum_{i \in N^{(k)}} \phi_i^{(k)} + \sum_{i \in R_a^{(k)}} F_i^{(k)^{-1}}(\beta_i|_{\beta_i^{(k)} = K^{(k)} + G^{(k)}(\lambda)}) = \phi^{(k)}, \quad (2.164)
$$

where $K^{(k)} = \min\limits_i F_i^{(k)}(\beta_i)$.

The combination of β that satisfies the above conditions is called the *equilibrium*. The existence of the equilibrium is assured in [KH88].

Comparison of both policies

We can observe a striking parallel between the conditions on the optimum and those on the equilibrium: If we replace every $F_i^{(k)}(\beta_i)$, $G^{(k)}(\lambda)$, and $K^{(k)}$ in the set of relations in (2.163) and (2.164) with $f_i^{(k)}(\beta_i)$, $g^{(k)}(\lambda)$, and $\alpha^{(k)}$, respectively, then the set becomes identical with the set of relations in (2.161) and (2.162). The parallel between the sets of relations (2.161), (2.162), and (2.163), (2.164) gives us other intuitive explanations of one in terms of the other. That is, the individually optimal policy would be realized by the overall optimal policy, if the values of the marginal node and communication delays were given as the expected node and communication delays, and vice versa.

We may therefore have three properties for the individually optimal policy like the three properties for the overall optimal policy obtained at section 1.3 of chapter 1. Using the three properties for the individually optimal policy, we may obtain an equilibrium load balancing algorithm. The formula of the equilibrium load balancing algorithm may have almost the same procedure as the optimum load balancing algorithm which is proposed at section 1.2 of chapter 1, except that the values of the marginal node and communication delays are replaced by the expected node and communication delays. The above relations (2.163) of the individually optimal policy is familiar to us than that of the overall optimal policy, because the expected node and communication delays may be obtained by watching real distributed computer systems.

Equilibrium load balancing algorithm

The equilibrium load balancing algorithm is shown. This algorithm (called a single-point algorithm of individually optimal policy) can obtain the equilibrium solution for an arbitrary set of parameter values.

o Single-point algorithm of individually optimal policy

1. Initialize.

   $l := 0$ (l: iteration number)

   $\boldsymbol{\beta}_{(l)} := \phi$

2. $l := l + 1$ and execute step 3.

3. Execute the following sub-algorithm for $k = 1, 2, .., m$ by using $\boldsymbol{\beta}_{(l)}^{(1)}, \boldsymbol{\beta}_{(l)}^{(2)}, .., \boldsymbol{\beta}_{(l)}^{(k-1)},$ $\boldsymbol{\beta}_{(l-1)}^{(k+1)}, .., \boldsymbol{\beta}_{(l-1)}^{(m)}$ as the values of $\boldsymbol{\beta}^{(1)}, .., \boldsymbol{\beta}^{(k-1)}, \boldsymbol{\beta}^{(k+1)}, .., \boldsymbol{\beta}^{(m)}$.

   Denote by $\boldsymbol{\beta}_{(l)}^{(k)}$, the value of $\boldsymbol{\beta}^{(k)}$ obtained by the sub-algorithm. Thus the resulting value of $\boldsymbol{\beta}$ at this step becomes as follows:

   $$(\boldsymbol{\beta}_{(l)}^{(1)}, \boldsymbol{\beta}_{(l)}^{(2)}, .., \boldsymbol{\beta}_{(l)}^{(k-1)}, \boldsymbol{\beta}_{(l)}^{(k)}, \boldsymbol{\beta}_{(l-1)}^{(k+1)}, .., \boldsymbol{\beta}_{(l-1)}^{(m)}).$$

   The final resulting value (at step $k = m$) of $\boldsymbol{\beta}$ is denoted by $\boldsymbol{\beta}_{(l)}$.

4. (Stopping rule) Compare $\boldsymbol{\beta}_{(l-1)}$ and $\boldsymbol{\beta}_{(l)}$.

   If $|\boldsymbol{\beta}_{(l)} - \boldsymbol{\beta}_{(l-1)}| < \varepsilon$ where $\varepsilon > 0$ is a properly chosen acceptance tolerance, then STOP.

   Otherwise, go to Step 2.

o Sub-algorithm that computes the value of $\boldsymbol{\beta}^{(k)}$ for class k jobs with $\boldsymbol{\beta}^{(1)}, .., \boldsymbol{\beta}^{(k-1)},$ $\boldsymbol{\beta}^{(k+1)}, .., \boldsymbol{\beta}^{(m)}$ being fixed.

1. Order nodes.

   Order nodes such that $F_1^{(k)}(\beta_1|_{\beta_1^{(k)}=\phi_1^{(k)}}) \leq F_2^{(k)}(\beta_2|_{\beta_2^{(k)}=\phi_2^{(k)}}) \leq \cdots \leq F_n^{(k)}(\beta_n|_{\beta_n^{(k)}=\phi_n^{(k)}})$.

   If $F_1^{(k)}(\beta_1|_{\beta_1^{(k)}=\phi_1^{(k)}}) + G^{(k)}(\boldsymbol{\lambda}|_{\lambda^{(k)}=0}) \geq F_n^{(k)}(\beta_n|_{\beta_n^{(k)}=\phi_n^{(k)}}),$

   then stop the sub-algorithm.

2. Determine $K^{(k)}$.

   Find $K^{(k)}$ such that $\lambda_{rec}^{(k)}(K^{(k)}) = \lambda_{sen}^{(k)}(K^{(k)})$.

   (by using, for example, a binary search),

where, given $K^{(k)}$, each value is calculated in the following order.

$$S^{(k)}(K^{(k)}) = \{i | F_i^{(k)}(\beta_i|_{\beta_i^{(k)}=\phi_i^{(k)}}) < K^{(k)}\},$$

$$\lambda_{rec}^{(k)}(K^{(k)}) = \sum_{i \in S^{(k)}(K^{(k)})} (F_i^{(k)^{-1}}(\beta_i|_{\beta_i^{(k)}=K^{(k)}}) - \phi_i^{(k)}),$$

$$R_d^{(k)}(K^{(k)}) = \{i | F_i^{(k)}(\beta_i|_{\beta_i^{(k)}=0}) \geq K^{(k)} + G^{(k)}(\lambda|_{\lambda^{(k)}=\lambda_{rec}^{(k)}(K^{(k)})})\},$$

$$R_a^{(k)}(K^{(k)}) = \{i | F_i^{(k)}(\beta_i|_{\beta_i^{(k)}=\phi_i^{(k)}}) > K^{(k)} + G^{(k)}(\lambda|_{\lambda^{(k)}=\lambda_{rec}^{(k)}(K^{(k)})}) > F_i^{(k)}(\beta_i|_{\beta_i^{(k)}=0})\},$$

$$\lambda_{sen}^{(k)}(K^{(k)}) = \sum_{i \in R_d^{(k)}(K^{(k)})} \phi_i^{(k)} +$$
$$\sum_{i \in R_a^{(k)}(K^{(k)})} (\phi_i^{(k)} - F_i^{(k)^{-1}}(\beta_i|_{\beta_i^{(k)}=K^{(k)}+G^{(k)}(\lambda|\lambda^{(k)}=\lambda_{rec}^{(k)}(K^{(k)}))})).$$

3. Determine optimal load.

$$\beta_i^{(k)} = F_i^{(k)^{-1}}(\beta_i|_{\beta_i^{(k)}=K^{(k)}}), \text{ for } i \in S^{(k)},$$

$$\beta_i^{(k)} = 0, \text{ for } i \in R_d^{(k)},$$

$$\beta_i^{(k)} = F_i^{(k)^{-1}}(\beta_i|_{\beta_i^{(k)}=K^{(k)}+G^{(k)}(\lambda)}), \text{ for } i \in R_a^{(k)}$$

$$\beta_i^{(k)} = \phi_i^{(k)}, \text{ for } i \in N^{(k)},$$

where $N^{(k)}(K^{(k)}) = \{i | K^{(k)} \leq F_i^{(k)}(\beta_i|_{\beta_i^{(k)}=\phi_i^{(k)}}) \leq$
$$K^{(k)} + G^{(k)}(\lambda|_{\lambda^{(k)}=\lambda_{rec}^{(k)}(K^{(k)})})\}.$$

Another sub-algorithm for multiple job classes of individually optimal policy that would be obtained directly from the relations (2.163) and (2.164) could be obtained. However, it is easy to see that the above sub-algorithm is more straightforward and efficient and has a better performance than the sub-algorithm directly obtained from the relations (2.163) and (2.164). Let us recall the results of Table 1.1.

2.4.2 Numerical experiment

We have examined numerically the effects of system parameters in several examples of the distributed computer system. Let us present some typical examples among them. We can consider three important parameters: the communication time of the network, the job service rate of each node, and the job arrival rate of each node. We have studied the effects of these system parameters in the two load balancing policies.

For the numerical experiments, the optimum load balancing algorithm proposed at section 1.3 of chapter 1 for the overall optimal policy and the equilibrium load balancing algorithm proposed above for the individually optimal policy are used.

Table 2.2: Examples of the set of parameter values of system models

		Job processing rate (jobs/sec)		Job arrival rate (jobs/sec)		
	Node	class 1	class 2	class 1	class 2	
Case 1	1	20.0	50.0	11.0	9.0	
	2	15.0	10.0	1.5	3.4	
Case 2	1	20.0	50.0	14.0	15.0	$(t^{(1)} = t^{(2)}$
	2	30.0	15.0	11.0	5.0	$= 0.06)$
Case 3	1	20.0	45.0	11.0	9.5	
	2	15.0	10.0	1.5	3.0	
	3	18.0	40.0	9.0	8.0	
	4	12.0	8.0	1.0	2.7	

The models used in the numerical examination

We consider the model of a distributed computer system that consists of an arbitrary number of host computers (nodes) connected via a single channel communications network. The system can process multiple class jobs. We model each node as a multi-class M/M/1 queuing model with processor-sharing discipline. The expected class k node delay of node i is given as follows.

$$F_i^{(k)}(\boldsymbol{\beta}_i) = \frac{\frac{1}{\mu_i^{(k)}}}{1 - (\frac{\beta_i^{(1)}}{\mu_i^{(1)}} + ... + \frac{\beta_i^{(m)}}{\mu_i^{(m)}})}, \tag{2.165}$$

where $\mu_i^{(k)}$, $i = 1, .., n$, $k = 1, .., m$, denotes the class k service rate at node i.

We consider a multi-class M/M/1 model with processor-sharing discipline for the single channel communications network. Expected class k communication delay is given as follows.

$$G^{(k)}(\boldsymbol{\lambda}) = \frac{t^{(k)}}{1 - (t^{(1)}\lambda^{(1)} + ... + t^{(m)}\lambda^{(m)})}, \tag{2.166}$$

where $t^{(k)}$, $k = 1, 2, .., m$, denotes the mean communication time (excluding the queuing time) for class k jobs.

As one example, a two node system model with only two job classes is treated for simplicity. For the example, we have the following parameter setting (see cases 1 and 2 of Table 2.2).

In this parameter setting, the model has a unique solution for optimal load. (For more general discussion on the uniqueness of the solution, see [KZ95].)

Other complicated models are also treated as examples. We will show one parameter setting of four node models (see case 3 of Table 2.2).

The effect of communication time

Figures 2.13 and 2.14 show the effect of communication time, $t^{(k)}$ ($t^{(1)} = t^{(2)}$), on two load balancing policies in case 1. This means each communication time is changed from 0 to a large value until no jobs are processed remotely in the two policies.

In Figures 2.13 (B) and 2.14 (B), it is observed that each class jobs are processed at remote nodes that have the large processing capacity for the class, when the communication time is short, i.e., almost all class 1 jobs are processed at node 2, but all class 2 jobs are processed at node 1 when the communication time is zero ($t^{(k)} = 0$). As the value of the communication time increases, the ratios of the remotely processed jobs gradually decrease, until all jobs are processed locally at each local node. We define *critical communication times* as follows: there exist remotely processed jobs if a communication time is shorter than the critical communication time and there exist no remotely processed jobs if a communication time is longer than the critical communication time. We can easily find the critical communication time of optimum and that of equilibrium are not always identical. The former is apparently much longer than the latter.

We can observe such an awkward phenomenon that an expected job response time (e.g., class 1 of node 2 in Figure 2.13 (A)) decreases first, then increases, and decreases again as the communication time increases. Expected job response times (e.g., classes 1 and 2 of node 2 in Figure 2.13 (A), classes 1 and 2 of node 1 in Figure 2.14 (A)) at some communication times are sometimes longer than the expected job response time when all jobs are processed at their local nodes.

Figure 2.15 shows that in the optimum, two groups of class 1 jobs that arrive at node 1 and are processed by different nodes (node 1 and node 2) according to the policy have distinct expected job response times. That is, $F_1^{(1)}(\boldsymbol{\beta}_1) \neq F_2^{(1)}(\boldsymbol{\beta}_2) + G^{(1)}(\boldsymbol{\lambda})$. In contrast, note that Figure 2.15 shows that in the equilibrium, the expected job response times of the two groups of class 1 jobs that arrive at node 1 and are processed by different nodes (node 1 and node 2) according to the policy are the same. That is, $F_1^{(1)}(\boldsymbol{\beta}_1) = F_2^{(1)}(\boldsymbol{\beta}_2) + G^{(1)}(\boldsymbol{\lambda})$.

We observe in Figures 2.14 (A) and 2.16 *anomalous* behavior in the equilibrium where the system-wide mean job response time at a certain communication time is longer than the system-wide mean job response time in the case where there exists no job processed remotely. In contrast, Figures 2.13 (A) and 2.16 show that the greater communication time leads to the larger system-wide mean job response time in the optimum which does not exceed the system-wide mean job response time in the case where there exists no job processed remotely.

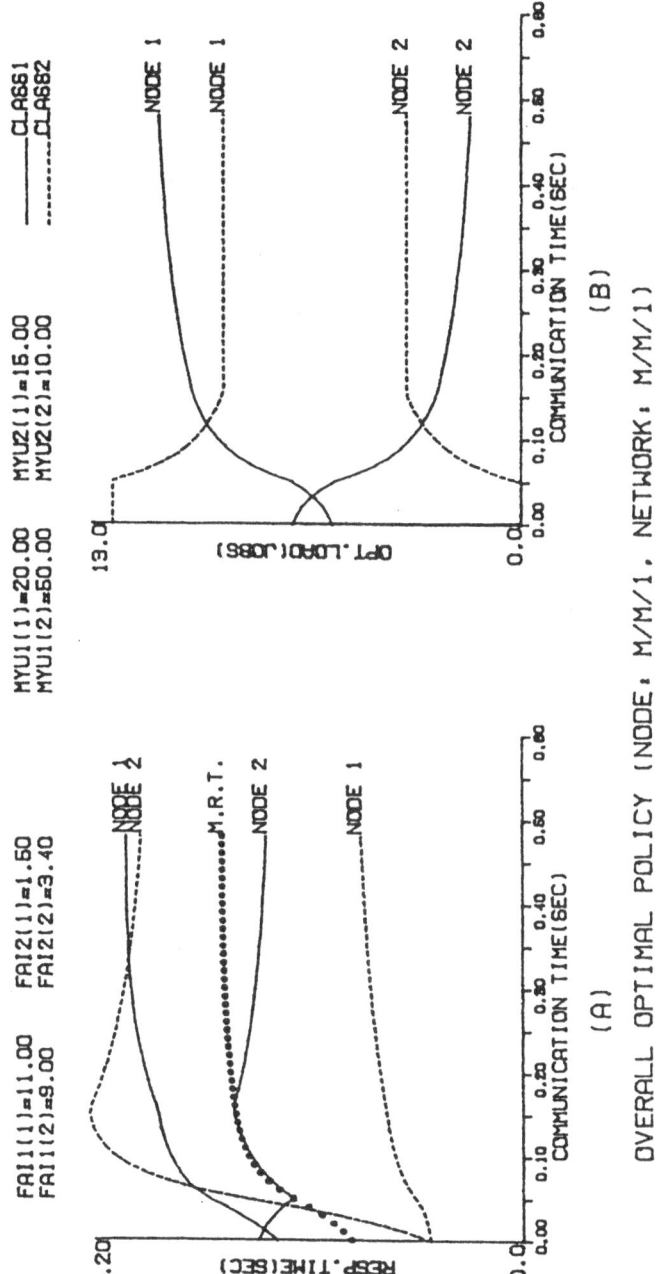

Figure 2.13: The effect of communication time on the optimum (case 1)

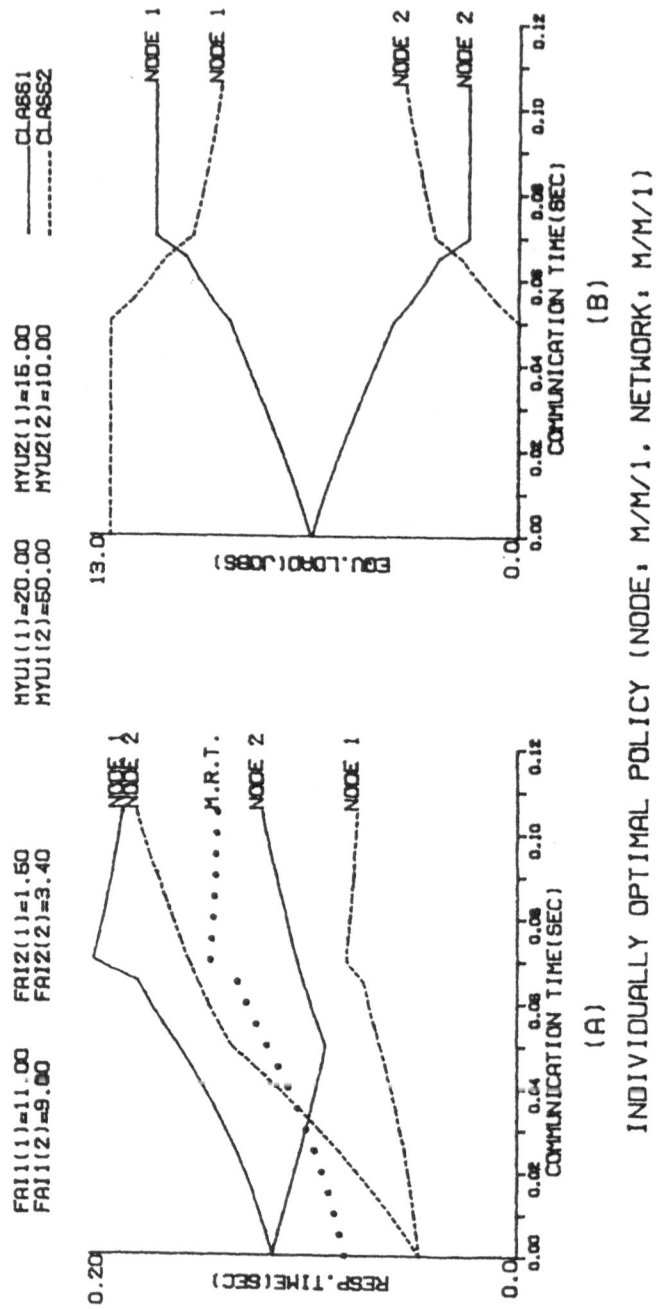

Figure 2.14: The effect of communication time on the equilibrium (case 1)

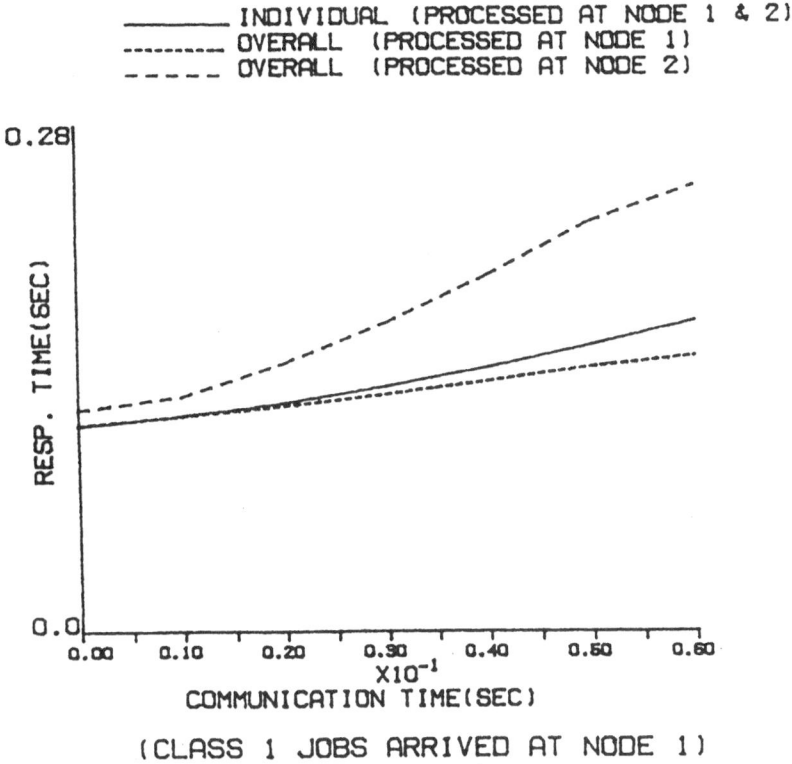

Figure 2.15: Unfairness in the optimum and fairness in the equilibrium (case 1)

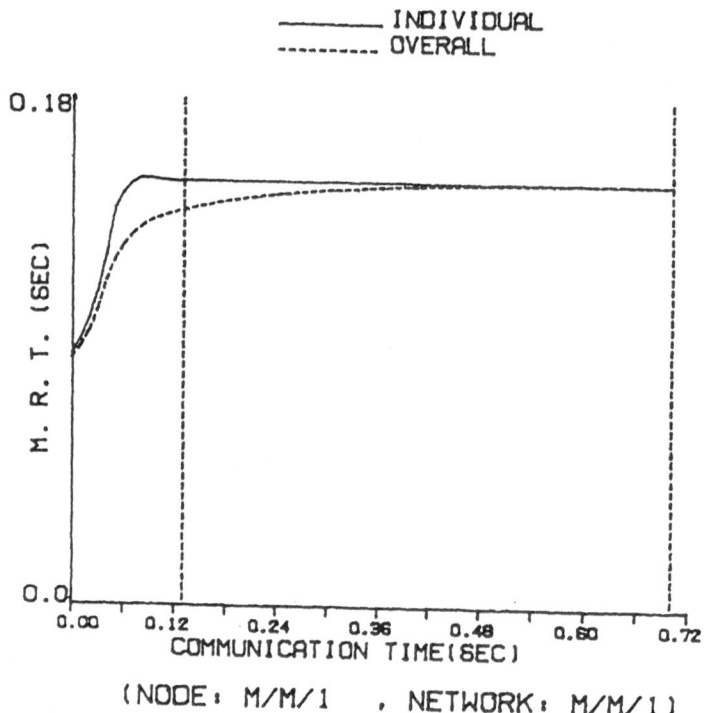

Figure 2.16: Comparison of sytem-wide mean job response times in the optimum and equilibrium (case 3)

The effect of arrival rates

Figure 2.17 shows that how the system in the optimum behaves as the class 1 arrival rate at node 2, $\phi_2^{(1)}$, decreases from 11.0 to 3.5. Figure 2.17 (B) shows that the remote processing ratio of the class 1 jobs that arrive at node 1 increases and the remote processing ratio of the class 2 jobs that arrive at node 1 also increases. Figure 2.17 (A) shows, however, that the smaller arrival rate leads to the smaller system-wide mean job response time and also leads to the smaller expected job response times for all classes and nodes.

The effect of service rates

Figure 2.18 shows that how the system in the optimum behaves as the class 1 service rate at node 2, $\mu_2^{(1)}$, increases from 30.0 to 80.0. Figure 2.18 (B) shows that the larger service rate, $\mu_2^{(1)}$, leads to the larger remote processing ratio of the class 1 jobs that arrive at node 1 and lead to the larger remote processing ratio of the class 2 jobs that arrive at node 1. Figure 2.18 (A) shows that the larger processing capacity of a class and a node, $\mu_2^{(1)}$, leads to the smaller system-wide mean job response time and also leads to smaller expected job response times for all classes and nodes.

2.4.3 Discussion

Consider the case where the communication time increases. At the short communication times, in both of the policies, most of jobs of each class are processed at remote nodes that have the high processing capacity for the class. We call this kind of phenomenon specialization. The critical communication time of the overall optimal policy is apparently much longer than that of the individually optimal policy. This means that the overall optimal policy has more remotely processed jobs than the individually optimal policy at the long values of communication time (recall Figure 2.16). All of the other experimental results show also that the critical communication time of the overall optimal policy is apparently much longer than that of the individually optimal policy. Therefore, we can see that in the overall optimal policy more jobs are processed remotely than in the individually optimal policy.

Unfairness in the overall optimal policy and fairness in the individually optimal policy are anticipated by Kameda and Hazeyama [KH88]. Let us consider two groups of jobs of the same class that arrive at the same node and that are processed by different nodes according to the policies (recall Figure 2.15). In the optimum under the overall optimal policy, the two groups of jobs show mutually different expected job response times, which seems *unfair*. On the other hand, in the equilibrium under the individually optimal policy, the two groups of jobs show mutually the same expected job response time, and each job

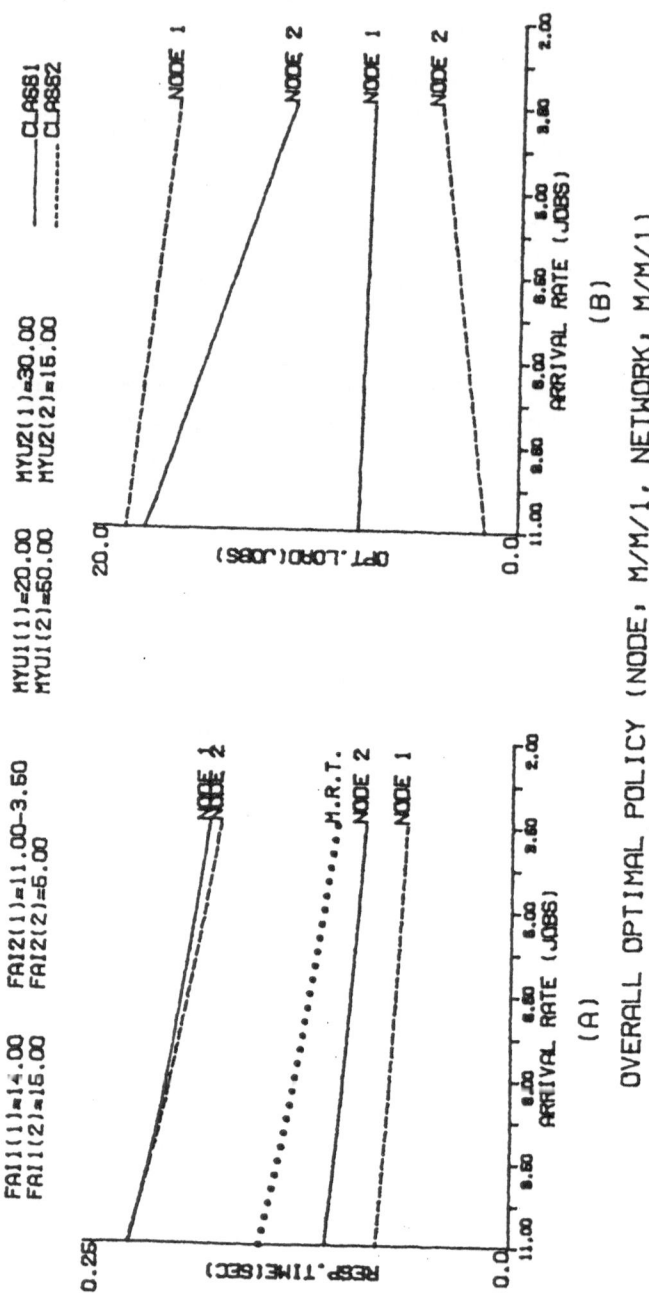

Figure 2.17: The effect of arrival rate on the optimum (case 2)

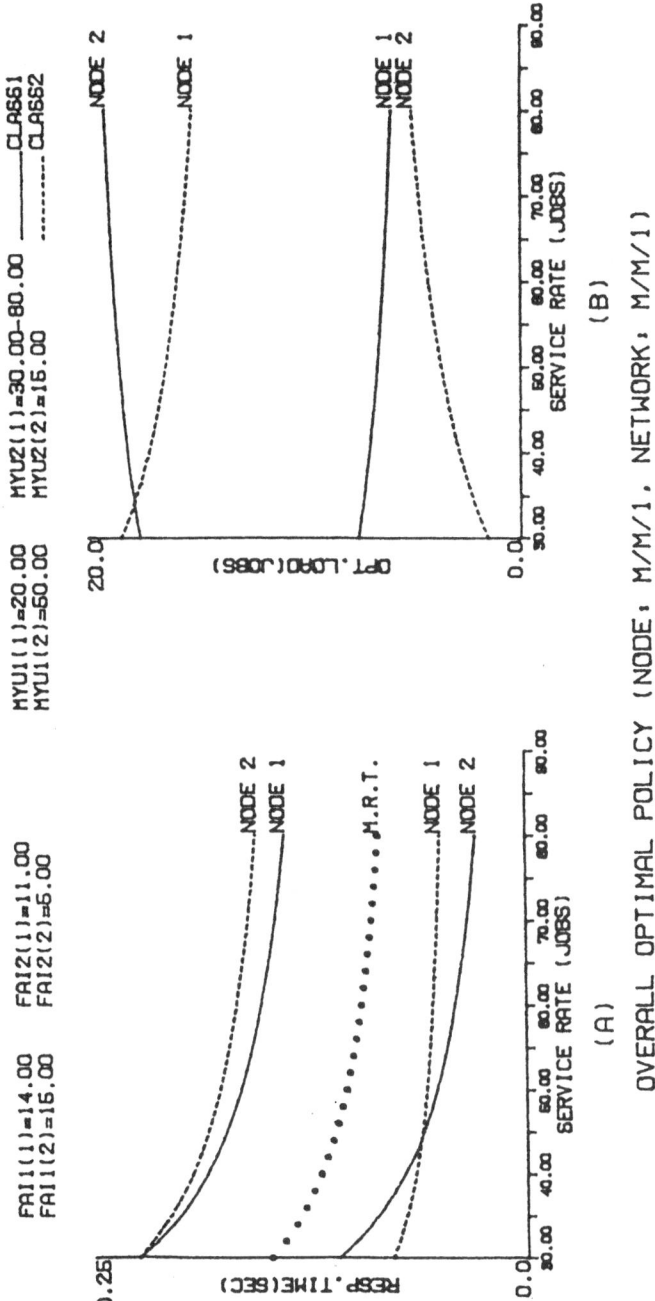

Figure 2.18: The effect of service rate on the optimum (case 2)

could anticipate no improvement in its expected response time even if it were processed by a node different from the processing node that had been determined by the policy. The equilibrium situation seems *fair* to all jobs in the sense that no job can expect any further improvement.

We can see that the smaller arrival rate and the larger processing capacity, for particular class k and node i, both have a similar effect on the system behavior in the optimum. The larger arrival rate and the smaller processing capacity also have the same effect on the system behavior in the optimum. We may observe the same tendencies in the equilibrium.

2.4.4 Conclusion

We studied the effects of system parameters on two contrasting policies for statically balancing the load in a distributed computer system with multiple job classes. One is called an overall optimal policy, whereby jobs are scheduled so that the system-wide mean job response time should be minimum. The other is called an individually optimal policy, whereby jobs are scheduled so that every job may feel that its own expected response time is minimum.

We studied the effects of system parameters numerically, i.e. the effects of various system parameters including the communication time, the job arrival rate, and the processing capacity, on the behavior of the optimum and of the equilibrium. In the results of numerical experiments, we observed several interesting phenomena. For the shorter communication times, it is observed that more of each class jobs are processed at remote nodes that have high service rates for the class. It seems that this kind of phenomenon, which we call specialization, is one of the major characteristics of load balancing in distributed computer systems. For the longer communication times, in the overall optimal policy, nodes exchange jobs more actively than in the individually optimal policy.

Two groups of jobs of the same class that arrive at the same node and that are processed by different nodes according to the policies are considered. In the optimum, the two groups of jobs may show mutually different expected response times, which seems unfair. On the other hand, in the equilibrium, the two groups of jobs may both show the same expected job response time, which seems fair. The results show that the objectives of two load balancing policies may conflict.

It is observed that decreasing the arrival rate, for particular class k and node i, has an effect similar to that of increasing the processing capacity. Increasing the arrival rate, for particular class k and node i, also has an effect similar to that of decreasing the processing capacity on the system behavior. We observed this *anomalous* behavior in the equilibrium where the system-wide mean job response time of the case where there exist some jobs processed remotely may be longer than that of the case where there exists no job processed remotely. This means that since minimizing the system-wide mean job response time is not a purpose of load balancing in the individually optimal policy, the policy may sometimes sacrifice the system-wide mean job response time in order to make each job see expected job response time minimized.

Based on the results of parametric analysis for the major system parameters in the two optimal load balancing policies, the questions of how a designed system behaves when the system parameters are changed and what kind of problems exist in the system are raised. The solutions of the two optimal static load balancing policies may also help us design the system.

Chapter 3

Load Balancing in Tree Hierarchy Network Configurations

3.1 Introduction

Of late, more and more computer networks are being established, and many of them have the tree topology. That is, we often have tree hierarchy organizations of computing machines, e.g., a super–power center machine is connected with remote mainframe computers each of which is, in turn, connected with server-type workstations each of which is connected with client-type workstations and personal computers, etc. Since the tree hierarchy network configuration is extensible and flexible, it is an important configuration in the literature. The star networks are special examples of the tree hierarchy networks.

In this chapter, we focus on the optimal static load balancing in a tree hierarchy computer network. Consider a set of heterogeneous host computers (nodes) interconnected by a tree hierarchy network. Jobs arrive at each node according to a time-invariant Poisson process. Each node determines whether each entire job should be processed locally or scheduled to another node for remote processing. In the latter case, there is a delay incurred as a result of forwarding the job and sending its response back over the network. Based on these assumptions, we formulate the optimal load balancing problem as a nonlinear optimization problem which minimizes the mean response time of a job in a tree hierarchy network. The optimal solution to this problem is referred to as the optimal load balancing strategy.

We derive theorems which give the necessary and sufficient conditions for the optimal solution to the tree hierarchy problem. On the basis of these theorems, it is shown that the tree hierarchy optimization problem can be solved by solving the star sub–optimization problem iteratively. Then we present a decomposition algorithm to solve the optimal static load balancing problem of a tree hierarchy network. A proof of the convergence of the proposed algorithm is given. In general, we note that the decomposition approaches are very effective to solve the optimization problems of large-scale and complicated systems [Cou77, LVM79].

The performance comparison of the proposed algorithm and two well-known algorithms, the FD (Flow Deviation) algorithm [FGK73] and the D-S (Dafermos–Sparrow) algorithm [DS69] which can be applied to solve the optimal load balancing problem, is provided. The algorithm performance is compared by measuring two criteria, the amount of the storage requirements and the amount of the computation time requirements. For the comparison of the storage requirements, our results show that the amounts of the storage required for the proposed algorithm and the FD algorithm are almost the same. However, the amount of the storage required for the D–S algorithm is much larger than that for the other two algorithms. For the comparison of the computation times, the results in our numerical experiments show that both the proposed algorithm and the D–S algorithm have much faster convergence in terms of CPU time than the FD algorithm.

We also study the effects of the link communication time and the node processing time on the following three system performance variables: link flow rates (i.e. the rate at which a node forwards jobs to another node for remote processing), node loads (i.e. the rate at which jobs are processed at a node), and the mean response time, of optimal static load balancing in tree hierarchy network configurations by parametric analysis. It is shown that the entire network is divided into several independent sub-tree networks with respect to the link flow rates and node loads, and that the communication time of a link and the processing time of a node have the effects only on the link flow rates and the loads of nodes that are in the same sub-tree network. Although the procedures of proofs are complicated, we obtain the following clear and simple relations.

- An increase in the communication time of a link causes a decrease in the link flow rates of its descendant nodes, its ancestor nodes and itself, but causes an increase in the link flow rate of other nodes in the same sub-tree network. Furthermore, it also causes an increase in the loads of its descendant nodes and itself, but causes a decrease in the loads of other nodes in the same sub-tree network.

- An increase in the processing time of a node causes an increase in the link flow rates of its ancestor nodes and itself, but causes a decrease in the link flow rates of other nodes in the same sub-tree network. Furthermore, it also causes a decrease in the load of itself, but causes an increase in the loads of other nodes in the same sub-tree network.

- In general, either an increase in the communication time of a link or an increase in the processing time of a node will cause an increase in the mean response time.

3.2 Model Description and Problem Formulation

We call G a *network* where G is a pair (N, L). Here N denotes a set whose elements are called *nodes* and L is a set whose elements are ordered pairs of elements of N. The ordered elements are called *links* which are represented by arrows from origin nodes to destination nodes. The direction of the arrow furnishes the link with an orientation.

By a *path* connecting the *ordered* pair $w = (x, y)$ of nodes we mean a sequence of links (p_1, p_2), $(p_2, p_3), ..., (p_{n-1}, p_n)$ where $p_1, p_2, ..., p_n$ are distinct nodes, $p_1 = x$, and $p_n = y$. We note that each link in a path has such an orientation that node p_i $(1 < i < n)$ is the destination node of link (p_{i-1}, p_i) and the origin node of link (p_i, p_{i+1}). In particular, each link is a path . If there is a path from node x to node y, then y is said to be *reachable* from x. A *cycle* is a sequence of links (p_1, p_2), $(p_2, p_3), ..., (p_{n-1}, p_n)$ where $p_1, p_2, p_3, ..., p_{n-1}$ are distinct nodes while the first node p_1 is the same as the last node p_n, $p_1 = p_n$. A network is *acyclic* if it has no cycles.

Before we define the tree hierarchy network, let us introduce the concepts of the ancestor, parent, descendant, child and root in a network. We call node j an *ancestor* of node i (or node i a *descendant* of node j) if node j is reachable from node i. We call node j a *parent* of node i (or node i a *child* of node j) if there is a link from node i to node j. A *root* node is a node that has no ancestor nodes. Now we define that a *tree hierarchy network* is an acyclic network which has a unique root node, each node of which has a unique parent if the node is not the root node (Figure 3.1). Then each node i has a unique link to its parent. We call such a link *link i*. In a tree hierarchy network, a node which has its parent and has no child is called a *terminal node*, and a node which has at least one child is called a *nonterminal node*. We will take the root node as node 0 in the remainder of this paper.

In this paper, nodes may be heterogeneous computer systems, that is, they may have different configurations, number of resources, and speed characteristics. However, they have sufficient processing capabilities, that is, a job may be processed from start to finish at any node in the network. Jobs arrive at node i $(i \in N)$ according to a time–invariant Poisson process. A job arriving at node i can be processed at node i or be transferred to one of the ancestor nodes of node i.

Note that in our model of a tree hierarchy network, we assume that a job arriving at node i can only be processed locally or be forwarded to the ancestor node of node i. This assumption is reasonable if the processing capacities of the ancestor nodes of a node i are larger than that of node i, and the processing capacities of the descendant nodes are smaller than that of node i. For example, in a tree hierarchy network a super–power center machine is the root node which is connected with remote mainframe computers each of

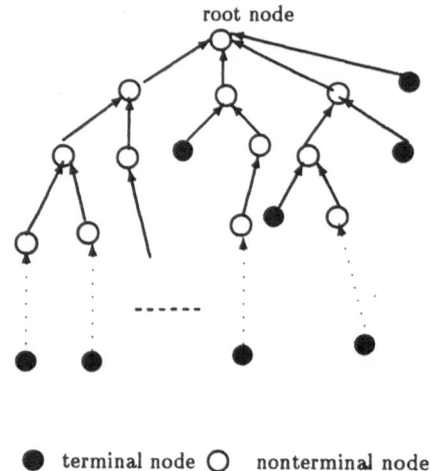

Figure 3.1: A tree hierarchy network configuration

which is, in turn, connected with server-type workstations each of which is connected
with client-type workstations and personal computers, etc. Notice that there are a lot
of distributed computing systems with the tree hierarchy network configuration. Note,
however, in some distributed computer systems with tree hierarchy configurations the
excessive jobs may be forwarded either to the ancestor or to the offspring nodes. This
model cannot treat such systems.

For the description of the tree hierarchy optimization problem, at first, let us introduce
the following notation and assumptions.

- B: the set of nonterminal nodes.

- T: the set of terminal nodes.

- L: the set of links, also we use this to denote the set of nodes that have a parent,
  i.e., $N - \{0\}$.

- V_j: the set of nodes that are children of node j, i.e.,

  $V_j = \{i \mid node\ i\ is\ the\ child\ of\ node\ j\}, \quad j \in B.$

- α_i: the Lagrange multiplier, $i \in N$.

- $\boldsymbol{\alpha}$: $[\alpha_i]$.

- β_i : the rate at which jobs are processed, also referred to as *load*, at node i, $i \in N$.

- $\boldsymbol{\beta}$: $[\beta_i]$.

- x_i : the job flow rate from child node i to its parent node on link i, $i \in L$.

- $\boldsymbol{x}$: $[x_i]$.

- ϕ_i: the external job arrival rate at node i, $i \in N$.

- ϕ_i': the total job flow into node i, i.e.,

$$\phi_i' = \begin{cases} \phi_i + \sum_{l \in V_i} x_l, & i \in B, \\ \phi_i, & i \in T. \end{cases}$$

- $F_i(\beta_i)$: the mean node delay of a job processed at node i, $i \in N$.

- $G_i(\boldsymbol{x})$: the mean communication delay for sending a job from node i to its parent node and sending the answer back to node i on link i $(i \in L)$. Assume that the mean communication time on link i depends solely upon the job flow rate x_i on this link. Then, we use $G_i(x_i)$ instead of $G_i(\boldsymbol{x})$.

- $D(\boldsymbol{\beta}, \boldsymbol{x})$: the mean delay (mean system response time) that a job spends in the network from the time of its arrival until the time of its departure. It consists of a node delay (queueing and processing delays) at the processing node in addition to a possible communication delay incurred due to job transfer.

In general, we assume that the mean node delay function $F_i(\beta_i)$ is a differentiable, increasing, and strictly convex function and the mean communication delay function $G_i(x_i)$ is a differentiable, nondecreasing, and convex function. The class of delay functions satisfying the assumptions is large, and it contains those of the M/M/c model, and the central server model [Kle76]. We express $D(\boldsymbol{\beta}, \boldsymbol{x})$ as the sum of the mean node delay at all nodes and the mean communication delay on all links by using the Little's result as follows,

$$D(\boldsymbol{\beta}, \boldsymbol{x}) = \frac{1}{\Phi} \{ \sum_{i \in N} \beta_i F_i(\beta_i) + \sum_{i \in L} x_i G_i(x_i) \},$$

where $\Phi = \sum_{i \in N} \phi_i$.

We have the optimization problem of minimizing the mean response time of a job which we can write by considering the structure of tree hierarchy configurations as follows:

$$\text{Minimize} \quad D(\boldsymbol{\beta}, \boldsymbol{x}) = \frac{1}{\Phi} \{ \sum_{i \in N} \beta_i F_i(\beta_i) + \sum_{i \in L} x_i G_i(x_i) \}, \tag{3.1}$$

with respect to the variables β_i, $i \in N$, and x_i, $i \in L$,

subject to

$$\phi_0' = \beta_0, \tag{3.2a}$$

$$\phi_i' = \beta_i + x_i, \quad i \in L, \tag{3.2b}$$

$$\text{where } \phi_i' = \begin{cases} \phi_i + \sum_{l \in V_i} x_l, & i \in B, \\ \phi_i, & i \in T, \end{cases} \tag{3.2c}$$

$$\beta_i \geq 0, \quad i \in N, \tag{3.2d}$$

$$x_i \geq 0, \quad i \in L. \tag{3.2e}$$

The first and second constraints (3.2a) and (3.2b) are the flow balance constraints. Each of them equates the flow into and out of each node in every node of the network. The third and fourth constraints (3.2d) and (3.2e) show the nonnegativity of β_i and x_i. We note that the flow into node i comes only from or through its child nodes and the flow out of node i goes only to or through its parent node.

3.3   Optimal Load Balancing

The load balancing strategy which minimizes the mean response time is obtained by solving the optimization problem (3.1).

In Tantawi and Towsley paper [TT85], they show that they can disjointedly divide the nodes in a single channel configured network if the communication network satisfies the triangle inequality property. This partitioning is important in showing the necessary and sufficient condition for optimal load balancing. In a tree hierarchy network, we make a disjoint partition of nodes without any assumption as follows. According to the magnitude of x_i ($i \in V_j$) relative to ϕ_i', we divide the nodes in the set V_j ($j \in B$) into three types, *idle source nodes*, *active source nodes*, and *neutral nodes* which we denote by Rd_j, Ra_j, and Nu_j, respectively. Node i is said to be a source if $\beta_i < \phi_i'$, i.e., it sends all or a part of its incoming jobs to one of its ancestor nodes. Node i is an idle source if $\beta_i = 0$ and an active source if $\beta_i > 0$. If node i processes all its incoming jobs locally ($\beta_i = \phi_i'$), it is a neutral node.

Furthermore, here we introduce two delay functions that we will use in the remainder of the paper. We define the *incremental node delay function* $f_i(\beta_i)$ and the *incremental communication delay function* $g_i(x_i)$ as follows,

$f_i(\beta_i) = \frac{d}{d\beta_i}\beta_i F_i(\beta_i)$,

$g_i(x_i) = \frac{d}{dx_i}x_i G_i(x_i)$.

Due to our assumption about $F_i(\beta_i)$ and $G_i(x_i)$, we note that $f_i(\beta_i)$ is an increasing function and $g_i(x_i)$ is a nondecreasing function. We can define the inverse of a function f^{-1} by

$$f^{-1}(x) = \begin{cases} a, & f(a) = x, \\ 0, & f(0) \geq x. \end{cases}$$

Theorem 3.1 *The set of values of $\boldsymbol{\beta}$ and $\boldsymbol{x}$ is an optimal solution to problem (3.1) if and only if the following set of relations holds for all $i \in V_j$, $j \in B$.*

$$f_i(\beta_i) \geq \alpha_i = \alpha_j + g_i(x_i), \qquad \beta_i = 0, \qquad (i \in Rd_j), \tag{3.3a}$$

$$f_i(\beta_i) = \alpha_i = \alpha_j + g_i(x_i), \quad 0 < \beta_i < \phi'_i, \quad (i \in Ra_j), \tag{3.3b}$$

$$f_i(\beta_i) = \alpha_i \leq \alpha_j + g_i(x_i), \qquad \beta_i = \phi'_i, \qquad (i \in Nu_j), \tag{3.3c}$$

$$\beta_i + x_i = \phi'_i, \tag{3.4a}$$

$$\beta_0 = \phi'_0, \tag{3.4b}$$

$$\beta_0 = f_0^{-1}(\alpha_0), \tag{3.5}$$

where
$$\phi'_i = \phi_i + \sum_{l \in V_i} x_l, \quad i \in B, \tag{3.6a}$$

$$= \phi_i, \quad i \in T, \tag{3.6b}$$

PROOF.

To obtain the optimal solution, we form the Lagrangian function as follows,

$$L(\boldsymbol{\beta}, \boldsymbol{x}, \boldsymbol{\alpha}) = \boldsymbol{\Phi}D(\boldsymbol{\beta}, \boldsymbol{x}) + \alpha_0(\phi'_0 - \beta_0) + \sum_{j \in B-\{0\}} \sum_{i \in V_j} \alpha_i(\phi'_i - \beta_i - x_i). \tag{3.7}$$

Since the objective function is convex by assumption and the feasible set is a convex set, a job flow pattern $(\boldsymbol{\beta}, \boldsymbol{x})$ is an optimal solution to the problem (3.1) if and only if it satisfies the following Kuhn-Tucker conditions (e.g., [Int71]),

$$\frac{\partial L}{\partial \beta_i} = f_i(\beta_i) - \alpha_i \geq 0, \tag{3.8a}$$

$$\beta_i \frac{\partial L}{\partial \beta_i} = \beta_i(f_i(\beta_i) - \alpha_i) = 0, \tag{3.8b}$$

$$\beta_i \geq 0, \tag{3.8c}$$

$$i \in N.$$

$$\frac{\partial L}{\partial x_i} = g_i(x_i) + \alpha_j - \alpha_i \geq 0, \tag{3.9a}$$

$$x_i \frac{\partial L}{\partial x_i} = x_i(g_i(x_i) + \alpha_j - \alpha_i) = 0, \tag{3.9b}$$

$$x_i \geq 0, \tag{3.9c}$$

$$i \in V_j, \quad j \in B.$$

$$\phi'_0 - \beta_0 = 0, \tag{3.10a}$$

$$\phi'_i - \beta_i + x_i = 0, \tag{3.10b}$$

$$i \in V_j, \quad j \in B.$$

By noting that $\beta_i + x_i = \phi'_i$, the above set of conditions is identical with the following set of relations,

$$f_i(\beta_i) \geq \alpha_i, \qquad \beta_i = 0, \tag{3.11a}$$

$$f_i(\beta_i) = \alpha_i, \quad 0 < \beta_i \leq \phi'_i, \tag{3.11b}$$

$$\text{for } i \in N.$$

$$\alpha_i = \alpha_j + g_i(x_i), \quad 0 \leq \beta_i < \phi'_i, \tag{3.12a}$$

$$\alpha_i \leq \alpha_j + g_i(x_i), \qquad \beta_i = \phi'_i, \tag{3.12b}$$

$$\text{for } i \in L.$$

$$\beta_i + x_i = \phi'_i, \text{ for } i \in L \tag{3.13a}$$

$$\beta_0 = \phi'_0. \tag{3.13b}$$

We can easily see the above set is equivalent to the set given in Theorem 3.1 by recalling the definition of $f^{-1}(\bullet)$. Q.E.D.

Remark: The above theorem shows that the values of β and x can be determined by using relations (3.3), (3.4) and (3.5) if the Lagrange multipliers can be determined. A limitation of the usefulness of the theorem may arise from the fact that it is not easy to determine all multipliers α_i ($i \in N$) by directly using Theorem 3.1. We proceed to find more properties about the optimal solution to problem (3.1).

3.4   Decomposability

Let us examine the structure of problem (3.1). Consider a particular node j ($\in B$). We note that the value of variables β_i, x_i ($i \in V_j$) and β_j can be varied with the values of other variables in β and x being fixed. We can do this without violating the constraints (3.2). If we keep such other variables constant, the minimization problem (3.1) reduces to the following subproblem,

$$\text{minimize } \frac{1}{\Phi}\{\beta_j F_j(\beta_j) + \sum_{i \in V_j}[x_i G_i(x_i) + \beta_i F_i(\beta_i)] + C\}, \tag{3.14}$$

with respect to the variables x_i, β_i, $i \in V_j$, and β_j,

subject to

$$\phi'_i = \beta_i + x_i, \quad i \in V_j, \tag{3.15a}$$

$$\phi'_j = \beta_j + x_j, \tag{3.15b}$$

$$(\text{i.e., } \phi_j + \textstyle\sum_{i \in V_j} x_i = \beta_j + x_j,)$$

$$\beta_j \geq 0, \quad x_i \geq 0, \quad \beta_i \geq 0, \quad i \in V_j. \tag{3.15c}$$

(C denotes a variable that is independent of x_i, β_i ($i \in V_j$) and β_j.)

This observation may lead us to an interesting decomposition technique to solve problem (3.1). In relation to this, we have the following theorem.

Theorem 3.2 *The set of values of β_i, x_i ($i \in V_j$) and β_j is an optimal solution to the subproblem (3.14) if and only if the following set of relations holds for all $i \in V_j$,*

$$f_i(\beta_i) \geq \alpha_i = \alpha_j + g_i(x_i), \qquad \beta_i = 0, \qquad (i \in Rd_j), \tag{3.16a}$$

$$f_i(\beta_i) = \alpha_i = \alpha_j + g_i(x_i), \quad 0 < \beta_i < \phi'_i, \quad (i \in Ra_j), \tag{3.16b}$$

$$f_i(\beta_i) = \alpha_i \leq \alpha_j + g_i(x_i), \qquad \beta_i = \phi'_i, \qquad (i \in Nu_j), \tag{3.16c}$$

$$\beta_i + x_i = \phi'_i, \quad i \in V_j, \tag{3.17a}$$

$$\beta_j + x_j = \phi'_j = \phi_j + \sum_{i \in V_j} x_i, \tag{3.17b}$$

$$\beta_j = f_j^{-1}(\alpha_j). \tag{3.18}$$

PROOF. We can show the above similarly as Theorem 3.1 by regarding node j as the root node and $\phi_j - x_j$ as ϕ_0. Only the difference from Theorem 3.1 lies in the possibility that $\phi_j - x_j$ can be negative in the subproblem. We can easily see that the proof of Theorem 3.1 can be extended to this generalization. Q.E.D.

Let us have the following definition,

$$h_i(\beta_i) = f_i(\beta_i) - g_i(\phi'_i - \beta_i).$$

Note that $h_i(\beta_i)$ is a monotonically increasing function.

Corollary 3.3 *The optimal solution to the subproblem (3.14) must satisfy the following relations for all $i \in V_j$,*

$$h_i(\beta_i) \geq \alpha_j, \qquad \beta_i = 0, \qquad (i \in Rd_j), \tag{3.19a}$$

$$h_i(\beta_i) = \alpha_j, \quad 0 < \beta_i < \phi'_i, \quad (i \in Ra_j), \tag{3.19b}$$

$$h_i(\beta_i) \leq \alpha_j, \qquad \beta_i = \phi'_i, \qquad (i \in Nu_j), \tag{3.19c}$$

$$\beta_j = f_j^{-1}(\alpha_j), \tag{3.20}$$

$$\beta_i + x_i = \phi'_i, \tag{3.21}$$

$$\sum_{i \in Rd_j} \phi'_i + \sum_{i \in Ra_j} (\phi'_i - h_i^{-1}(\alpha_j)) = f_j^{-1}(\alpha_j) - \phi_j + x_j. \tag{3.22}$$

PROOF. Relations (3.19) are derived from relations (3.16). Eq.(3.22) is derived from eqs.(3.17b) and (3.20) and relation (3.19b). Q.E.D.

Now we proceed to show that the set of relations given in Corollary 3.3 uniquely determines the values of β_i, x_i $(i \in V_j)$ and β_j. We show the following Corollaries 3.4 and 3.5.

Corollary 3.4 *The set of relations (3.19) is equivalent to the following set of relations for all $i \in V_j$,*

$$h_i(0) \geq \alpha_j, \quad \beta_i = 0, \quad (i \in Rd_j), \tag{3.23a}$$

$$h_i(\phi'_i) > \alpha_j > h_i(0), \quad 0 < \beta_i < \phi'_i, \quad (i \in Ra_j), \tag{3.23b}$$

$$h_i(\phi'_i) \leq \alpha_j, \quad \beta_i = \phi'_i, \quad (i \in Nu_j). \tag{3.23c}$$

Note the following definitions in an optimal solution to subproblem (3.14):

$$Rd_j = \{i|\ \beta_i = 0,\ i \in V_j\}, \quad \text{(idle sources)},$$

$$Ra_j = \{i|\ 0 < \beta_i < \phi'_i,\ i \in V_j\}, \quad \text{(active sources)},$$

$$Nu_j = \{i|\ \beta_i = \phi'_i,\ i \in V_j\}, \quad \text{(neutral)}.$$

Corollary 3.5 *If β_i $(i \in V_j)$ is an optimal solution to the subproblem (3.14), then we have,*

$$\beta_i = 0, \quad i \in Rd_j, \tag{3.24a}$$

$$\beta_i = h_i^{-1}(\alpha_j), \quad i \in Ra_j, \tag{3.24b}$$

$$\beta_i = \phi'_i, \quad i \in Nu_j. \tag{3.24c}$$

Let us introduce the following definitions in the order shown below for an arbitrary α:

$$Rd_j(\alpha) = \{i|\ h_i(0) \geq \alpha,\ i \in V_j\},$$

$$Ra_j(\alpha) = \{i|\ h_i(\phi'_i) > \alpha > h_i(0),\ i \in V_j\},$$

$$Nu_j(\alpha) = \{i|\ h_i(\phi'_i) \leq \alpha,\ i \in V_j\},$$

$$\lambda_{S_j}(\alpha) = f_j^{-1}(\alpha) - \phi_j + x_j,$$

$$\lambda_{R_j}(\alpha) = \sum_{i \in Rd_j(\alpha)} \phi'_i + \sum_{i \in Ra_j(\alpha)} (\phi'_i - h_i^{-1}(\alpha)),$$

$$\beta_i(\alpha) = \begin{cases} 0, & i \in Rd_j(\alpha), \\ h_i^{-1}(\alpha), & i \in Ra_j(\alpha), \\ \phi_i', & i \in Nu_j(\alpha), \end{cases}$$

$$x_i(\alpha) = \phi_i' - \beta_i(\alpha),$$

$$\beta_j(\alpha) = \phi_j - x_j + \sum_{i \in Ra_j(\alpha)} x_i(\alpha) + \sum_{i \in Rd_j(\alpha)} \phi_i'.$$

Then, if an optimal α is given, $Rd_j = Rd_j(\alpha)$, $Ra_j = Ra_j(\alpha)$, and $Nu_j = Nu_j(\alpha)$, and

$$\lambda_{S_j}(\alpha) = \lambda_{R_j}(\alpha). \quad \text{(see eq.(3.22))} \tag{3.25}$$

Lemma 3.6 *As α increases, $\lambda_{S_j}(\alpha)$ first stays to be $x_j - \phi_j$ while $\alpha \le f_j(0)$ and then increases monotonically while $\alpha > f_j(0)$. On the other hand, as α increases, $\lambda_{R_j}(\alpha)$ first stays to be $\sum_{i \in V_j} \phi_i'$ while $\alpha \le \min_{i \in V_j}\{h_i(0)\}$ and then decreases monotonically while $\min_{i \in V_j}\{h_i(0)\} < \alpha < \max_{i \in V_j}\{h_i(\phi_i')\}$ and then stays to be zero while $\alpha \ge \max_{i \in V_j}\{h_i(\phi_i')\}$.*
PROOF. It is easy to prove this from the above definitions and the properties on $f_i(\beta_i)$ and $g_i(x_i)$. Q.E.D.

Naturally we can assume that $x_j - \phi_j \le \sum_{i \in V_j} \phi_i'$, i.e., the rate at which node j sends jobs to its parent cannot exceed the sum of the external job arrival rates at node i ($i \in V_j$), and node j. Then we have the following property.

Lemma 3.7 *The set of relations given in Corollary 3.3 determines uniquely the set of values of β_i, x_i ($i \in V_j$) and β_j, which is the optimal solution to subproblem (3.14). It determines also the value of α_j uniquely except in the following two cases,*

    case 1. $x_j - \phi_j = \sum_{i \in V_j} \phi_j'$. Then $\alpha \le \min\{f_j(0), \min_{i \in V_j}\{h_i(0)\}\}$,

    case 2. $x_j - \phi_j = 0$ and $f_j(0) > \max_{i \in V_j}\{h_i(\phi_i')\}$. Then $\max_{i \in V_j}\{h_i(\phi_i')\} \le \alpha \le f_j(0)$.
PROOF. From Lemma 3.6 and by noting that $x_j - \phi_j \le \sum_{i \in V_j} \phi_i'$, we can see that there exists such an α that satisfies $\lambda_{S_j}(\alpha) = \lambda_{R_j}(\alpha)$ (eq.(3.25)). Such an α is unique except in cases 1 and 2, and then $\beta_i(\alpha)$, $x_i(\alpha)$ ($i \in V_j$) and $\beta_j(\alpha)$ are also determined uniquely through the definitions.

Let us examine the two exceptional cases.

Case 1. $x_j - \phi_j = \sum_{i \in V_j} \phi_i'$. We have from Lemma 3.6 that eq.(3.25) holds for every α that satisfies $\alpha \le \min\{f_j(0), \min_{i \in V_j}\{h_i(0)\}\}$.
In this case, node i ($i \in V_j$) and node j are all idle, and thus

$$x_i = \phi_i', \beta_i = 0 \ (i \in V_j) \text{ and } \beta_j = 0.$$

Case 2. $x_j - \phi_j = 0$ and $f_j(0) > \max_{i \in V_j}\{h_i(\phi_i')\}$. We have from Lemma 3.6 that eq.(3.25) holds for every α that satisfies $\max_{i \in V_j}\{h_i(\phi_i')\} \le \alpha \le f_j(0)$.

In this case, node j is idle and all of node i ($i \in V_j$) are neutral and thus,

$$x_i = 0, \; \beta_i = \phi_i' \; (i \in V_j) \text{ and } \beta_j = 0.$$

In the above all cases, we can see that we have a unique set of values of β_i, x_i ($i \in V_j$) and β_j that satisfies the relations given in Corollary 3.3. Thus from Corollary 3.3, we see that the unique set of values is the optimal solution to the subproblem (3.14). Q.E.D.

Remark: From the above consideration, we easily see that the uniquely determined α_j must be within the range shown below.

Case 3. $x_j - \phi_j > 0$. Then $\min\{f_j(0), \min_{i \in V_j}\{h_i(0)\}\} \le \alpha_j \le \max_{i \in V_j}\{h_i(\phi_i')\}$.

Case 4. $x_j - \phi_j < 0$.

Case 4.1. $f_j(\phi_j - x_j) > \max_{i \in V_j}\{h_i(\phi_i')\}$. Then $\alpha_j = f_j(\phi_j - x_j)$.

Case 4.2. $f_j(\phi_j - x_j) \le \max_{i \in V_j}\{h_i(\phi_i')\}$. Then $f_j(\phi_j - x_j) \le \alpha_j \le \max_{i \in V_j}\{h_i(\phi_i')\}$.

We then have the following property.

Theorem 3.8 *The set of values of β and x is an optimal solution of problem (3.1) if and only if the set of values of β_i, x_i ($i \in V_j$) and β_j is an optimal solution to subproblem (3.14) for all $j \in B$ at the same time.*

PROOF. Since problem (3.1) is not an unconstrained optimization problem, careful examination is necessary as shown below.

The necessity is clear since such a set of values of β, x, and α that satisfies the relations of Theorem 3.1 satisfies the relations of Theorem 3.2 for all $j \in B$ all at the same time.

The sufficiency is clear if we can show that the values of α_i ($i \in V_j$) and α_j are determined uniquely and consistently for all j, $j \in B$, by the relations given in Theorem 3.2. It is clear in all the cases other than cases 1 and 2 given in Lemma 3.7. In these cases, α_j is uniquely determined and thus one of relations (3.16) determines α_i ($i \in V_j$) uniquely. In case 1, we can choose, for example, $\alpha_j = \min\{f_j(0), \min_{i \in V_j}\{h_i(0)\}\}$. In this case, every node i ($i \in V_j$) is idle and α_i is uniquely determined by relation (3.16a). In case 2, we can choose the value of α_j such that, for example, $\alpha_j = f_j(0) \ge \max_{i \in V_j}\{h_i(\phi_i')\}$. In this case, every node i ($i \in V_j$) is neutral and α_i is uniquely determined by relation (3.16c). In this way, α_i ($i \in V_j$) and α_j can be determined uniquely and consistently for all $j \in B$, and we can show the sufficiency. Q.E.D

Note the similarity between the set of relations given in Corollary 3.3 and that of Theorem 3.1 given by Tantawi and Towsley [TT84]. In the case where $x_j - \phi_j \ge 0$, the former is identical with the latter, and we can use the Tantawi and Towsley algorithm

[TT84] or the Kim and Kameda algorithm [KK92a] to obtain the optimal solution to the subproblem (3.14).

Consider the case where $x_j - \phi_j < 0$, i.e., the parent node j is forced to send a certain rate of jobs to ancestor nodes. We can provide a modified version of the Tantawi and Towsley or Kim and Kameda algorithm [TT84, KK92a] to include this case also. In the next section, we present the design and principles of the decomposition algorithm based on the theories given in this section.

3.5 Proposed Algorithm

The general structure of the algorithm should now be clear. That is to solve the tree hierarchy optimization problem (3.1) by solving the much simpler star subproblem (3.14) iteratively.

Two algorithms, the Tantawi and Towsley (T&T) algorithm [TT84] and the Kim and Kameda (K&K) algorithm [KK92a], have been developed to solve the star optimization subproblem (3.14) when the external job arrival rates at the central node and the terminate nodes are nonnegative. The main process of the K&K algorithm is to determine the Lagrange multiplier α_j directly by a simple one-dimensional search method and then to determine the optimal load in the star network. The T&T algorithm determines the node partitions before calculating α_j exactly. Here we apply the K&K algorithm to solve the star subproblem (3.14) if the external job arrival rate at the central node is nonnegative, i.e., $\phi_j - x_j \geq 0$. If the external job arrival rate at the central node is negative, i.e., $\phi_j - x_j < 0$, according to Corollary 3.3, we can develop an extension to the K&K algorithm. We can apply the gold section search to determine α_j from eq.(3.22). After determining the multiplier α_j, we can obtain the optimal solution to the subproblem (3.14) in the same way as the K&K algorithm.

Before we introduce the proposed algorithm, we define the set of feasible solutions FS as follows,

$FS = \{(\boldsymbol{\beta}, \boldsymbol{x}) \mid (\boldsymbol{\beta}, \boldsymbol{x}) \ satisfies \ the \ constraints \ (3.2)\}$. We have that the set FS is a convex polyhedron. We also note that the set FS is closed according to the constraints (3.2) which the job flow pattern $(\boldsymbol{\beta}, \boldsymbol{x})$ in the set FS must satisfy.

The following is the proposed algorithm.

- Decomposition Algorithm

1. Initialization.

    $r = 0$. (r : iteration number)

    Find $(\boldsymbol{\beta}^{(0)}, \boldsymbol{x}^{(0)})$ ($\in FS$) as an initial feasible solution to the problem (3.1).

2. Solve the star subproblem (3.14) iteratively to obtain $(\boldsymbol{\beta}^{(r)}, \boldsymbol{x}^{(r)})$.

   Increase the iteration number r by 1.

   Do the following with each node j ($j \in B$) taken downward in the tree hierarchy network. Take node j as the central node and node i ($i \in V_j$) as a satellite node in the star subproblem (3.14). Let $\beta_j^{(r)}$, $\phi_j - x_j^{(r)}$, $\beta_i^{(r)}$, $x_i^{(r)}$ and $\phi_i''^{(r)}$ ($i \in V_j$) be β_j, $\phi_j - x_j$, β_i, x_i and ϕ_i' in the star subproblem (3.14), respectively, with $\phi_j - x_j^{(r)}$ and $\phi_i''^{(r)}$ being given, where $\phi_i''^{(r)} = \begin{cases} \phi_i + \sum_{l \in V_i} x_l^{(r-1)}, & i \in B, \\ \phi_i, & i \in T. \end{cases}$ Apply the sub-algorithm given below to solve the subproblem (3.14) for node j.

3. Stopping rule.

   If $|D(\boldsymbol{\beta}^{(r)}, \boldsymbol{x}^{(r)}) - D(\boldsymbol{\beta}^{(r-1)}, \boldsymbol{x}^{(r-1)})| / D(\boldsymbol{\beta}^{(r)}, \boldsymbol{x}^{(r)}) < \varepsilon$ then STOP, where ε is a proper acceptance tolerance, otherwise, goto step 2.

- Sub-algorithm that solves the subproblem (3.14) with x_j and ϕ_i' ($i \in V_j$) being given.

1. Compare. $O(n)$

   If $\phi_j - x_j \geq 0$ and $\max_{i \in V_j}\{h_i(\phi_i')\} \leq f_j(\phi_j - x_j)$, then no load balancing is required. In this case, let

   $\beta_i = \phi_i'$, $x_i = 0$, $\forall i \in V_j$, $\beta_j = \phi_j - x_j$, and

   return to the main algorithm.

   Otherwise, proceed to step 2.

2. Determine α_j. $O(n)$

   Find such α_j that satisfies

   $$\lambda_{S_j}(\alpha_j) = \lambda_{R_j}(\alpha_j), \quad (\text{see eq.}(3.25))$$

   by applying a golden section search, where

   $$\lambda_{R_j}(\alpha_j) = \sum_{i \in Rd_j(\alpha_j)} \phi_i' + \sum_{i \in Ra_j(\alpha_j)} [\phi_i' - h_i^{-1}(\phi_i')],$$
   $$\lambda_{S_j}(\alpha_j) = f_j^{-1}(\alpha_j) - \phi_j + x_j.$$

   The range of the value of α_j is given for each of cases 1, 3, and 4.2 examined in Lemma 3.7 and in the remark following Lemma 3.7.

   Proceed to step 3.

3. Determine the optimal load. $O(n)$

$$\beta_i = 0, \quad \forall i \in Rd_j(\alpha_j),$$

$$\beta_i = h_i^{-1}(\alpha_j), \quad \forall i \in Ra_j(\alpha_j),$$

$$\beta_i = \phi_i', \quad \forall i \in Nu_j(\alpha_j),$$

$$x_i = \phi_i' - \beta_i, \quad \forall i \in V_j,$$

$$\beta_j = \phi_j + \sum_{i \in Ra_j(\alpha_j)} x_i + \sum_{i \in Rd_j(\alpha_j)} x_i - x_j.$$

4. Return to the main algorithm.

In the proposed algorithm, the first step finds an initial feasible solution (β^0, x^0).

The second step solves a star subproblem (3.14) iteratively by using the sub-algorithm. The main process of the sub-algorithm is to determine the Lagrange multiplier α_j by the golden section search and then to determine the optimal load in the star network.

The last step of the proposed algorithm determines whether the algorithm stops or not.

The convergence of the algorithm is shown in the following theorem.

Theorem 3.9 *The sequence $\{(\beta^{(r)}, x^{(r)})\}$ in the proposed algorithm converges to the optimal solution of the optimization problem (3.1).*

PROOF.

The proof of convergence of the proposed algorithm can be shown in similar way as that given by Dafermos and Sparrow [DS69].

We note that, under the constraints (3.2), β is determined if x is given. We, therefore, denote the job flow pattern and the mean delay by x and $D(x)$ instead of (β, x) and $D(\beta, x)$, respectively.

Let us call the subalgorithm for node j $(j \in B)$ "equilibration operator" $E^{(j)}$.

$$E^{(j)} : FS \longrightarrow FS, \quad j \in B,$$

where FS is the set of all feasible solutions x.

Then we have an equilibration operator,

$$E = E^{(b-1)} \circ E^{(b-2)} \circ \dots \circ E^{(1)} \circ E^{(0)}$$

where b is the number of elements of B, and nodes are numbered in such a way that, if node i is the parent of node j, $i < j$ $(i, j \in N)$.

E is a map

$$E : FS \longrightarrow FS.$$

Define that $E_{(r)}$ means r times iterative application of E. Denote by x^* an optimal solution x.

Like Dafermos and Sparrow [DS69], we have the following.

If E has the following properties, then the proposed algorithm can have an optimal solution to problem (3.1).

(3.1) $E\boldsymbol{x} = \boldsymbol{x}$ for some $\boldsymbol{x} \in FS$ implies that $\boldsymbol{x}$ satisfies the relations given in Corollary 3.3 for all $j \in B$, so that $\boldsymbol{x} = \boldsymbol{x}^*$.

(3.2) E is a continuous mapping from FS to FS.

(3.3) $D(\boldsymbol{x}) \geq D(E\boldsymbol{x})$ for all $\boldsymbol{x} \in FS$.

(3.4) $D(\boldsymbol{x}) = D(E\boldsymbol{x})$ for some $\boldsymbol{x} \in FS$ implies that $E\boldsymbol{x} = \boldsymbol{x}$.

This theorem can be proven if we show that equilibration operator E satisfies the above four properties (3.1), (3.2), (3.3), and (3.4). We give the proof as follows.

(1) $E\boldsymbol{x} = \boldsymbol{x}$ implies $E^{(j)}\boldsymbol{x} = \boldsymbol{x}$ for all $j \in B$. Then we have property (3.1).

(2) By carefully examining the subalgorithm and by noting that $f_i(\beta_i)$ and $h_i(\beta_i)$ are continuous functions and are increasing with increase in β_i, we see that there exists $\delta'(\varepsilon') > 0$ for $\varepsilon' > 0$, such that

$$|\alpha_j(\boldsymbol{x}') - \alpha_j(\boldsymbol{x})| < \varepsilon' \text{ if } |\boldsymbol{x}' - \boldsymbol{x}| < \delta'(\varepsilon').$$

And also there exists $\delta''(\varepsilon) > 0$ for $\varepsilon > 0$ such that

$$\|E^{(i)}\boldsymbol{x}' - E^{(i)}\boldsymbol{x}\| < \varepsilon \text{ if } |\alpha_j(\boldsymbol{x}') - \alpha_j(\boldsymbol{x})| < \delta''(\varepsilon).$$

Therefore, we see that there exists $\delta(\varepsilon) > 0$ for $\varepsilon > 0$ such that

$$\|E^{(i)}\boldsymbol{x}' - E^{(i)}\boldsymbol{x}\| < \varepsilon \text{ if } |\boldsymbol{x}' - \boldsymbol{x}| < \delta(\varepsilon),$$

where $\delta = \delta' \circ \delta''$.

That is, $E^{(i)}$ is a continuous mapping from FS to FS. Therefore, by noting the definitions of E, we see that E is a continuous mapping from FS to FS.

(3) Consider an arbitrary $\boldsymbol{x} \in FS$. Let us define

$$\boldsymbol{x}^{(j)} = E^{(j)} \circ E^{(j-1)} \circ ... \circ E^{(0)}\boldsymbol{x}, \quad j = 0, 1, 2, ..., b-1.$$

Then we have

$$\boldsymbol{x}^{(j)} = E^{(j)}\boldsymbol{x}^{(j-1)}, \quad j = 0, 1, 2, ..., b-1,$$

where we let $\boldsymbol{x}^{(-1)} = \boldsymbol{x}$.

By carefully examining the subalgorithm, we easily see

$$D(\boldsymbol{x}^{(j-1)}) \geq D(E^{(j)}\boldsymbol{x}^{(j-1)}), \quad j = 0, 1, 2, ..., b-1.$$

Therefore, we have

$$D(\boldsymbol{x}) \geq D(\boldsymbol{x}^{(0)}) \geq \cdots \geq D(\boldsymbol{x}^{(b-1)}),$$

where $E\boldsymbol{x} = \boldsymbol{x}^{(b-1)}$.

Thus we have $D(\boldsymbol{x}) \geq D(E\boldsymbol{x})$.

(4) If $D(\boldsymbol{x}) = D(E\boldsymbol{x})$, from property (3.3) above,

$$D(\boldsymbol{x}) = D(\boldsymbol{x}^{(0)}) = \cdots = D(\boldsymbol{x}^{(b-1)}).$$

That is, $D(E^{(j)}\boldsymbol{x}^{(j)}) = D(\boldsymbol{x}^{(j)})$ for all $j \in B$.

Note that $\boldsymbol{x}$ consists of $\boldsymbol{x}_j$, $j = 0, 1, ..., b-1$, where $\boldsymbol{x}_j$ consists of x_i, $i \in V_j$. Then $D(E^{(j)}\boldsymbol{x}^{(j)}) = D(\boldsymbol{x}^{(j)})$, which means $\boldsymbol{x}^{(j)}$ satisfies the optimal conditions given in Corollary

3.3, and $E^{(j)}\boldsymbol{x}^{(j)} = \boldsymbol{x}^{(j)}$. Therefore, we have

$\quad \boldsymbol{Ex} = E^{(b-1)} \circ E^{(b-2)} \circ ...E^{(1)} \circ E^{(0)}\boldsymbol{x} = \boldsymbol{x}$.

$\quad$ Q.E.D.

3.6 Comparison of Algorithm Performance

We compare the decomposition algorithm proposed above (we will call it the DC algorithm in the remainder of the paper) with the two other algorithms, the FD algorithm and the D-S algorithm. We examine two performance measures: the storage requirements and the computation time requirements. The storage requirements show the capability of an algorithm solving large-scale problems. The computation time requirements show how fast an algorithm can converge to an accuracy level.

3.6.1 Comparison of Storage Requirements

For the DC algorithm and the FD algorithm, it is only necessary to calculate the elements in the vector $\boldsymbol{\beta}$ in each iteration. Suppose that the number of elements in $\boldsymbol{\beta}$ is n $(n = |N|)$. Hence, the amount of storage requirements of these two algorithms are $O(n)$. In the case of the D-S algorithm, not only the elements of $\boldsymbol{\beta}$ but also the origin-destination job flow rate should be obtained. According to the special structure of the tree hierarchy, the average length of the paths from an origin node to its destination nodes is $\log(n)$. Therefore, the D-S algorithm requires the amount of storage of $O(n \log(n))$. We summarize the storage requirements of the three algorithms in Table 3.1.

Table 3.1: Storage requirements of algorithms

Algorithm	Storage Requirement
FD	$O(n)$
D-S	$O(n \log(n))$
DC	$O(n)$

3.6.2 Comparison of Computation Time Requirements

We program the FD algorithm, the D-S algorithm and the DC algorithm in FORTRAN and run these three algorithms in a SPARC workstation. The FD algorithm is programmed according to the L. Fratta, M. Gerla and L. Kleinrock paper [FGK73]. The

D–S algorithm has been improved by using the column generation technique for the path generation [LNT73] which can save some storage requirements especially for the large scale problem. We program the modified version of D–S algorithm embodying the column generation method [LNT73]. For the FD algorithm, the stopping rule is based on having the marginal improvements in the objective function below a given acceptance tolerance (See page 114 of [FGK73]). In the D–S algorithm and the DC algorithm, the stopping rules are based on the acceptance tolerances for the total improvements in the objective functions in order to obtain the close objective values to the one of the FD algorithm. The same initial feasible solution is given for the three algorithms.

In the numerical experiments, we assume that the node model is an M/M/1 and the communication link model also is an M/M/1. So the node delay function $F_i(\beta_i)$ and the communication link delay function $G_i(x_i)$ have the following forms,

$$F_i(\beta_i) = \frac{1}{c_i - \beta_i},$$

$$G_i(x_i) = \frac{1}{c_i' - x_i},$$

where c_i and c_i' are the processing capacities of node i and link i respectively.

Consider the set of nodes whose paths to the root node contain the same number of links. We say that such nodes belong to the same layer. The root node alone constitutes a layer. Three cases in the numerical experiments are considered as follows,

Case 1: 13 nodes constitute a tree hierarchy network in 3 layers.

Case 2: 25 nodes constitute a tree hierarchy network in 3 layers.

Case 3: 27 nodes constitute a tree hierarchy network in 9 layers.

We examine case 1 as an example of small-scale networks. We examine case 2 to see the effects of adding more nodes to the network such as case 1. We examine case 3 to see the effects of adding more layers to the network such as case 1.

The configurations and the parameters of the networks in the three cases are shown in Figures 3.2, 3.3, 3.4 and Tables 3.2, 3.3 and 3.4.

The comparison is given in terms of the CPU processing time on a SUN–4 workstation and the number of iterations to reach a given acceptance tolerance on the stopping rule. ε denotes the acceptance tolerance for the relative error of $D(\beta, x)$. Tables 3.5, 3.6, and 3.7 show the results of comparing the three algorithms for the above three cases respectively.

The numerical results in Tables 3.5, 3.6, and 3.7 show that the FD algorithm has slower convergence than the D–S algorithm and the DC algorithm. We consider the effect of the size of the acceptance tolerance ε on the computation time requirements. It is found that the CPU processing time and the number of iterations that the FD algorithm requires increase rapidly as the acceptance tolerance ε decreases (i.e., the accuracy level

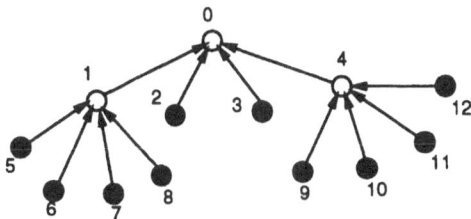

Figure 3.2: Configuration of the network in case 1

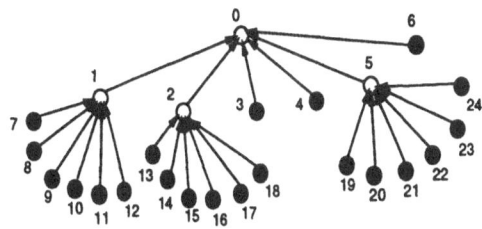

Figure 3.3: Configuration of the network in case 2

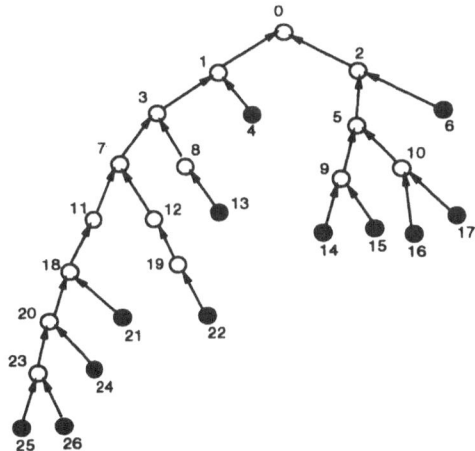

Figure 3.4: Configuration of the network in case 3

Table 3.2: Parameters of the network in case 1

Parent	Node	Processing Capacity of node (jobs/sec.)	Processing Capacity of link (jobs/sec.)	Job arrival rate (jobs/sec.)
	0	200	Null	50
0	1	10	10	1
	2	10	12.5	4
	3	10	12.5	5
	4	10	20	4
1	5	5	100	2
	6	5	66.667	3.3
	7	3.333	66.667	2.5
	8	3.333	100	1.8
4	9	10	1.8182	3
	10	3.333	1.8182	1.8
	11	3.333	5	1
	12	5	5	2

Table 3.3: Parameters of the network in case 2

Parent	Node	Processing Capacity of node (jobs/sec.)	Processing Capacity of link (jobs/sec.)	Job arrival rate (jobs/sec.)
	0	100	Null	10
0	1	20	10	2
	2	20	10	9
	3	5	10	4
	4	10	5	8
	5	10	5	7
	6	10	5	6
1	7	20	10	10
	8	50	10	30
	9	66.667	5	20
	10	100	10	30
	11	20	50	15
	12	12.5	10	10
2	13	10	5	7
	14	10	10	2
	15	5	10	2
	16	100	50	10
	17	50	10	10
	18	50	10	10
5	19	10	10	5
	20	50	5	20
	21	10	10	5
	22	10	5	5
	23	5	10	2
	24	33.333	10	6

Table 3.4: Parameters of the network in case 3

Parent	Node	Processing Capacity of node (jobs/sec.)	Processing Capacity of link (jobs/sec.)	Job arrival rate (jobs/sec.)
	0	200	Null	30
0	1	20	50	6
	2	20	50	7
1	3	20	50	8
	4	12.5	10	9
2	5	50	12.5	6
	6	30	10	10
3	7	100	50	11
	8	12.5	20	10
5	9	10	10	2
	10	10	2	5
7	11	25	30	8
	12	20	10	9
8	13	50	10	4
9	14	10	50	6
	15	10	20	3
10	16	20	10	10
	17	50	25	12
11	18	12.5	25	11
12	19	20	20	10
18	20	20	25	10
	21	50	30	20
19	22	50	25	21
20	23	10	10	2
	24	10	10	5
23	25	20	10	7
	26	20	10	9

Table 3.5: Comparison of computation time requirements for case 1

ε	Algorithm	CPU time (sec.) (Iteration number)	Response time (in seconds)
10^{-2}	FD	1.09 (30)	0.0792618285
	D-S	1.95 (3)	0.0791115645
	DC	0.97 (4)	0.0791118110
10^{-3}	FD	8.90 (297)	0.0791226371
	D-S	1.95 (3)	0.0791115645
	DC	1.24 (5)	0.0790849446
10^{-4}	FD	179.65 (5903)	0.0790852869
	D-S	2.46 (4)	0.0791115645
	DC	1.51 (6)	0.0790798758
10^{-5}	FD	2296.18 (67514)	0.0790796276
	D-S	2.46 (4)	0.0791110689
	DC	1.76 (7)	0.0790791268
10^{-6}	FD	17890.38 (505751)	0.0790790001
	D-S	3.01 (5)	0.0791110302
	DC	2.29 (9)	0.0790790005

Table 3.6: Comparison of Computation Time Requirements for Case 2

ε	Algorithm	CPU time (sec.) (Iteration number)	Response time (in seconds)
	FD	3.49 (55)	0.0886088626
10^{-2}	D-S	1.43 (2)	0.0882478261
	DC	0.53 (2)	0.0882634530
	FD	49.02 (783)	0.0883058770
10^{-3}	D-S	1.69 (3)	0.0882477096
	DC	0.77 (3)	0.0882484131
	FD	532.11 (8379)	0.0882549158
10^{-4}	D-S	1.69 (3)	0.0882477096
	DC	1.03 (4)	0.0882477424
	FD	6517.37 (93533)	0.0882483927
10^{-5}	D-S	1.70 (3)	0.0882477096
	DC	1.03 (4)	0.0882477424
	FD	66026.85 (926608)	0.0882477121
10^{-6}	D-S	1.85 (4)	0.0882477095
	DC	1.30 (5)	0.0882477111

Table 3.7: Comparison of Computation Time Requirements for Case 3

ε	Algorithm	CPU time (sec.) (Iteration number)	Response time (in seconds)
10^{-2}	FD	1.51 (23)	0.0700432467
	D-S	1.29 (2)	0.0698515387
	DC	0.82 (3)	0.0699039977
10^{-3}	FD	19.57 (297)	0.0698896075
	D-S	1.29 (2)	0.0698515387
	DC	1.10 (4)	0.0698592506
10^{-4}	FD	279.78 (4218)	0.0698568518
	D-S	1.29 (2)	0.0698515387
	DC	1.38 (5)	0.0698532243
10^{-5}	FD	3632.15 (50520)	0.0698520665
	D-S	1.29 (2)	0.0698515387
	DC	1.93 (7)	0.0698518231
10^{-6}	FD	33005.53 (446471)	0.0698515453
	D-S	1.35 (3)	0.0698515387
	DC	2.22 (8)	0.0698518231

increases). On the other hand, the CPU processing time and the iteration number of the D–S algorithm and the DC algorithm increase rather slowly as the acceptance tolerance ε decreases. We also note that the FD algorithm is as good as the other two algorithms for the larger ε (e.g., $\varepsilon > 10^{-2}$ in our numerical experiments), but is much poorer than the others for the smaller ε (e.g., $\varepsilon < 10^{-4}$ in our numerical experiments) in terms of the CPU processing time.

The above results are expected because both the DC algorithm and the D–S algorithm take the advantage of the network structure of problem. Notice that the main process of the sub-algorithm in the DC algorithm is determining a Lagrange multiplier by applying an effective golden section search. And at each step of the D–S algorithm, only a simple problem that balances loads between two paths in the network is solved. Some other numerical experiments wherein we examined different system models were also conducted. They showed results similar to those of the above examples.

3.7 Effects of Link Communication Time and Node Processing Time on Optimal Load Balancing

The link communication time, i.e., the traffic–independent mean service time of each communication channel, and the node processing time, i.e., the mean service time of each host computer, are critical system parameters that affect the system performance. In designing a distributed system with the tree hierarchy network configuration, it is important to know the effects of link communication time and node processing time on the system performance variables. By using the knowledge of the effects of link communication time and node processing time, we also can make a parametric adjustment to improve the system performance.

Some related works have been conducted for specifically configured distributed systems. Zhang, Kameda, and Shimizu [ZKS91] have studied the effects of link communication time, node processing time and job arrival rates on optimal static load balancing in star network configurations. They [ZKS92] also have studied the effects of link communication time, node processing time and job arrival rates on optimal static load balancing in single channel network configurations.

Here we study the effects of the link communication time and the node processing time on the following three system performance variables: link flow rates (i.e. the rate at which a node forwards jobs to another node for remote processing), node loads (i.e. the rate at which jobs are processed at a node), and the mean response time, of optimal static load balancing in tree hierarchy network configurations by parametric analysis. It is shown that the entire network is divided into several independent sub-tree networks with respect to the link flow rates and node loads, and that the communication time of a link and the processing time of a node have the effects only on the link flow rates and the loads of nodes that are in the same sub-tree network. Although the procedures of proofs are complicated, we obtain the following clear and simple relations.

- An increase in the communication time of a link causes a decrease in the link flow rates of its descendant nodes, its ancestor nodes and itself, but causes an increase in the link flow rate of other nodes in the same sub-tree network. Further, it also causes an increase in the loads of its descendant nodes and itself, but causes a decrease in the loads of other nodes in the same sub-tree network.

- An increase in the processing time of a node causes an increase in the link flow rates of its ancestor nodes and itself, but causes a decrease in the link flow rates of other nodes in the same sub-tree network. Further, it also causes a decrease in the load of itself, but causes an increase in the loads of other nodes in the same sub-tree network.

- In general, either an increase in the communication time of a link or an increase in the processing time of a node will cause an increase in the mean response time.

3.7.1   Concept of Sub-Tree Networks

The notation and assumptions used here are the same as that in section 3.2. Furthermore, let Ra, Rd, and Nu be the sets of active source nodes, idle source nodes, and neutral nodes, respectively, in the entire tree hierarchy network. That is $Ra = \bigcup_{i \in B} Ra_i$, $Rd = \bigcup_{i \in B} Rd_i$, $Nu = \bigcup_{i \in B} Nu_i$.

Here we introduce an important concept of *sub-tree networks* in the optimal load balancing. Noting that the link flow rates of neutral nodes are zero, we can regard that the links of neutral nodes are cut. Then an entire tree hierarchy network can be divided into $m + 1$ ($m = |Nu|$) independent *sub-tree networks* with respect to the link flow rates and node loads, each of which consists of a *sub-root j* ($j \in Nu$ or j is the root node) and the links which have the positive link flow rates. We call a sub-tree network *sub-tree network j* if its sub-root is node j. In a sub-tree network, there are also terminal nodes and nonterminal nodes. Let $\tilde{T}_j$ and $\tilde{B}_j$ be the the set of the terminal nodes and the set of the nonterminal nodes in sub-tree network j, respectively. Furthermore, from each sub-tree network, we take away such idle source nodes all of whose descendants are also idle source nodes, because they do not affect (or are affected by) other nodes in link flow rates or in node loads. We note that the terminal nodes in sub-tree networks must be active source nodes.

By $q \Rightarrow i$, we mean that node q is a descendant of node i, and nodes q and i are in the same sub-tree network. By $q \stackrel{\circ}{=} i$, we mean both that $q = i$ (i.e., node q is the same as i), and that node q is in a sub-tree network. (We note that an idle source node is taken away from a sub-tree network if all its descendants are also idle source nodes in the sub-tree network. If node q is the node that is taken away from a sub-tree network and $q = i$, we do not have $q \stackrel{\circ}{=} i$.) Let $q \stackrel{s}{\Rightarrow} i$ if $q \Rightarrow i$, or $q \stackrel{\circ}{=} i$. In contrast to the definitions in section 2, we say that node i is a *substantial* ancestor of node q (or node q is a *substantial* descendant of node i) if $q \Rightarrow i$. For a node in a sub-tree network, its substantial ancestors, its substantial descendants and itself are its *substantial direct line* nodes. By $q \not\Rightarrow i$, we mean that node q is not the substantial descendant node of node i. By $q \stackrel{s}{\not\Rightarrow} i$, we mean that $q \neq i$, or node q does not belong to any sub-tree networks. Let $q \stackrel{s}{\not\Rightarrow} i$ if $q \not\Rightarrow i$ and $q \stackrel{s}{\not\Rightarrow} i$. We call node i a *substantial collateral* node of node q if $i \stackrel{s}{\not\Rightarrow} q$, $q \stackrel{s}{\not\Rightarrow} i$, but i and q have a common substantial ancestor.

3.7.2 Parametric Analysis

In this section we analyse the effects of link communication time and node processing in a tree hierarchy network by parametric analysis. Consider the *communication time*, i.e., the traffic–independent mean service time of each communication channel, t_i, on link i ($i \in L$) and the *processing time*, i.e., the mean service time of each host computer, u_i, on node i ($i \in N$) as important system parameters. Note that t_i is the mean communication time (without queueing time) for forwarding and sending back a job through link i. We use the notation $G_i(x_i, t_i)$ and $g_i(x_i, t_i)$ for the mean communication delay and the incremental communication delay, respectively, in order to make the parameter t_i explicit. We also use the notation $F_i(\beta_i, u_i)$ and $f_i(\beta_i, u_i)$ for the mean node delay and the incremental node delay, respectively, in order to make the parameter u_i explicit.

Assume that the incremental communication delay on link i $g_i(x_i, t_i)$ increases with t_i. And $\dfrac{\partial g_i(x_i, t_i)}{\partial t_q} > 0$ if $q = i$; $\dfrac{\partial g_i(x_i, t_i)}{\partial t_q} = 0$ if $q \neq i$. Also assume that the the incremental node delay on node i, $f_i(\beta_i, u_i)$, increases with u_i. And $\dfrac{\partial f_i(\beta_i, u_i)}{\partial u_q} > 0$ if $q = i$; $\dfrac{\partial f_i(\beta_i, u_i)}{\partial u_q} = 0$ if $q \neq i$.

Denote $[t_i]$, $[u_i]$ by t and u. Let p denote (t, u). Link flow rate x_i ($i \in L$), node load β_i ($i \in N$), and Lagrange multiplier α_i ($i \in N$) in the optimal solution are determined by given p, and are written as $x_i(p)$, $\beta_i(p)$, and $\alpha_i(p)$.

Note that as the communication time of a certain link or the processing time of a certain node increase, it would cause some nodes to process more jobs; on the other hand, it would also cause some other nodes to send more jobs to their parent nodes in a tree hierarchy network. This means that the increase in communication time of the link or the increase in processing time of the node has the effects on the node loads and the link flow rates, and then on the mean response time in the network. In the limit, large link communication time or node processing time may lead to an infinitely long response time if the loads on some nodes exceed the processing capacities of these nodes. In previous section, the optimal solution is characterized by the partition of nodes into different types (active source, idle source, and neutral). Here we consider an interval of the link communication time or and interval of the node processing time during which the node partition remains the same. In following sections, we first show the effects of the link communication time on the system performance variables. Then we show the effects of the node processing time on the system performance variables.

3.7.3   Effects of Link Communication Time

From relations (3.3c) and (3.5), we note that $f_j(\beta_j(p), u_j) = \alpha_j(p)$ if $j \in Nu$ or if j is the root node. First we have the following lemma.

Lemma 3.10 *The following relations hold for the incremental node delay $\alpha_j(t)$ in a tree hierarchy network,*

$$\frac{\partial \alpha_j}{\partial t_q} < 0, \quad if \; q \Rightarrow j,$$

$$= 0, \quad else, \tag{3.26}$$

where node j is a sub-tree root (i.e., $j \in Nu$, or node j is the root node ($j = 0$) in a tree hierarchy network).

PROOF.

According to relations (3.3) in Theorem 3.1, we have

$$f_l(\beta_l(t)) = \alpha_i + g_l(x_l(t), t_l) \quad l \in Ra_i, \; i \in B,$$

$$\alpha_l = f_l(\beta_l(t)) \qquad\qquad \beta_l > 0, \tag{3.27}$$

$$\alpha_l = \alpha_i + g_l(x_l(t), t_l) \qquad l \in Rd_i, \; i \in B.$$

Note that the changing communication time of a link in a sub-tree network does not affect other nodes which do not belong to the same sub-tree network both in link flow rates and in node loads. Without loss of generality, we assume node l belongs to sub-tree network j.

First, we consider the case that node l is a terminal node ($l \in \tilde{T}_j$). Note that terminal nodes in each sub-tree network must be active nodes. Using $\beta_l = \phi'_l - x_l$ and $\frac{\partial \phi'_l}{\partial t_q} = 0$, we have,

$$\frac{\partial \alpha_i}{\partial t_q} = -(\frac{\partial f_l}{\partial \beta_l} + \frac{\partial g_l}{\partial x_l})\frac{\partial x_l}{\partial t_q} - \frac{\partial g_l}{\partial t_q}, \; l \in \tilde{T}_j \cap Ra_i. \tag{3.28}$$

Rearranging (3.28), we have

$$\frac{\partial x_l}{\partial t_q} = -Y_l - A_l \frac{\partial \alpha_i}{\partial t_q}, \quad l \in \tilde{T}_j \cap Ra_i \tag{3.29}$$

where,

$$Y_l = \frac{1}{U_l}\frac{\partial g_l}{\partial t_q}, \quad A_l = \frac{1}{U_l}, \quad U_l = \frac{\partial f_l}{\partial \beta_l} + \frac{\partial g_l}{\partial x_l}.$$

Because $\frac{\partial f_l}{\partial \beta_l} > 0$ and $\frac{\partial g_l}{\partial x_l} > 0$, we have $A_l > 0$, $U_l > 0$, and

$$Y_l = \begin{cases} \frac{1}{U_l}\frac{\partial g_l}{\partial t_q} > 0, & q = l, \\ 0, & \text{else.} \end{cases}$$

More generally, we consider that node l is either a terminal node or a nonterminal node ($l \in \tilde{T}_j \cup \tilde{B}_j$). Note that node l may be an active node ($l \in Ra$) or an idle node ($l \in Rd$), and the child nodes of l may also be active nodes or idle nodes. Using (3.27) and (3.29), we have the following recursive formula.

$$\frac{\partial x_l}{\partial t_q} = -Y_l - A_l \frac{\partial \alpha_i}{\partial t_q}, \quad l \in Ra_i, \tag{3.30}$$

where

$$
\begin{aligned}
Y_l &= \frac{1}{U_l}[\frac{\partial f_l}{\partial \beta_l}(\sum_{r \in Ra_l} Y_r + \sum_{r \in Rd_l} Z_r) \\
&\quad + \frac{\partial g_l}{\partial t_q}(1 + \frac{\partial f_l}{\partial \beta_l}(\sum_{r \in Ra_l} A_r + \sum_{r \in Rd_l} B_r))], \\
A_l &= \frac{1}{U_l}[1 + \frac{\partial f_l}{\partial \beta_l}(\sum_{r \in Ra_l} A_r + \sum_{r \in Rd_l} B_r)], \\
U_l &= \frac{\partial f_l}{\partial \beta_l} + \frac{\partial g_l}{\partial x_l}[1 + \frac{\partial f_l}{\partial \beta_l}(\sum_{r \in Ra_l} A_r + \sum_{r \in Rd_l} B_r)].
\end{aligned}
\tag{3.31}
$$

$$\frac{\partial x_l}{\partial t_q} = -Z_l - B_l \frac{\partial \alpha_i}{\partial t_q}, \quad l \in Rd_i, \tag{3.32}$$

where

$$
\begin{aligned}
Z_l &= \frac{1}{V_l}[(\sum_{r \in Ra_l} Y_r + \sum_{r \in Rd_l} Z_r) \\
&\quad + \frac{\partial g_l}{\partial t_q}(\sum_{r \in Ra_l} A_r + \sum_{r \in Rd_l} B_r)], \\
B_l &= \frac{1}{V_l}(\sum_{r \in Ra_l} A_r + \sum_{r \in Rd_l} B_r), \\
V_l &= 1 + \frac{\partial g_l}{\partial x_l}(\sum_{r \in Ra_l} A_r + \sum_{r \in Rd_l} B_r).
\end{aligned}
\tag{3.33}
$$

As we have done when l is a terminal node, we also have the following relations.

$U_l > 0$, $A_l > 0$, and

$$Y_l = \begin{cases} \frac{1}{U_l}\frac{\partial f_l}{\partial \beta_l} Y_r > 0, & q \overset{*}{\Rightarrow} r,\; r \in Ra_l, \\ \frac{1}{U_l}\frac{\partial f_l}{\partial \beta_l} Z_r > 0, & q \overset{*}{\Rightarrow} r,\; r \in Rd_l, \\ \frac{1}{U_l}\frac{\partial g_l}{\partial t_q}(1 + \frac{\partial f_l}{\partial \beta_l}(\sum_{r \in Ra_l} A_r + \sum_{r \in Rd_l} B_r)) > 0, & q = l, \\ 0, & \text{else.} \end{cases} \tag{3.34}$$

$V_l > 0$, $B_l > 0$, and

$$Z_l = \begin{cases} \frac{1}{V_l} Y_r > 0, & q \overset{*}{\Rightarrow} r,\; r \in Ra_l, \\ \frac{1}{V_l} Z_r > 0, & q \overset{*}{\Rightarrow} r,\; r \in Rd_l, \\ \frac{1}{V_l}\frac{\partial g_l}{\partial t_q}(\sum_{r \in Ra_l} A_r + \sum_{r \in Rd_l} B_r) > 0, & q = l, \\ 0, & \text{else.} \end{cases} \tag{3.35}$$

Note that for $l \in \tilde{T}_j$, we naturally have $A_r = 0$, $Y_r = 0$; and $B_r = 0$, $Z_r = 0$ both in (3.34) and (3.35).

When $j \in Nu$ or node j is the root, by noting that $\beta_j = \phi'_j = \sum_{i \in V_j} x_i$, from (3.27), we have

$$\frac{\partial \alpha_j}{\partial t_q} = \frac{\partial f_j}{\partial \beta_j}\frac{\partial \beta_j}{\partial t_q} = \frac{\partial f_j}{\partial \beta_j}(\sum_{i \in Ra_j}\frac{\partial x_i}{\partial t_q} + \sum_{i \in Rd_j}\frac{\partial x_i}{\partial t_q}). \tag{3.36}$$

Substituting (3.30) and (3.32) into (3.36), we have

$$\frac{\partial \alpha_j}{\partial t_q} = -A_j(\sum_{i \in Ra_j} Y_i + \sum_{i \in Rd_j} Z_i), \tag{3.37}$$

where

$$A_j = \frac{\frac{\partial f_j}{\partial \beta_j}}{1 + \frac{\partial f_j}{\partial \beta_j}(\sum_{i \in Ra_j} A_i + \sum_{i \in Rd_j} B_i)}. \tag{3.38}$$

According to (3.37), by using (3.34) and (3.35), we obtain the relations (3.26). Q.E.D.

The effects of link communication time on link flow rates

The effects of the link communication time on the link flow rates in a tree hierarchy network are specified in Theorem 3.11.

Theorem 3.11 *The following relations hold for the link flow rate $x_l(t)$ $(l \in L)$ in a tree hierarchy network,*

$$\frac{\partial x_l}{\partial t_q} \quad < \quad 0, \quad \text{if } q \overset{*}{\Rightarrow} l, \text{ or } l \Rightarrow q, \tag{3.39a}$$

$$> \quad 0, \quad \text{if } q \not\Rightarrow l,\; l \not\Rightarrow q,\; q \Rightarrow j, \text{ and } l \Rightarrow j, \tag{3.39b}$$

$$= \quad 0, \quad \text{else,} \tag{3.39c}$$

where node j is a sub-tree root.

PROOF.

We consider the effects of the communication time of an arbitrary link q, t_q, on link flow rate of link l ($l \in L$). Without loss of generality, we assume that node l belongs to sub-tree network 0 ($j = 0$), and that the substantial ancestor nodes of l are i_{m-1}, i_{m-2}, ..., i_1, i_0 ($i_0 = 0$) (i.e., $l \in Ra_{i_{m-1}} \cup Rd_{i_{m-1}}$, $i_{m-1} \in Ra_{i_{m-2}} \cup Rd_{i_{m-2}}, \cdots, i_1 \in Ra_{i_0} \cup Rd_{i_0}$ ($i_0 = 0$)). Note that node l may be an active node or an idle node (i.e., $l \in Ra \cup Rd$). We prove the theorem in case 1 that node l is an idle node and all its substantial ancestor nodes are also idle nodes (i.e., $l \in Rd_{i_{m-1}}$, $i_{m-k} \in Rd_{i_{m-k-1}}$, $k = 1, 2, ..., m - 1$) in detail, and then make remarks on the other cases in which some substantial ancestors of node l are active nodes.

Case 1 $l \in Rd_{i_{m-1}}$, $i_{m-1} \in Rd_{i_{m-2}}$, $i_{m-2} \in Rd_{i_{m-3}}$, $\cdots$, $i_2 \in Rd_{i_1}$, $i_1 \in Rd_0$.

According to (3.33), we note that

$$1 - \frac{\partial g_{i_{m-k}}}{\partial x_{i_{m-k}}} B_{i_{m-k}} = \frac{1}{V_{i_{m-k}}}, \quad k = 1, 2, ..., m - 1. \tag{3.40}$$

From (3.27), we have

$$\frac{\partial \alpha_{i_{m-k}}}{\partial t_q} = \frac{\partial \alpha_{i_{m-k-1}}}{\partial t_q} + \frac{\partial g_{i_{m-k}}}{\partial x_{i_{m-k}}} \frac{\partial x_{i_{m-k}}}{\partial t_q} + \frac{\partial g_{i_{m-k}}}{\partial t_q}. \tag{3.41}$$

From (3.41), using (3.32) and (3.40), we have the useful recursive relation as follows,

$$\frac{\partial \alpha_{i_{m-k}}}{\partial t_q} = \frac{1}{V_{i_{m-k}}} \frac{\partial \alpha_{i_{m-k-1}}}{\partial t_q} + \frac{\partial g_{i_{m-k}}}{\partial t_q} - \frac{\partial g_{i_{m-k}}}{\partial x_{i_{m-k}}} Z_{i_{m-k}}, \quad k = 1, 2, ..., m - 1. \tag{3.42}$$

We also note that node q may be in the following sub-cases: case 1.1 $q = i_{m-k}$ ($1 \le k < m$) (i.e., node q is a substantial ancestor node of node l); case 1.2 $q \overset{*}{\Rightarrow} l$ (i.e., node q is a substantial descendant node of node l or $q = l$); and case 1.3 $q \overset{*}{\not\Rightarrow} l$, $q \ne i_{m-k}$ ($0 < k < m$), $l \Rightarrow 0$, $q \Rightarrow 0$ (i.e., node q is a substantial collateral node of node l). We consider each case separately.

Case 1.1 $q = i_{m-k}$ ($1 \le k < m$), i.e., q is a substantial ancestor node of l.

From (3.32) and (3.42), using (3.35), we have

$$\frac{\partial x_l}{\partial t_q} = -B_l \frac{1}{V_{i_{m-1}}} \cdots \frac{1}{V_{i_{m-k}}} \left(\frac{\partial g_{i_{m-k}}}{\partial t_q} - \frac{\partial g_{i_{m-k-1}}}{\partial x_{i_{m-k-1}}} Z_{i_{m-k-1}} \right.$$
$$\left. - \cdots - \frac{1}{V_{i_{m-k-1}}} \cdots \frac{1}{V_{i_2}} \frac{\partial g_{i_1}}{\partial x_{i_1}} Z_{i_1} - \frac{1}{V_{i_{m-k-1}}} \cdots \frac{1}{V_{i_1}} A_0 Z_{i_1} \right). \tag{3.43}$$

Using (3.33), we have following expansion formula,

$$\frac{\partial g_{i_{m-k}}}{\partial t_q} = \frac{\partial g_{i_{m-k}}}{\partial t_q} \frac{V_{i_{m-k-1}}}{V_{i_{m-k-1}}}$$

$$= \frac{\partial g_{i_{m-k}}}{\partial t_q} \frac{1}{V_{i_{m-k-1}}} \frac{V_{i_{m-k-2}}}{V_{i_{m-k-2}}}$$

$$+ \frac{\partial g_{i_{m-k}}}{\partial t_q} \frac{1}{V_{i_{m-k-1}}} \frac{\partial g_{i_{m-k-1}}}{\partial x_{i_{m-k-1}}} \left(\sum_{u \in Ra_{i_{m-k-1}}} A_u + \sum_{u \in Rd_{i_{m-k-1}}} B_u \right)$$

$$= H_1 + H_2 + \cdots + H_{m-k}, \tag{3.44}$$

where

$$H_1 = \frac{\partial g_{i_{m-k}}}{\partial t_q} \frac{1}{V_{i_{m-k-1}}} \frac{1}{V_{i_{m-k-2}}} \cdots \frac{1}{V_{i_1}},$$

$$H_2 = \frac{\partial g_{i_{m-k}}}{\partial t_q} \frac{1}{V_{i_{m-k-1}}} \frac{1}{V_{i_{m-k-2}}} \cdots \frac{1}{V_{i_1}} \frac{\partial g_{i_1}}{\partial x_{i_1}} \left(\sum_{u \in Ra_{i_1}} A_u + \sum_{u \in Rd_{i_1}} B_u \right),$$

$$\cdots \cdots \tag{3.45}$$

$$H_{m-k} = \frac{\partial g_{i_{m-k}}}{\partial t_q} \frac{1}{V_{i_{m-k-1}}} \frac{\partial g_{i_{m-k-1}}}{\partial x_{i_{m-k-1}}} \left(\sum_{u \in Ra_{i_{m-k-1}}} A_u + \sum_{u \in Rd_{i_{m-k-1}}} B_u \right).$$

By substituting (3.44) into (3.43), we have

$$\frac{\partial x_l}{\partial t_q} = -B_l \frac{1}{V_{i_{m-1}}} \cdots \frac{1}{V_{i_{m-k}}}$$

$$\times \left\{ \left(H_{m-k} - \frac{\partial g_{i_{m-k-1}}}{\partial x_{i_{m-k-1}}} Z_{i_{m-k-1}} \right) + \left(H_{m-k-1} - \frac{1}{V_{i_{m-k-1}}} \frac{\partial g_{i_{m-k-2}}}{\partial x_{i_{m-k-2}}} Z_{i_{m-k-2}} \right) \right.$$

$$+ \cdots + \left(H_2 - \frac{1}{V_{i_{m-k-1}}} \cdots \frac{1}{V_{i_2}} \frac{\partial g_{i_1}}{\partial x_{i_1}} Z_{i_1} \right) + \left. \left(H_1 - \frac{1}{V_{i_{m-k-1}}} \cdots \frac{1}{V_{i_1}} A_0 Z_{i_1} \right) \right\} \tag{3.46}$$

From (3.45), we have

$$H_1 - \frac{1}{V_{i_{m-k-1}}} \cdots \frac{1}{V_{i_1}} A_0 Z_{i_1} = \frac{1}{V_{i_{m-k-1}}} \cdots \frac{1}{V_{i_1}} \left(\frac{\partial g_{i_{m-k}}}{\partial t_q} - A_0 Z_{i_1} \right) \tag{3.47}$$

From (3.35), we have

$$Z_{i_1} = \frac{1}{V_{i_1}} \cdots \frac{1}{V_{i_{m-k}}} \frac{\partial g_{i_{m-k}}}{\partial t_q} \left(\sum_{u \in Ra_{i_{m-k}}} A_u + \sum_{u \in Rd_{i_{m-k}}} B_u \right) \quad \text{if } q = i_{m-k}. \tag{3.48}$$

From (3.38) and (3.48), we have

$$H_1 - \frac{1}{V_{i_{m-k-1}}} \cdots \frac{1}{V_{i_1}} A_0 Z_{i_1} =$$

$$\frac{\frac{1}{V_{i_{m-k-1}}} \cdots \frac{1}{V_{i_1}}}{1 + \frac{\partial f_0}{\partial \beta_0} \left(\sum_{u \in Ra_0} A_u + \sum_{u \in Rd_0} B_u \right)} \left\{ \frac{\partial g_{i_{m-k}}}{\partial t_q} \left(1 + \frac{\partial f_0}{\partial \beta_0} \left(\sum_{u \in Ra_0} A_u + \sum_{u \in Rd_0} B_u \right) \right) \right.$$

$$\left. - \frac{\partial f_0}{\partial \beta_0} \frac{1}{V_{i_{m-k}}} \cdots \frac{1}{V_{i_1}} \frac{\partial g_{i_{m-k}}}{\partial t_q} \left(\sum_{u \in Ra_{i_{m-k}}} A_u + \sum_{u \in Rd_{i_{m-k}}} B_u \right) \right\} \tag{3.49}$$

By noting (3.37) that $B_{i_{m-k}} = \frac{1}{V_{i_{m-k}}} \left(\sum_{u \in Ra_{i_{m-k}}} A_u + \sum_{u \in Rd_{i_{m-k}}} B_u \right)$, $(1 \leq k < m)$, it is easy to obtain that

$$H_1 - \frac{1}{V_{i_{m-k-1}}} \cdots \frac{1}{V_{i_1}} A_0 Z_{i_1} > 0. \tag{3.50}$$

Similarly, we can obtain that

$$H_t - \frac{1}{V_{i_{m-k-1}}} \cdots \frac{1}{V_{i_t}} \frac{\partial g_{i_{t-1}}}{\partial x_{i_{t-1}}} Z_{i_{t-1}} > 0, \quad t=2,3,...,m\text{-}k\text{-}1,$$

$$H_{m-k} - \frac{\partial g_{i_{m-k-1}}}{\partial x_{i_{m-k-1}}} Z_{i_{m-k-1}} > 0. \tag{3,51}$$

From (3.50) and (3.51), we have

$$\frac{\partial x_l}{\partial t_q} < 0, \quad \text{if } q = i_{m-k}, \, (1 \le k < m). \tag{3.52}$$

Case 1.2 $q \overset{*}{\Rightarrow} l$, i.e., q is a substantial descendant node of l or $q = l$.
According to (3.32) and (3.42), noting (3.35), we have

$$\frac{\partial x_l}{\partial t_q} = -Z_l + B_l \frac{\partial g_{i_{m-1}}}{\partial x_{i_{m-1}}} Z_{i_{m-1}} + B_l \frac{1}{V_{i_{m-1}}} \frac{\partial g_{i_{m-2}}}{\partial x_{i_{m-2}}} Z_{i_{m-2}}$$
$$+ \cdots + B_l \frac{1}{V_{i_{m-1}}} \cdots \frac{1}{V_{i_2}} \frac{\partial g_{i_1}}{\partial x_{i_1}} Z_{i_1} + B_l \frac{1}{V_{i_{m-1}}} \cdots \frac{1}{V_{i_1}} A_0 Z_{i_1}. \tag{3.53}$$

We use a similar method to that in case 1.1 and obtain that

$$\frac{\partial x_l}{\partial t_q} < 0. \quad \text{if } q \overset{*}{\Rightarrow} l. \tag{3.54}$$

Case 1.3 $q \overset{*}{\not\Rightarrow} l$, $q \ne i_{m-k}$ $(0 < k < m)$, $l \Rightarrow 0$, $q \Rightarrow 0$, i.e., q is a substantial collateral node of l.
According to (3.32) and (3.42), noting (3.35), we have

$$\frac{\partial x_l}{\partial t_q} = B_l \frac{\partial g_{i_{m-1}}}{\partial x_{i_{m-1}}} Z_{i_{m-1}} + B_l \frac{1}{V_{i_{m-1}}} \frac{\partial g_{i_{m-2}}}{\partial x_{i_{m-2}}} Z_{i_{m-2}} + \cdots$$
$$+ B_l \frac{1}{V_{i_{m-1}}} \cdots \frac{1}{V_{i_2}} \frac{\partial g_{i_1}}{\partial x_{i_1}} Z_{i_1} + B_l \frac{1}{V_{i_{m-1}}} \frac{1}{V_{i_{m-2}}} \cdots \frac{1}{V_{i_1}} A_0 \Big(\sum_{u \in Ra_0} Y_u + \sum_{u \in Rd_0, u \ne i_1} Z_u \Big)$$
$$> 0, \quad \text{if } q \overset{*}{\not\Rightarrow} l, \, q \ne i_{m-k} \, (0 < k < m), \, l \Rightarrow 0, \, q \Rightarrow 0. \tag{3.55}$$

Remark: For the other cases in which some substantial ancestors of node l are active nodes, we can use $U_{i_{m-k}}$, $Y_{i_{m-k}}$ and $A_{i_{m-k}}$ for node i_{m-k} if node i_{m-k} is an active source node instead of using $V_{i_{m-k}}$, $Z_{i_{m-k}}$ and $B_{i_{m-k}}$ in the procedure of the proof for case 1. For brevity, we will skip this part of the proof, but it has a similar result.

To summarize, we have the relations in Theorem 3.11. Q.E.D.

Remark: Relation (3.39a) states that an increase in the communication time of a link causes a decrease in the link flow rates of its substantial direct line nodes. Relation (3.39b) states that an increase in the communication time of a link causes an increase in the link

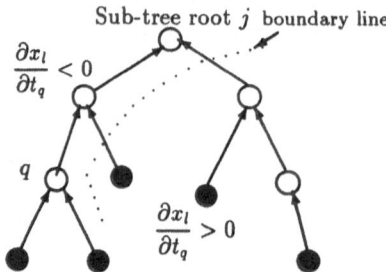

Figure 3.5: Effects of the communication time of link q on link flow rates

flow rates of its substantial collateral nodes. Otherwise, an increase in the communication time of a link has no effects on the link flow rates.

We illustrate the effects of the communication time of a certain link q ($q \in L$), t_q, on the link flow rates in a sub-tree network j by using Figure 3.5. We note that a node (l, $l \in L$) which belongs to the left side of the boundary line is the direct line node of node q in the sub-tree network. A node (l, $l \in L$) which belongs to the right side of the boundary line is a substantial collateral node of node q. Theorem 3.11 implies that an increase in t_q causes a decrease in the link flow rates of nodes that belong to the left side of the boundary line, but causes an increase in the link flow rates of nodes that belong to the right side of the boundary line.

The effects of link communication time on node loads

Theorem 3.12 *The following relations hold for the node load $\beta_l(t)$ ($l \in N$) in a tree hierarchy network,*

$$\frac{\partial \beta_l}{\partial t_q} \ > \ 0, \quad l \overset{*}{\Rightarrow} q, \text{ and } l \in Ra, \tag{3.56a}$$

$$< \ 0, \quad q \Rightarrow l, \text{ and } l \in Ra, \tag{3.56b}$$

$$< \ 0, \quad q \not\Rightarrow l, l \not\Rightarrow q, q \Rightarrow j, l \Rightarrow j, \text{ and } l \in Ra, \tag{3.56c}$$

$$< \ 0, \quad l = j, q \Rightarrow l, \tag{3.56d}$$

$$= \ 0, \quad else, \tag{3.56e}$$

where node j is a sub-tree root.
PROOF.

We consider exhaustively the following three cases: case 1. node l is a neutral node ($l \in Nu$) or the root node; case 2. node l is an active node ($l \in Ra$); case 3. node l is an idle node ($l \in Rd$).

Case 1. l is a neutral node or the root node, i.e., $l = j$.

From Lemma 3.10, by noting $\frac{\partial \alpha_l}{\partial t_q} = \frac{\partial f_l}{\partial \beta_l} \frac{\partial \beta_l}{\partial t_q}$ ($l \in Rd_i$) and $\frac{\partial f_l}{\partial \beta_l} > 0$, we have

$$
\begin{aligned}
\frac{\partial \beta_l}{\partial t_q} &< 0, \quad \text{if } q \Rightarrow l, \\
\frac{\partial \beta_l}{\partial t_q} &= 0, \quad \text{else,}
\end{aligned}
\tag{3.57}
$$

which corresponds to eq.(3.56d).

Case 2. l is the active source node.

Without loss of generality, we assume node l belongs to sub-tree network 0 ($j = 0$). Note that $\frac{\partial \alpha_l}{\partial t_q} = \frac{\partial f_l}{\partial \beta_l} \frac{\partial \beta_l}{\partial t_q}$ ($l \in Ra_i$) and $\frac{\partial f_l}{\partial \beta_l} > 0$. To prove the theorem, we first prove the relations that $\frac{\partial \alpha_l}{\partial t_q}$ holds, and then show the relations that $\frac{\partial \beta_l}{\partial t_q}$ holds.

In general, we assume that $l \in Ra_{i_{m-1}}$, $i_{m-1} \in Ra_{i_{m-2}} \cup Rd_{i_{m-2}}$, $i_{m-2} \in Ra_{i_{m-3}} \cup Rd_{i_{m-3}}$, ..., $i_2 \in Ra_{i_1} \cup Rd_{i_1}$, $i_1 \in Ra_{i_0} \cup Rd_{i_0}$ ($i_0 = 0$). Here we only show the theorem on the sub-case 2.1 that $i_{m-1} \in Ra_{i_{m-2}}$, $i_{m-2} \in Ra_{i_{m-3}}$, ..., $i_2 \in Ra_{i_1}$, $i_1 \in Ra_{i_0}$ (i.e., all substantial ancestor nodes of node l are active nodes). For other sub-cases in which some substantial ancestors of node l are idle nodes, we will make a remark.

Case 2.1 $i_{m-1} \in Ra_{i_{m-2}}$, ..., $i_2 \in Ra_{i_1}$, $i_1 \in Ra_{i_0}$ ($i_0 = 0$).

According to (3.31), we have

$$
1 - \frac{\partial g_l}{\partial x_l} A_l = \frac{1}{U_l} \frac{\partial f_l}{\partial \beta_l}
\tag{3.58}
$$

From (3.30) and (3.27), using (3.58), we have

$$
\frac{\partial \alpha_l}{\partial t_q} = \frac{\partial \alpha_{i_{m-1}}}{\partial t_q} \frac{1}{S_l} - \frac{\partial g_l}{\partial x_l} Y_l + \frac{\partial g_l}{\partial t_q},
\tag{3.59}
$$

where

$$
\frac{1}{S_l} = \frac{1}{U_l} \frac{\partial f_l}{\partial \beta_l}.
$$

Similarly, we also have,

$$
\frac{\partial \alpha_{i_{m-k}}}{\partial t_q} = \frac{\partial \alpha_{i_{m-k-1}}}{\partial t_q} \frac{1}{S_{i_{m-k}}} - \frac{\partial g_{i_{m-k}}}{\partial x_{i_{m-k}}} Y_{i_{m-k}} + \frac{\partial g_{i_{m-k}}}{\partial t_q}, \quad k = 1, 2, ..., m-1,
\tag{3.60}
$$

where

$$
\frac{1}{S_{i_{m-k}}} = \frac{1}{U_{i_{m-k}}} \frac{\partial f_{i_{m-k}}}{\partial \beta_{i_{m-k}}}.
$$

Note that the increase of the communication time of a link q in a sub-tree network does not affect the loads of other nodes which belong to other sub-tree networks. We consider the following four sub-cases separately: case 2.1.1. q is a substantial ancestor node of l; case 2.1.2 $q = l$; case 2.1.3. q is a substantial descendant node of l; and case 2.1.4. q is a substantial collateral node of l.

Case 2.1.1 $q = i_{m-k}$, ($0 < k \le m$), i.e., q is a substantial ancestor node of l

From (3.59), (3.60) and (3.34), we have

$$\frac{\partial \alpha_l}{\partial t_q} = \frac{1}{S_l}\frac{1}{S_{i_{m-1}}} \cdots \frac{1}{S_{i_2}}\frac{1}{S_{i_1}}\frac{\partial \alpha_0}{\partial t_q} - \frac{1}{S_l}\frac{1}{S_{i_{m-1}}} \cdots \frac{1}{S_{i_2}}\frac{\partial g_{i_1}}{\partial x_{i_1}}Y_{i_1}$$

$$- \cdots - \frac{1}{S_l}\frac{1}{S_{i_{m-1}}} \cdots \frac{1}{S_{i_{m-k+1}}}\frac{\partial g_{i_{m-k}}}{\partial x_{i_{m-k}}}Y_{i_{m-k}} + \frac{1}{S_l}\frac{1}{S_{i_{m-1}}} \cdots \frac{1}{S_{i_{m-k+1}}}\frac{\partial g_{i_{m-k}}}{\partial t_q} \quad (3.61)$$

From (3.61), using (3.37) and (3.34), we have

$$\frac{\partial \alpha_l}{\partial t_q} = \frac{1}{S_l}\frac{1}{S_{i_{m-1}}} \cdots \frac{1}{S_{i_{m-k+1}}}\left(\frac{\partial g_{i_{m-k}}}{\partial t_q} - \frac{1}{S_{i_{m-k}}} \cdots \frac{1}{S_{i_1}}A_0 Y_{i_1}\right.$$

$$\left. - \frac{1}{S_{i_{m-k}}} \cdots \frac{1}{S_{i_2}}\frac{\partial g_{i_1}}{\partial x_{i_1}}Y_{i_1} - \cdots - \frac{1}{S_{i_{m-k}}}\frac{\partial g_{i_{m-k-1}}}{\partial x_{i_{m-k-1}}}Y_{i_{m-k-1}} - \frac{\partial g_{i_{m-k}}}{\partial x_{i_{m-k}}}Y_{i_{m-k}}\right). \quad (3.62)$$

According to (3.31), we let

$$\frac{\partial g_{i_{m-k}}}{\partial t_q} = \frac{\partial g_{i_{m-k}}}{\partial t_q}\frac{U_{i_{m-k}}}{U_{i_{m-k}}}$$

$$= \frac{\partial g_{i_{m-k}}}{\partial t_q}\frac{1}{S_{i_{m-k}}}\frac{U_{i_{m-k-1}}}{U_{i_{m-k-1}}} + \frac{\partial g_{i_{m-k}}}{\partial t_q}\frac{\partial g_{i_{m-k}}}{\partial x_{i_{m-k}}}A_{i_{m-k}} \quad (3.63)$$

$$= \cdots \cdots$$

$$= H_0 + H_1 + \cdots H_{m-k-1} + H_{m-k},$$

where
$$H_0 = \frac{\partial g_{i_{m-k}}}{\partial t_q}\frac{1}{S_{i_{m-k}}} \cdots \frac{1}{S_{i_1}},$$

$$H_t = \frac{\partial g_{i_{m-k}}}{\partial t_q}\frac{1}{S_{i_{m-k}}} \cdots \frac{1}{S_{i_{t+1}}}\frac{\partial g_{i_t}}{\partial x_{i_t}}A_{i_t}, \quad t = 1, ..., m-k-1, \quad (3.64)$$

$$H_{m-k} = \frac{\partial g_{i_{m-k}}}{\partial t_q}\frac{\partial g_{i_{m-k}}}{\partial x_{i_{m-k}}}A_{i_{m-k}}.$$

Note that,
$$H_0 - \frac{1}{S_{i_{m-k}}} \cdots \frac{1}{S_{i_1}}A_0 Y_{i_1}$$

$$= \frac{1}{S_{i_{m-k}}} \cdots \frac{1}{S_{i_1}}\left(\frac{\partial g_{i_{m-k}}}{\partial t_q} - A_0 Y_{i_1}\right) \quad (3.65)$$

From (3.34) and (3.31), here we have

$$Y_{i_{m-t}} = \frac{1}{S_{i_{m-t}}} \cdots \frac{1}{S_{i_{m-k-1}}}Y_{i_{m-k}}$$

$$= \frac{1}{S_{i_{m-t}}} \cdots \frac{1}{S_{i_{m-k-1}}}\frac{\partial g_{i_{m-k}}}{\partial t_q}A_{i_{m-k}}, \quad q = i_{m-k}, \ k < t < m. \quad (3.66)$$

Using (3.37) and (3.66), it is easy to find that

$$H_0 - \frac{1}{S_{i_{m-k}}} \cdots \frac{1}{S_{i_1}}A_0 Y_{i_1} > 0. \quad (3.67)$$

Similarly, we can have

$$H_{m-t} - \frac{1}{S_{i_{m-k}}} \cdots \frac{1}{S_{i_{m-t+1}}} \frac{\partial g_{i_{m-t}}}{\partial x_{i_{m-t}}} Y_{i_{m-t}} > 0, \quad k < t < m. \tag{3.68}$$

$$H_{m-k} - \frac{\partial g_{i_{m-k}}}{\partial x_{i_{m-k}}} Y_{i_{m-k}} = 0. \tag{3.69}$$

From (3.67), (3.68) and (3.69), we have

$$\frac{\partial \alpha_l}{\partial t_q} > 0, \quad \text{if } q = i_{m-k}, \, (0 < k < m). \tag{3.70}$$

Case 2.1.2 $q = l$

From (3.59), (3.60), using (3.34), we have

$$\frac{\partial \alpha_l}{\partial t_q} = \frac{\partial g_l}{\partial t_q} - \frac{1}{S_l} \frac{1}{S_{i_{m-1}}} \cdots \frac{1}{S_{i_2}} \frac{1}{S_{i_1}} A_0 Y_{i_1}$$
$$- \frac{1}{S_l} \frac{1}{S_{i_{m-1}}} \cdots \frac{1}{S_{i_2}} \frac{\partial g_{i_1}}{\partial x_{i_1}} Y_{i_1} - \cdots - \frac{1}{S_l} \frac{\partial g_{i_{m-1}}}{\partial x_{i_{m-1}}} Y_{i_{m-1}} - \frac{\partial g_l}{\partial x_l} Y_l \tag{3.71}$$

Applying a similar method to that in that in case 2.1.1, we can obtain

$$\frac{\partial \alpha_l}{\partial t_q} > 0, \quad \text{if } q = l. \tag{3.72}$$

Case 2.1.3 $q \Rightarrow l$, i.e., q is a substantial descendant node of l.

From (3.59), (3.60) and (3.34), we have

$$\frac{\partial \alpha_l}{\partial t_q} = \frac{1}{S_l} \frac{1}{S_{i_{m-1}}} \cdots \frac{1}{S_{i_2}} \frac{1}{S_{i_1}} \frac{\partial \alpha_0}{\partial t_q}$$
$$- \frac{1}{S_l} \frac{1}{S_{i_{m-1}}} \cdots \frac{1}{S_{i_2}} \frac{\partial g_{i_1}}{\partial x_{i_1}} Y_{i_1} - \cdots - \frac{1}{S_l} \frac{\partial g_{i_{m-1}}}{\partial x_{i_{m-1}}} Y_{i_{m-1}} - \frac{\partial g_l}{\partial x_l} Y_l. \tag{3.73}$$

Using (3.37) and (3.34), it is easy to see

$$\frac{\partial \alpha_l}{\partial t_q} < 0, \quad \text{if } q \Rightarrow l. \tag{3.74}$$

Case 2.1.4 $q \not\Rightarrow l$, $q \neq i_{m-k} \ (0 < k < m)$, $l \Rightarrow 0$, $q \Rightarrow 0$, i.e., q is a substantial collateral node of l.

From (3.59), (3.60), (3.34) and (3.37), it is easy to have

$$\frac{\partial \alpha_l}{\partial t_q} = -\frac{1}{S_l} \frac{\partial g_{i_{m-1}}}{\partial x_{i_{m-1}}} Y_{i_{m-1}} - \frac{1}{S_l} \frac{1}{S_{i_{m-1}}} \frac{\partial g_{i_{m-2}}}{\partial x_{i_{m-2}}} Y_{i_{m-2}}$$
$$- \cdots - \frac{1}{S_l} \frac{1}{S_{i_{m-1}}} \cdots \frac{1}{S_{i_2}} \frac{\partial g_{i_1}}{\partial x_{i_1}} Y_{i_1}$$
$$- \frac{1}{S_l} \frac{1}{S_{i_{m-1}}} \cdots \frac{1}{S_{i_2}} \frac{1}{S_{i_1}} A_0 \left(\sum_{u \in Ra_0, u \neq i_1} Y_u + \sum_{u \in Rd_0} Z_u \right)$$
$$< 0, \quad \text{if } q \not\Rightarrow l, \, q \neq i_{m-k} \ (0 < k < m), \, l \Rightarrow 0, \, q \Rightarrow 0. \tag{3.75}$$

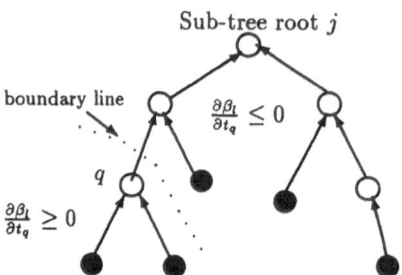

Figure 3.6: Effects of the communication time of link q on node loads

Remark: For other sub-cases in which some substantial ancestors are idle nodes, we use $B_{i_{m-k}}$, $Z_{i_{m-k}}$ and $V_{i_{m-k}}$ for node i_{m-k} if node i_{m-k} is an idle source node instead of using $A_{i_{m-k}}$, $Y_{i_{m-k}}$ and $U_{i_{m-k}}$ in the procedure of the proof on case 2.1. For brevity, we skip this part of the proof, but it has a similar relation to that for case 2.1.

Summarizing the results in case 2, we have the relations shown in (3.56a), (3.56b) and (3.56c).

Case 3. l is the idle source node (i.e., $\beta_l = 0$).

We have

$$\frac{\partial \beta_l}{\partial t_q} = 0. \tag{3.76}$$

According to the results in cases 1, 2, and 3, we complete the proof. Q.E.D.

Remark: Theorem 3.12 shows that the link communication time has effects only on the loads on active source nodes and sub-tree root nodes. Relation (3.56a) states that an increase in the communication time of a link causes an increase in the loads of its substantial descendant nodes and itself. Relations (3.56b), (3.56c) and (3.56d) state that an increase in the communication time of a link causes a decrease in the loads of its substantial ancestor nodes and its substantial collateral nodes. An increase in the communication time of a link in a sub-tree network, as stated by (3.56e), has no effects on the loads of the nodes that do not belong to the same sub-tree network.

By using Figure 3.6, we illustrate the effects of the communication time of a certain link q ($q \in L$), t_q, on the node loads in a sub-tree network j. Theorem 3.12 implies that an increase in t_q causes an increase in the loads of active source nodes (l) which belong to the left side of the boundary line, and causes a decrease in the loads of active source nodes (l) and the sub-tree root j which belong to the right side of the boundary line.

The effects of link communication time on mean response time

Theorem 3.13 *The following relations hold for the mean response time, $D(\boldsymbol{\beta}(t), \boldsymbol{x}(t))$, in a tree hierarchy network,*

$$\frac{\partial D(\boldsymbol{\beta}, \boldsymbol{x})}{\partial t_q} \quad > \quad 0, \quad if \ q \in Ra \cup Rd, \tag{3.77a}$$

$$= \quad 0, \quad else. \tag{3.77b}$$

PROOF.

From the definitions of $D(\boldsymbol{\beta}(t), \boldsymbol{x}(t))$, we have

$$
\begin{aligned}
D(\boldsymbol{\beta}(t), \boldsymbol{x}(t)) = \\
\frac{1}{\Phi}\{\beta_0(t) F_0(\beta_0(t)) + \sum_{i \in B}[\sum_{l \in Ra_i}(\beta_l(t) F_l(\beta_l(t)) + x_l G_l(x_l(t), t_l)) \\
+ \sum_{l \in Nu_i} \beta_l(t) F_l(\beta_l(t)) + \sum_{l \in Rd_i} x_l(t) G_l(x_l(t), t_l)]\}.
\end{aligned}
\tag{3.78}
$$

Without loss of generality, we suppose that node q belongs to sub-tree network 0. Note $\frac{\partial \beta_l}{\partial t_q}$ and $\frac{\partial x_l}{\partial t_q}$ will be zero if node l does not belong to sub-tree network 0. Also note that $\frac{\partial G_l}{\partial t_q} > 0$ if $q = l$; $\frac{\partial G_l}{\partial t_q} = 0$ if $q \neq l$. Using the definition of f_i and g_i, we have,

$$
\begin{aligned}
\frac{\partial D(\boldsymbol{\beta}, \boldsymbol{x})}{\partial t_q} &= \frac{1}{\Phi}\{f_0 \frac{\partial \beta_0}{\partial t_q} + \sum_{i \in \tilde{B}_0}[\sum_{l \in Ra_i}(f_l \frac{\partial \beta_l}{\partial t_q} + g_l \frac{\partial x_l}{\partial t_q}) + \sum_{l \in Rd_i} g_l \frac{\partial x_l}{\partial t_q}] \\
&\quad + \sum_{i \in \tilde{B}_0} \sum_{l \in Ra_i \cup Rd_i} x_l \frac{\partial G_l}{\partial t_q}\}.
\end{aligned}
\tag{3.79}
$$

Note $x_l = \phi'_l - \beta_l$, and $\alpha_i = \alpha_l - g_l$ ($l \in Ra_i \cup Rd_i$). According to Theorem 3.1, we have

$$
\begin{aligned}
\frac{\partial D(\boldsymbol{\beta}, \boldsymbol{x})}{\partial t_q} &= \frac{1}{\Phi}\{\alpha_0 \frac{\partial \beta_0}{\partial t_q} + \sum_{i \in \tilde{B}_0}[\sum_{l \in Ra_i}(\alpha_i \frac{\partial \beta_l}{\partial t_q} + g_l \frac{\partial \phi'_l}{\partial t_q}) + \sum_{l \in Rd_i} g_l \frac{\partial \phi'_l}{\partial t_q}] \\
&\quad + \sum_{i \in \tilde{B}_0} \sum_{l \in Ra_i \cup Rd_i} x_l \frac{\partial G_l}{\partial t_q}\}.
\end{aligned}
\tag{3.80}
$$

Note that

$$
\begin{aligned}
\frac{\partial \beta_0}{\partial t_q} &= \sum_{l \in Ra_0} \frac{\partial x_l}{\partial t_q} + \sum_{l \in Rd_0} \frac{\partial \phi'_l}{\partial t_q}, \\
\frac{\partial x_l}{\partial t_q} &= \frac{\partial \phi'_l}{\partial t_q} - \frac{\partial \beta_l}{\partial t_q}, \\
\frac{\partial \phi'_l}{\partial t_q} &= \begin{cases} \sum_{u \in Ra_l} \frac{\partial x_u}{\partial t_q} + \sum_{u \in Rd_l} \frac{\partial \phi'_u}{\partial t_q}, & \text{if } l \in \tilde{B}_0, \\ 0, & \text{if } l \in \tilde{T}_0. \end{cases}
\end{aligned}
\tag{3.81}
$$

Substituting (3.81) into (3.80), using that $\alpha_i = \alpha_l - g_l$ ($l \in Ra_i \cup Rd_i$), we can easily obtain

$$\frac{\partial D(\beta, x)}{\partial t_q} = \frac{1}{\Phi} \sum_{i \in \tilde{B}_0} \sum_{l \in Ra_i \cup Rd_i} x_l \frac{\partial G_l}{\partial t_q}, \quad \text{if } q \text{ belongs to sub-tree network } 0. \qquad (3.82)$$

For node q belongs to another sub-tree network i ($i \in Nu$), we have the same results. Q.E.D.

3.7.4   Effects of Node Processing Time

To analyse the effects of node processing time on the system performance variables, the following lemma is obtained.

Lemma 3.14 *The following relations hold for the incremental node delay $\alpha_j(u)$ in a tree hierarchy network,*

$$\frac{\partial \alpha_j}{\partial u_q} > 0, \quad \text{if } q \Rightarrow j \text{ and } q \in Ra, \text{ or } q = j, \qquad (3.83a)$$

$$= 0, \quad \text{else}, \qquad (3.83b)$$

where node j is a sub-tree root (i.e., $j \in Nu$, or node j is the root node ($j = 0$) in a tree hierarchy network).

PROOF.

According to the relations (3.3) in Theorem 3.1, we have

$$f_l(\beta_l(u), u_l) = \alpha_i + g_l(x_l(u)) \quad \text{if } l \in Ra_i,$$

$$\alpha_l = f_l(\beta_l(u), u_l) \qquad\qquad \text{if } \beta_l > 0,$$

$$\alpha_l = \alpha_i + g_l(x_l(u)) \qquad\quad \text{if } l \in Rd_i, \qquad (3.84)$$

$$i \in \tilde{B}.$$

Note that the change of the processing time of a node in a sub-tree network does not affect the link flow rates and node loads of nodes which belong to the other sub-tree networks. That is $\frac{\partial x_i}{\partial u_q} = 0$ and $\frac{\partial \beta_i}{\partial u_q} = 0$ if nodes i and q do not belong to the same sub-tree network. Without loss of generality, we assume node l belongs to sub-tree network j (i.e., $l \in \tilde{T}_j \cup \tilde{B}_j$) and its parent node is node i.

First, we consider the case that node l is a terminal node ($l \in \tilde{T}_j$). Note that terminal nodes in each sub-tree network must be active nodes. From (3.84), noting $\beta_l = \phi'_l - x_l$ and $\frac{\partial \phi'_l}{\partial u_q} = 0$ for $l \in \tilde{T}_j$, we have,

$$\frac{\partial \alpha_i}{\partial u_q} = -\left(\frac{\partial f_l}{\partial \beta_l} + \frac{\partial g_l}{\partial x_l}\right)\frac{\partial x_l}{\partial u_q} + \frac{\partial f_l}{\partial u_q}, \quad (l \in Ra_i). \qquad (3.85)$$

Rearranging (3.85), we have,

$$\frac{\partial x_l}{\partial u_q} = Y_l - A_l \frac{\partial \alpha_i}{\partial u_q}, \tag{3.86}$$

where,

$$Y_l = \frac{1}{T_l}\frac{\partial f_l}{\partial u_q}, \quad A_l = \frac{1}{T_l}, \quad T_l = \frac{\partial f_l}{\partial \beta_l} + \frac{\partial g_l}{\partial x_l}.$$

Because $\frac{\partial f_l}{\partial \beta_l} > 0$ and $\frac{\partial g_l}{\partial x_l} > 0$, we have $T_l > 0$, $A_l > 0$, and

$$Y_l = \begin{cases} \frac{1}{T_l}\frac{\partial f_l}{\partial u_q} > 0, & \text{if } q = l, \\ \\ 0, & \text{else.} \end{cases}$$

More generally, we consider that node l is an arbitrary node except the sub-tree root ($l \in \tilde{T}_j \cup \tilde{B}_j - \{j\}$). Note that node l may be an active node ($l \in Ra$) or an idle node ($l \in Rd$), and its substantial child nodes may also be active nodes or idle nodes. Using (3.84) and (3.86), we have the following recursive formula.

$$\frac{\partial x_l}{\partial u_q} = Y_l - A_l\frac{\partial \alpha_i}{\partial u_q}, \quad l \in Ra_i, \tag{3.87}$$

where

$$Y_l = \frac{1}{T_l}[\frac{\partial f_l}{\partial u_q} + \frac{\partial f_l}{\partial \beta_l}(\sum_{r \in Ra_l} Y_r + \sum_{r \in Rd_l} Z_r)]$$

$$A_l = \frac{1}{T_l}[1 + \frac{\partial f_l}{\partial \beta_l}(\sum_{r \in Ra_l} A_r + \sum_{r \in Rd_l} B_r)], \tag{3.88}$$

$$T_l = \frac{\partial f_l}{\partial \beta_l} + \frac{\partial g_l}{\partial x_l}[1 + \frac{\partial f_l}{\partial \beta_l}(\sum_{r \in Ra_l} A_r + \sum_{r \in Rd_l} B_r)].$$

$$\frac{\partial x_l}{\partial u_q} = Z_l - B_l\frac{\partial \alpha_i}{\partial u_q}, \quad l \in Rd_i, \tag{3.89}$$

where

$$Z_l = \frac{1}{U_l}(\sum_{r \in Ra_l} Y_r + \sum_{r \in Rd_l} Z_r),$$

$$B_l = \frac{1}{U_l}(\sum_{r \in Ra_l} A_r + \sum_{r \in Rd_l} B_r), \tag{3.90}$$

$$U_l = 1 + \frac{\partial g_l}{\partial x_l}(\sum_{r \in Ra_l} A_r + \sum_{r \in Rd_l} B_r).$$

As we have done when l is a terminal node, we also have the following relations. $T_l > 0$, $A_l > 0$, and

$$Y_l = \begin{cases} \frac{1}{T_l}\frac{\partial f_l}{\partial \beta_l}(\sum_{r \in Ra_l} Y_r + \sum_{r \in Rd_l} Z_r) > 0, & \text{if } q \Rightarrow l, \text{ and } q \in Ra, \\ \\ \frac{1}{T_l}\frac{\partial f_l}{\partial u_q} > 0, & \text{if } q = l, \\ \\ 0, & \text{else.} \end{cases} \tag{3.91}$$

$U_l > 0, \quad B_l > 0$, and

$$
Z_l = \begin{cases} \dfrac{1}{U_l}(\sum_{r \in Ra_l} Y_r + \sum_{r \in Rd_l} Z_r) > 0 & \text{if } q \Rightarrow l, \text{ and } q \in Ra, \\[4mm] 0 & \text{else.} \end{cases} \tag{3.92}
$$

Note that for $l \in \tilde{T}_j$, we naturally have $Y_r = 0$, and $Z_r = 0$ both in (3.91) and (3.92).

For node l is the sub-tree root (i.e., $l = j$, $j \in Nu$ or node j is the root) by noting that $\beta_j = \phi_j' = \sum_{r \in V_j} x_r$, from (3.84), we have

$$
\begin{aligned}
\frac{\partial \alpha_j}{\partial u_q} &= \frac{\partial f_j}{\partial \beta_j} \frac{\partial \beta_j}{\partial u_q} + \frac{\partial f_j}{\partial u_q}, \\
&= \frac{\partial f_j}{\partial \beta_j}(\sum_{r \in Ra_j} \frac{\partial x_r}{\partial u_q} + \sum_{r \in Rd_j} \frac{\partial x_r}{\partial u_q}) + \frac{\partial f_j}{\partial u_q}.
\end{aligned} \tag{3.93}
$$

Substituting (3.87) and (3.89) into (3.93), we have

$$
\frac{\partial \alpha_j}{\partial u_q} = A_j[\frac{\partial f_j}{\partial \beta_j}(\sum_{r \in Ra_j} Y_r + \sum_{r \in Rd_j} Z_r) + \frac{\partial f_j}{\partial u_q}] \tag{3.94}
$$

where

$$
A_j = \frac{1}{1 + \frac{\partial f_j}{\partial \beta_j}(\sum_{r \in Ra_j} A_r + \sum_{r \in Rd_j} B_r)}. \tag{3.95}
$$

Note $A_j > 0$. From (3.94), by using (3.91) and (3.92), we obtain the relations (3.77). Q.E.D.

The effects of node processing time on link flow rates

The effects of the node processing time on the link flow rates in a tree hierarchy network are specified in Theorem 3.15.

Theorem 3.15 *The following relations hold for the link flow* $x_l(u)$,

$$
\begin{aligned}
\frac{\partial x_l}{\partial u_q} \quad &< \quad 0, \quad \text{if } l \Rightarrow q, \ q \in Ra \text{ or } q = j, &\tag{3.96a} \\
&> \quad 0, \quad \text{if } q \overset{*}{\Rightarrow} l, \ q \in Ra, &\tag{3.96b} \\
&< \quad 0, \quad \text{if } q \not\Rightarrow l, \ l \not\Rightarrow q, \ l \Rightarrow j, \ q \Rightarrow j, \ q \in Ra, &\tag{3.96c} \\
&= \quad 0, \quad \text{else,} &\tag{3.96d}
\end{aligned}
$$

where node j *is a sub-tree root.*
PROOF.

We consider the effects of the processing time of an arbitrary node q, u_q, on link flow rate of link l ($l \in L$). Without loss of generality, we assume that node l belongs to sub-tree network 0 ($j = 0$). If l belongs to another sub-tree network j ($j \in Nu$), we note that the procedure of the proof is the same as that for sub-tree network 0. Note that node l may be an active node or an idle node (i.e., $l \in Ra \cup Rd$) and its substantial ancestor nodes except the sub-tree root also may be active nodes or idle nodes. We further assume that the substantial ancestor nodes of l are $i_{m-1}, i_{m-2}, ..., i_1, i_0$ ($i_0 = 0$) (i.e., $l \in Ra_{i_{m-1}} \cup Rd_{i_{m-1}}$, $i_{m-1} \in Ra_{i_{m-2}} \cup Rd_{i_{m-2}}, \cdots, i_1 \in Ra_{i_0} \cup Rd_{i_0}$ ($i_0 = 0$)). We prove the theorem for case 1 that node l is an active node and all its substantial ancestor nodes are also active nodes (i.e., $l \in Ra_{i_{m-1}}$, $i_{m-k} \in Ra_{i_{m-k-1}}$, $k = 1, 2, ..., m - 1$) in detail; and then make a remark on the other cases in which some substantial ancestors of node l are idle nodes.

Case 1 $l \in Ra_{i_{m-1}}$, $i_{m-1} \in Ra_{i_{m-2}}$, $i_{m-2} \in Ra_{i_{m-3}}, \cdots, i_2 \in Ra_{i_1}$, $i_1 \in Ra_0$.

According to (3.88), we note that

$$1 - \frac{\partial g_{i_{m-k}}}{\partial x_{i_{m-k}}} A_{i_{m-k}} = \frac{\partial f_{i_{m-k}}}{\partial \beta_{i_{m-k}}} \frac{1}{T_{i_{m-k}}}. \tag{3.97}$$

From (3.84), using (3.87) and (3.97), we have the following useful recursive relation,

$$\begin{aligned}
\frac{\partial \alpha_{i_{m-k}}}{\partial u_q} &= \frac{\partial \alpha_{i_{m-k-1}}}{\partial u_q} + \frac{\partial g_{i_{m-k}}}{\partial x_{i_{m-k}}} (Y_{i_{m-k}} - A_{i_{m-k}} \frac{\partial \alpha_{i_{m-k-1}}}{\partial u_q}) \\
&= (1 - \frac{\partial g_{i_{m-k}}}{\partial x_{i_{m-k}}} A_{i_{m-k}}) \frac{\partial \alpha_{i_{m-k-1}}}{\partial u_q} + \frac{\partial g_{i_{m-k}}}{\partial x_{i_{m-k}}} Y_{i_{m-k}} \\
&= \frac{1}{S_{i_{m-k}}} \frac{\partial \alpha_{i_{m-k-1}}}{\partial u_q} + \frac{\partial g_{i_{m-k}}}{\partial x_{i_{m-k}}} Y_{i_{m-k}}, \tag{3.98} \\
&(k = 1, 2, ..., m - 1),
\end{aligned}$$

where

$$\frac{1}{S_{i_{m-k}}} = \frac{\partial f_{i_{m-k}}}{\partial \beta_{i_{m-k}}} \frac{1}{T_{i_{m-k}}}.$$

Notice that the increase of the processing time of a node q in a sub-tree network does not affect the loads of other nodes that belong to other sub-tree networks. We note that node q may be in the following sub-cases: case 1.1 $q = i_{m-k}$ ($1 \le k \le m$) (i.e., node q is a substantial ancestor node of node l); case 1.2 $q \xrightarrow{s} l$ (i.e., node q is a substantial descendant node of node l or $q = l$); and case 1.3 $q \not\xrightarrow{s} l$, $q \ne i_{m-k}$ ($1 \le k \le m$), $l \Rightarrow 0$, $q \Rightarrow 0$ (i.e., node q is a substantial collateral node of node l). We consider each case separately.

Case 1.1 $q = i_{m-k}$ ($1 \le k \le m$), i.e., node q is a substantial ancestor node of node l. From (3.87) and (3.98), noting (3.91) and (3.92), we have

$$\frac{\partial x_l}{\partial u_q} = -A_l \frac{1}{S_{i_{m-1}}} \cdots \frac{1}{S_{i_{m-k+1}}} \frac{\partial \alpha_{i_{m-k}}}{\partial u_q}$$

$$
\begin{aligned}
= & -A_l \frac{1}{S_{i_{m-1}}} \cdots \frac{1}{S_{i_{m-k+1}}} \left(\frac{\partial g_{i_{m-k}}}{\partial x_{i_{m-k}}} Y_{i_{m-k}} + \frac{1}{S_{i_{m-k}}} \frac{\partial g_{i_{m-k-1}}}{\partial x_{i_{m-k-1}}} Y_{i_{m-k-1}} \right. \\
& \left. + \cdots + \frac{1}{S_{i_{m-k}}} \cdots \frac{1}{S_{i_2}} \frac{\partial g_{i_1}}{\partial x_{i_1}} Y_{i_1} + \frac{1}{S_{i_{m-k}}} \cdots \frac{1}{S_{i_2}} \frac{1}{S_{i_1}} \frac{\partial \alpha_{i_0}}{\partial u_q} \right)
\end{aligned}
\tag{3.99}
$$

Noting (3.91) and (3.92), we have

$$
\frac{\partial \alpha_{i_0}}{\partial u_q} = A_0 \frac{\partial f_0}{\partial \beta_0} Y_{i_1},
$$

$$
Y_{i_{m-t}} = \frac{1}{S_{i_{m-k-1}}} \cdots \frac{1}{S_{i_{m-t+1}}} \frac{1}{S_{i_{m-t}}} Y_{i_{m-k}}, \quad (t = k+1, ..., m-1),
$$

$$
Y_{i_{m-k}} = \frac{1}{T_{i_{m-k}}} \frac{\partial f_{i_{m-k}}}{\partial u_q} > 0.
\tag{3.100}
$$

Then we have,

$$
\frac{\partial x_l}{\partial u_q} < 0, \quad \text{if } q = i_{m-k} \ (1 < k < m).
\tag{3.101}
$$

Case 1.2 $q \overset{\rightarrow}{\Rightarrow} l$, i.e., node q is a substantial descendant node of node l or $q = l$.
From (3.91) and (3.92), we have

$$
\frac{\partial x_l}{\partial u_q} = 0, \quad \text{if } q \in Rd.
\tag{3.102}
$$

If $q \in Ra$, we have the following.
From (3.87) and (3.98), using (3.91),

$$
\begin{aligned}
\frac{\partial x_l}{\partial u_q} = & Y_l - A_l \Big(\frac{\partial g_{i_{m-1}}}{\partial x_{i_{m-1}}} Y_{i_{m-1}} + \frac{1}{S_{i_{m-1}}} \frac{\partial g_{i_{m-2}}}{\partial x_{i_{m-2}}} Y_{i_{m-2}} \\
& + \cdots + \frac{1}{S_{i_{m-1}}} \cdots \frac{1}{S_{i_2}} \frac{\partial g_{i_1}}{\partial x_{i_1}} Y_{i_1} + \frac{1}{S_{i_{m-1}}} \cdots \frac{1}{S_{i_2}} \frac{1}{S_{i_1}} \frac{\partial \alpha_0}{\partial u_q} \Big).
\end{aligned}
\tag{3.103}
$$

From (3.94) and (3.88), using (3.91) and (3.92), we have

$$
\frac{\partial \alpha_0}{\partial u_q} = A_0 \frac{\partial f_0}{\partial \beta_0} Y_{i_1},
$$

$$
Y_{i_{m-t}} = \frac{1}{S_{i_{m-1}}} \cdots \frac{1}{S_{i_{m-t}}} Y_l, \quad (t = 1, 2, ..., m-1).
\tag{3.104}
$$

From (3.88), we have

$$
\begin{aligned}
Y_l = & Y_l \frac{T_{i_{m-1}}}{T_{i_{m-1}}} \\
= & Y_l \frac{1}{T_{i_{m-1}}} \Big[\frac{\partial f_{i_{m-1}}}{\partial \beta_{i_{m-1}}} + \frac{\partial g_{i_{m-1}}}{\partial x_{i_{m-1}}} \Big(1 + \frac{\partial f_{i_{m-1}}}{\partial \beta_{i_{m-1}}} \Big(\sum_{u \in Ra_{i_{m-1}}} A_u + \sum_{u \in Rd_{i_{m-1}}} B_u \Big) \Big) \Big] \\
= & Y_l \frac{1}{S_{i_{m-1}}} \frac{T_{i_{m-2}}}{T_{i_{m-2}}} + Y_l \frac{\partial g_{i_{m-1}}}{\partial x_{i_{m-1}}} A_{i_{m-1}} \\
= & \cdots \cdots \\
= & H_0 + H_1 + \cdots + H_{m-1}
\end{aligned}
\tag{3.105}
$$

where

$$
\begin{aligned}
H_0 &= Y_l \frac{1}{S_{i_{m-1}}} \cdots \frac{1}{S_{i_2}} \frac{1}{S_{i_1}}, \\
H_1 &= Y_l \frac{1}{S_{i_{m-1}}} \cdots \frac{1}{S_{i_2}} \frac{\partial g_{i_1}}{\partial x_{i_1}} A_{i_1}, \\
&\quad \cdots \cdots \\
H_{m-2} &= Y_l \frac{1}{S_{i_{m-1}}} \frac{\partial g_{i_{m-2}}}{\partial x_{i_{m-2}}} A_{i_{m-2}}, \\
H_{m-1} &= Y_l \frac{\partial g_{i_{m-1}}}{\partial x_{i_{m-1}}} A_{i_{m-1}}.
\end{aligned}
\tag{3.106}
$$

By substituting (3.105) into (3.103), we have

$$
\begin{aligned}
\frac{\partial x_l}{\partial u_q} &= \left(H_0 - A_l \frac{1}{S_{i_{m-1}}} \cdots \frac{1}{S_{i_1}} \frac{\partial \alpha_0}{\partial u_q} \right) + \left(H_1 - A_l \frac{1}{S_{i_{m-1}}} \cdots \frac{1}{S_{i_2}} \frac{\partial g_{i_1}}{\partial x_{i_1}} Y_{i_1} \right) \\
&\quad + \cdots + \left(H_{m-2} - A_l \frac{1}{S_{i_{m-1}}} \frac{\partial g_{i_{m-2}}}{\partial x_{i_{m-2}}} Y_{i_{m-2}} \right) \\
&\quad + \left(H_{m-1} - A_l \frac{\partial g_{i_{m-1}}}{\partial x_{i_{m-1}}} Y_{i_{m-1}} \right).
\end{aligned}
\tag{3.107}
$$

From (3.104) and (3.106), we have

$$
\begin{aligned}
H_0 &- A_l \frac{1}{S_{i_{m-1}}} \cdots \frac{1}{S_{i_1}} \frac{\partial \alpha_0}{\partial u_q} \\
&= \frac{1}{S_{i_{m-1}}} \cdots \frac{1}{S_{i_1}} \left(Y_l - A_l A_0 \frac{\partial f_0}{\partial \beta_0} Y_{i_1} \right) \\
&= \frac{1}{S_{i_{m-1}}} \cdots \frac{1}{S_{i_1}} Y_l \left(1 - A_l A_0 \frac{\partial f_0}{\partial \beta_0} \frac{1}{S_{i_{m-1}}} \cdots \frac{1}{S_{i_1}} \right)
\end{aligned}
\tag{3.108}
$$

From (3.88), (3.90), (3.91), (3.92) and (3.95), it is easy to find that

$$
1 - A_l A_0 \frac{\partial f_0}{\partial \beta_0} \frac{1}{S_{i_{m-1}}} \cdots \frac{1}{S_{i_1}} > 0.
\tag{3.109}
$$

Then,

$$
H_0 - A_l \frac{1}{S_{i_{m-1}}} \cdots \frac{1}{S_{i_1}} \frac{\partial \alpha_0}{\partial u_q} > 0.
\tag{3.110}
$$

From (3.104) and (3.106), it also has

$$
\begin{aligned}
H_{m-k} &- A_l \frac{1}{S_{i_{m-1}}} \cdots \frac{1}{S_{i_{m-k+1}}} \frac{\partial g_{i_{m-k}}}{\partial x_{i_{m-k}}} Y_{i_{m-k}} = \\
&\frac{1}{S_{i_{m-1}}} \cdots \frac{1}{S_{i_{m-k+1}}} Y_l \frac{\partial g_{i_{m-k}}}{\partial x_{i_{m-k}}} \left(A_{i_{m-k}} - A_l \frac{1}{S_{i_{m-1}}} \cdots \frac{1}{S_{i_{m-k}}} \right) \\
&k = 2, 3, ..., m - 1.
\end{aligned}
\tag{3.111}
$$

From (3.88), (3.91), (3.90) and (3.92), it is easy to find that

$$A_{i_{m-k}} - A_l \frac{1}{S_{i_{m-1}}} \cdots \frac{1}{S_{i_{m-k}}} > 0. \tag{3.112}$$

Then

$$H_{m-k} - A_l \frac{1}{S_{i_{m-1}}} \cdots \frac{1}{S_{i_{m-k+1}}} \frac{\partial g_{i_{m-k}}}{\partial x_{i_{m-k}}} Y_{i_{m-k}} > 0 \quad k = 2, 3, ..., m-1. \tag{3.113}$$

From (3.104), we have

$$H_{m-1} - A_l \frac{\partial g_{i_{m-1}}}{\partial x_{i_{m-1}}} Y_{i_{m-1}}$$

$$= Y_l \frac{\partial g_{i_{m-1}}}{\partial x_{i_{m-1}}} (A_{i_{m-1}} - A_l \frac{1}{S_{i_{m-1}}})$$

$$= Y_l \frac{\partial g_{i_{m-1}}}{\partial x_{i_{m-1}}} \frac{1}{T_{i_{m-1}}} [1 + \frac{\partial f_{i_{m-1}}}{\partial \beta_{i_{m-1}}} (\sum_{u \in Ra_{i_{m-1}}, u \neq l} A_u + \sum_{u \in Rd_{i_{m-1}}} B_u)]$$

$$> 0. \tag{3.114}$$

From (3.102), (3.110), (3.113) and (3.114), we obtain

$$\frac{\partial x_l}{\partial u_q} > 0, \quad \text{if } q \overset{\rightarrow}{\Rightarrow} l, q \in Ra,$$

$$= 0, \quad \text{if } q \overset{\rightarrow}{\Rightarrow} l, q \in Rd. \tag{3.115}$$

Case 1.3 $q \overset{\rightarrow}{\not\Rightarrow} l, q \neq i_{m-k}$ $(1 \leq k \leq m), l \Rightarrow 0, q \Rightarrow 0$, i.e., node q is a substantial collateral node of node l.

From (3.87), (3.98) and (3.95), using (3.91) and (3.92), we have

$$\frac{\partial x_l}{\partial u_q} = -A_l \frac{\partial g_{i_{m-1}}}{\partial x_{i_{m-1}}} Y_{i_{m-1}} - A_l \frac{1}{S_{i_{m-1}}} \frac{\partial g_{i_{m-2}}}{\partial x_{i_{m-2}}} Y_{i_{m-2}}$$

$$- \cdots - A_l \frac{1}{S_{i_{m-1}}} \cdots \frac{1}{S_{i_2}} \frac{\partial g_{i_1}}{\partial x_{i_1}} Y_{i_1}$$

$$- A_l \frac{1}{S_{i_{m-1}}} \cdots \frac{1}{S_{i_1}} A_0 \frac{\partial f_0}{\partial \beta_0} (\sum_{u \in Ra_0, u \neq i_1} Y_u + \sum_{u \in Rd_0} Z_u)$$

$$< 0, \quad \text{if } q \in Ra. \tag{3.116}$$

Remark: For the other cases in which some substantial ancestors of node l are idle nodes, we can use $U_{i_{m-k}}$, $Z_{i_{m-k}}$ and $B_{i_{m-k}}$ instead of using $S_{i_{m-k}}$, $Y_{i_{m-k}}$ and $A_{i_{m-k}}$ for node i_{m-k} if it is the idle node in the procedure of the proof on case 1. For brevity, we skip this part of the proof, but it has a similar result.

To summarize, we have the relations in Theorem 3.15. Q.E.D.

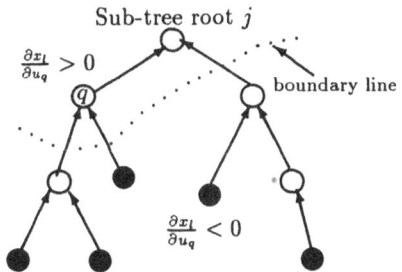

q is an active source node

Figure 3.7: Effects of the processing time of node q on link flow rates

Theorem 3.15 may be interpreted as follows. Relation (3.96b) states that an increase in the processing time of an active source node causes an increase in the link flow rates of its substantial ancestor nodes and itself; relations (3.96a) and (3.96c) state that a decrease in the processing time of an active source node or the sub-tree root node causes a decrease in the link flow rates of its substantial descendant nodes and its substantial collateral nodes. Otherwise, any change in the processing time of a node does not affect the link flow rates in the entire network.

Figure 3.7 illustrates the effects on the processing time of a certain active node q ($q \in Ra$), u_q, in the link flow rates in a sub-tree network j. Notice that a node (l, $l \in L$) that belongs to the left side of the boundary line is either a substantial ancestor node of node q or node q itself; a node (l, $l \in L$) that belongs to the right side of the boundary line is a substantial descendant node or a substantial collateral node of q. Theorem 3.15 implies that an increase in u_q causes an increase in the link flow rates of nodes that belong to the left side of the boundary line, but causes an decrease in the link flow rates of nodes that belong to the right side of the boundary line.

The effects of node processing time on node loads

Theorem 3.16 *The following relations hold for the node load* $\beta_l(\boldsymbol{u})$,

$$\frac{\partial \beta_l}{\partial u_q} \quad > \quad 0, \quad \text{if } l \Rightarrow q, \; q \in Ra \text{ or } q = j, \quad\quad (3.117a)$$

$$< \quad 0, \quad \text{if } q = l, \; q \in Ra \text{ or } q = j, \quad\quad (3.117b)$$

$$> \quad 0, \quad \text{if } q \Rightarrow l, \; q \in Ra, \quad\quad (3.117c)$$

$$> \quad 0, \quad \text{if } q \not\Rightarrow l, \; l \not\Rightarrow q, \; l \Rightarrow j, \; q \Rightarrow j, \; q \in Ra, \quad\quad (3.117d)$$

$$= \quad 0, \quad \text{else}, \quad\quad (3.117e)$$

where node j is a sub-tree root.

PROOF.

We consider exhaustively the following three cases: case 1. node l is a neutral node $(l \in Nu)$ or the root node; case 2. node l is an active node $(l \in Ra)$; case 3. node l is an idle node $(l \in Rd)$.

Case 1. l is a neutral node or the root node, i.e., $l = j$.

From (3.93) and (3.94), we have

$$
\begin{aligned}
\frac{\partial f_l}{\partial \beta_l}\frac{\partial \beta_l}{\partial u_q} &= \frac{\partial \alpha_l}{\partial u_q} - \frac{\partial f_l}{\partial u_q} \\
&= A_l\frac{\partial f_l}{\partial \beta_l}(\sum_{i \in Ra_l} Y_i + \sum_{i \in Rd_l} Z_i) + (A_l - 1)\frac{\partial f_l}{\partial u_q}
\end{aligned}
\tag{3.118}
$$

From (3.95), note that

$$
A_l - 1 = -A_l\frac{\partial f_l}{\partial \beta_l}(\sum_{i \in Ra_l} A_i + \sum_{i \in Rd_l} B_i) < 0.
\tag{3.119}
$$

Note $\frac{\partial f_l}{\partial \beta_l} > 0$. From (3.118), using (3.91) and (3.92), we obtain

$$
\begin{aligned}
\frac{\partial \beta_l}{\partial u_q} &> 0, \quad \text{if } q \Rightarrow l, q \in Ra, \\
&< 0, \quad \text{if } q = l, \\
&= 0, \quad \text{else.}
\end{aligned}
\tag{3.120}
$$

Case 2. l is an active node, $(l \in Ra)$.

Without loss of generality, we assume that node l belong to sub-tree network 0 (i.e., $j = 0$). From (3.84), we note that

$$
\frac{\partial f_l}{\partial \beta_l}\frac{\partial \beta_l}{\partial u_q} = \frac{\partial \alpha_l}{\partial u_q} - \frac{\partial f_l}{\partial u_q} \quad \text{if } l \in Ra.
\tag{3.121}
$$

To prove the theorem in this case, we first prove the relations that $\frac{\partial \alpha_l}{\partial u_q}$ holds, then show the relations that $\frac{\partial \beta_l}{\partial u_q}$ holds.

In general, we assume that $l \in Ra_{i_{m-1}}$, $i_{m-1} \in Ra_{i_{m-2}} \cup Rd_{i_{m-2}}$, $i_{m-2} \in Ra_{i_{m-3}} \cup Rd_{i_{m-3}}$, ..., $i_2 \in Ra_{i_1} \cup Rd_{i_1}$, $i_1 \in Ra_{i_0} \cup Rd_{i_0}$ $(i_0 = 0)$. Here we carry out the proof for sub-case 2.1 that $i_{m-1} \in Ra_{i_{m-2}}$, $i_{m-2} \in Ra_{i_{m-3}}$, ..., $i_2 \in Ra_{i_1}$, $i_1 \in Ra_{i_0}$ (i.e., all substantial ancestor nodes of node l are active nodes). The proof for other sub-cases in which some substantial ancestors of node l are idle nodes is essentially the same.

Case 2.1 $i_{m-1} \in Ra_{i_{m-2}}$, ..., $i_2 \in Ra_{i_1}$, $i_1 \in Ra_{i_0}$ $(i_0 = 0)$.

According to (3.88), we have

$$
1 - \frac{\partial g_l}{\partial x_l}A_l = \frac{1}{T_l}\frac{\partial f_l}{\partial \beta_l}
\tag{3.122}
$$

From (3.84) and (3.87), using (3.122), we have

$$\frac{\partial \alpha_l}{\partial u_q} = \frac{\partial \alpha_{i_{m-1}}}{\partial u_q} \frac{1}{S_{i_l}} + \frac{\partial g_l}{\partial x_l} Y_{l},$$ (3.123)

where

$$\frac{1}{S_{i_l}} = \frac{1}{T_l} \frac{\partial f_l}{\partial \beta_l}.$$

Similarly, we also have,

$$\frac{\partial \alpha_{i_{m-k}}}{\partial u_q} = \frac{1}{S_{i_{m-k}}} \frac{\partial \alpha_{i_{m-k-1}}}{\partial u_q} + \frac{\partial g_{i_{m-k}}}{\partial x_{i_{m-k}}} Y_{i_{m-k}}, \quad k = 1, 2, ..., m-1,$$ (3.124)

where

$$\frac{1}{S_{i_{m-k}}} = \frac{1}{T_{i_{m-k}}} \frac{\partial f_{i_{m-k}}}{\partial \beta_{i_{m-k}}}.$$

Note that the increase of the processing time of a node q in a sub-tree network does not affect the loads of other nodes which belong to other sub-tree networks. We consider the following four sub-cases separately: case 2.1.1. q is a substantial ancestor node of l; case 2.1.2 $q = l$; case 2.1.3. q is a substantial descendant node of l; and case 2.1.4. q is a substantial collateral node of l.

Case 2.1.1 $q = i_{m-k}$, $(k = 1, 2, ..., m)$, i.e., q is a substantial ancestor node of l

From (3.121), (3.123) and (3.124), using (3.91), we have

$$\frac{\partial f_l}{\partial \beta_l} \frac{\partial \beta_l}{\partial u_q} = \frac{1}{S_l} \frac{1}{S_{i_{m-1}}} \cdots \frac{1}{S_{i_{m-k+1}}} \frac{\partial \alpha_{i_{m-k}}}{\partial u_q}.$$ (3.125)

From (3.124), we have

$$\frac{\partial \alpha_{i_{m-k}}}{\partial u_q} = \frac{\partial g_{i_{m-k}}}{\partial x_{i_{m-k}}} Y_{i_{m-k}} + \frac{1}{S_{i_{m-k}}} \frac{\partial g_{i_{m-k-1}}}{\partial x_{i_{m-k-1}}} Y_{i_{m-k-1}}$$
$$+ \cdots + \frac{1}{S_{i_{m-k}}} \cdots \frac{1}{S_{i_2}} \frac{\partial g_{i_1}}{\partial x_{i_1}} Y_{i_1} + \frac{1}{S_{i_{m-k}}} \cdots \frac{1}{S_{i_1}} \frac{\partial \alpha_0}{\partial u_q}.$$ (3.126)

From (3.91) and Lemma 3.12, we obtain

$$\frac{\partial \beta_l}{\partial u_q} > 0, \quad \text{if } q = i_{m-k}, (k = 1, 2, ..., m-1).$$ (3.127)

Case 2.1.2 $q = l$

From (3.123) and (3.124), using (3.91), we have

$$\frac{\partial \alpha_l}{\partial u_q} = \frac{1}{S_l} \frac{1}{S_{i_{m-1}}} \cdots \frac{1}{S_{i_1}} \frac{\partial \alpha_0}{\partial u_q} + \frac{1}{S_l} \frac{1}{S_{i_{m-1}}} \cdots \frac{1}{S_{i_2}} \frac{\partial g_{i_1}}{\partial x_{i_1}} Y_{i_1}$$
$$+ \cdots + \frac{1}{S_l} \frac{\partial g_{i_{m-1}}}{\partial x_{i_{m-1}}} Y_{i_{m-1}} + \frac{\partial g_l}{\partial x_l} Y_{l}.$$ (3.128)

Substituting (3.128) into (3.121), using (3.94), we have

$$\frac{\partial f_l}{\partial \beta_l}\frac{\partial \beta_l}{\partial u_q} = \frac{1}{S_l}\frac{1}{S_{i_{m-1}}}\cdots\frac{1}{S_{i_1}}A_0\frac{\partial f_0}{\partial \beta_0}Y_{i_1} + \frac{1}{S_l}\frac{1}{S_{i_{m-1}}}\cdots\frac{1}{S_{i_2}}\frac{\partial g_{i_1}}{\partial x_{i_1}}Y_{i_1}$$
$$+\cdots+\frac{1}{S_l}\frac{\partial g_{i_{m-1}}}{\partial x_{i_{m-1}}}Y_{i_{m-1}} + \frac{\partial g_l}{\partial x_l}Y_l - \frac{\partial f_l}{\partial u_q}. \tag{3.129}$$

Applying a similar method to that for case 1.2 in the proof of Theorem 3.15 we can obtain

$$\frac{\partial \beta_l}{\partial u_q} < 0, \quad \text{if } q = l. \tag{3.130}$$

Case 2.1.3 $q \Rightarrow l$, i.e., q is a substantial descendant node of l

Note that $\frac{\partial f_l}{\partial u_q} = 0$. From (3.123) and (3.124), using (3.91), we have

$$\frac{\partial f_l}{\partial \beta_l}\frac{\partial \beta_l}{\partial u_q} = \frac{\partial \alpha_l}{\partial u_q}$$
$$= \frac{1}{S_l}\frac{1}{S_{i_{m-1}}}\cdots\frac{1}{S_{i_1}}\frac{\partial \alpha_0}{\partial u_q} + \frac{1}{S_l}\frac{1}{S_{i_{m-1}}}\cdots\frac{1}{S_{i_2}}\frac{\partial g_{i_1}}{\partial x_{i_1}}Y_{i_1}$$
$$+\cdots+\frac{1}{S_l}\frac{\partial g_{i_{m-1}}}{\partial x_{i_{m-1}}}Y_{i_{m-1}} + \frac{\partial g_l}{\partial x_l}Y_l \tag{3.131}$$

Using (3.94) and (3.91), we obtain

$$\frac{\partial \beta_l}{\partial u_q} > 0, \quad \text{if } q \Rightarrow l,\, q \in Ra. \tag{3.132}$$

Case 2.1.4 $q \not\Rightarrow l$, $q \neq i_{m-k}$ $(0 < k < m)$, $l \Rightarrow 0$, $q \Rightarrow 0$, i.e., q is a substantial collateral node of l

Note that $\frac{\partial f_l}{\partial u_q} = 0$ in this case. From (3.123) and (3.124), using (3.91), we have

$$\frac{\partial f_l}{\partial \beta_l}\frac{\partial \beta_l}{\partial u_q} = \frac{\partial \alpha_l}{\partial u_q}$$
$$= \frac{1}{S_l}\frac{1}{S_{i_{m-1}}}\cdots\frac{1}{S_{i_1}}\frac{\partial \alpha_0}{\partial u_q} + \frac{1}{S_l}\frac{1}{S_{i_{m-1}}}\cdots\frac{1}{S_{i_2}}\frac{\partial g_{i_1}}{\partial x_{i_1}}Y_{i_1}$$
$$+\cdots+\frac{1}{S_l}\frac{\partial g_{i_{m-1}}}{\partial x_{i_{m-1}}}Y_{i_{m-1}} \tag{3.133}$$

Using (3.94) and (3.91), we obtain

$$\frac{\partial \beta_l}{\partial u_q} > 0, \quad \text{if } q \not\Rightarrow l,\, q \neq i_{m-k}\ (0 < k < m),\, l \Rightarrow 0,\, q \Rightarrow 0,\, q \in Ra. \tag{3.134}$$

Case 3. l is the idle node.

It is clear that

$$\frac{\partial \beta_l}{\partial u_q} = 0. \tag{3.135}$$

From the results in cases 1, 2 and 3, we obtain the relations in the theorem. Q.E.D.

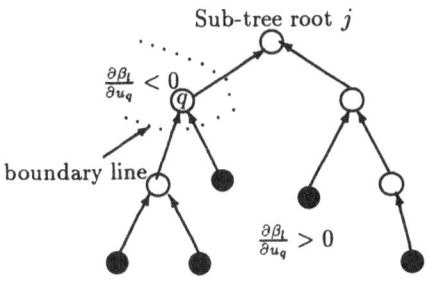

Figure 3.8: Effects of the processing time of node q on node loads

Relation (3.117b) states that an increase in the processing time of an active source node or the sub-tree root node causes a decrease in the load of itself. But an increase in the processing time of an active source node or the sub-tree root node causes an increase in the loads of its substantial descendant nodes, its substantial ancestor nodes and its substantial collateral nodes, as stated by (3.117a), (3.117c) and (3.117d). Otherwise, any change in the processing time of a node does not affect the node loads in the entire network.

By using Figure 3.8, we illustrate the effects of the processing time of a certain active node q $(q \in Ra)$, u_q, on the node loads in a sub-tree network j. Theorem 3.16 implies that an increase in u_q causes a decrease in the load of node q itself, but causes an increase in the loads of other nodes (l) in the sub-tree network.

The effects of node processing time on mean response time

We have the following theorem showing the effects of processing time on mean system response time.

Theorem 3.17 *The following relations hold for the mean response time, $D(\boldsymbol{\beta}(\boldsymbol{u}), \boldsymbol{x}(\boldsymbol{u}))$, in a tree hierarchy network,*

$$\frac{\partial D(\boldsymbol{\beta}, \boldsymbol{x})}{\partial u_q} \quad > \quad 0, \quad if\, q \in Ra \cup Nu, \; or\, q = 0 \tag{3.136a}$$

$$= \quad 0, \quad else. \tag{3.136b}$$

PROOF.

From the definitions of $D(\boldsymbol{\beta}, \boldsymbol{x})$, we have

$$D(\boldsymbol{\beta}, \boldsymbol{x}) \quad = \quad \frac{1}{\Phi}\{\beta_0 F_0(\beta_0) + \sum_{i \in B}[\sum_{l \in Ra_i} (\beta_l F_l(\beta_l) + x_l G_l(x_l))$$

$$+ \sum_{l \in Nu_i} \beta_l F_l(\beta_l) + \sum_{l \in Rd_i} x_l G_l(x_l)]\}. \tag{3.137}$$

Without loss of generality, we suppose that node q belongs to sub-tree 0. Note that $\frac{\partial \beta_l}{\partial u_q}$ and $\frac{\partial x_l}{\partial u_q}$ will be zero if node l does not belong to sub-tree 0. Also note that $\frac{\partial F_l}{\partial u_q} > 0$ if $q = l$; and $\frac{\partial F_l}{\partial u_q} = 0$ if $q \neq l$. From (3.137) we have

$$\frac{\partial D(\boldsymbol{\beta}, \boldsymbol{x})}{\partial u_q} = \frac{1}{\Phi}\{f_0 \frac{\partial \beta_0}{\partial u_q} + \sum_{i \in \check{B}_0}[\sum_{l \in Ra_i}(f_l \frac{\partial \beta_l}{\partial u_q} + g_l \frac{\partial x_l}{\partial u_q}) $$
$$+ \sum_{l \in Rd_i} g_l \frac{\partial x_l}{\partial u_q}] + \beta_0 \frac{\partial F_0}{\partial u_q} + \sum_{i \in \check{B}_0} \sum_{l \in Ra_i \cup Nu_i} \beta_l \frac{\partial F_l}{\partial u_q}\}. \tag{3.138}$$

Note $x_l = \phi_l' - \beta_l$, and $\alpha_i = \alpha_l - g_l$ ($l \in Ra_i \cup Rd_i$). According to Theorem 3.1, we have

$$\frac{\partial D(\boldsymbol{\beta}, \boldsymbol{x})}{\partial u_q} = \frac{1}{\Phi}\{\alpha_0 \frac{\partial \beta_0}{\partial u_q} + \sum_{i \in \check{B}_0}[\sum_{l \in Ra_i}(\alpha_i \frac{\partial \beta_l}{\partial u_q} + g_l \frac{\partial \phi_l'}{\partial u_q}) $$
$$+ \sum_{l \in Rd_i} g_l \frac{\partial \phi_l'}{\partial u_q}] + \beta_0 \frac{\partial F_0}{\partial u_q} + \sum_{i \in \check{B}_0} \sum_{l \in Ra_i} x_l \frac{\partial F_l}{\partial u_q}\}. \tag{3.139}$$

Note that

$$\frac{\partial \beta_0}{\partial u_q} = \sum_{l \in Ra_0} \frac{\partial x_l}{\partial u_q} + \sum_{l \in Rd_0} \frac{\partial \phi_l'}{\partial u_q},$$
$$\frac{\partial x_l}{\partial u_q} = \frac{\partial \phi_l'}{\partial u_q} - \frac{\partial \beta_l}{\partial u_q}, \tag{3.140}$$
$$\frac{\partial \phi_l'}{\partial u_q} = \begin{cases} \sum_{u \in Ra_l} \frac{\partial x_u}{\partial u_q} + \sum_{u \in Rd_l} \frac{\partial \phi_u'}{\partial u_q}, & \text{if } l \in \check{B}_0, \\ 0, & \text{if } l \in \tilde{T}_0. \end{cases}$$

Substituting (3.140) into (3.139), using $\alpha_i = \alpha_l - g_l$ ($l \in Ra_i \cup Rd_i$), we can easily obtain

$$\frac{\partial D(\boldsymbol{\beta}, \boldsymbol{x})}{\partial u_q} = \frac{1}{\Phi}\{\beta_0 \frac{\partial F_0}{\partial u_q} + \sum_{i \in \check{B}_0} \sum_{l \in Ra_i} x_l \frac{\partial F_l}{\partial u_q}\}. \tag{3.141}$$

From (3.141), it is obvious that $\frac{\partial D(\boldsymbol{\beta}, \boldsymbol{x})}{\partial u_q} > 0$ if $q = 0$ or $q \in Ra$; otherwise $\frac{\partial D(\boldsymbol{\beta}, \boldsymbol{x})}{\partial u_q} = 0$. For node q belongs to another sub-tree i ($i \in Nu$), we have a similar result. Q.E.D.

3.8   Conclusion

In this chapter, we have studied the optimal static load balancing problem in a distributed computer system with the tree hierarchy configurations. The problem is formulated as a nonlinear optimization problem. We have studied the necessary and sufficient conditions for the optimal solution of the problem, and we have shown that we can decompose the tree hierarchy optimization problem into a collection of much simpler star subproblems. Then, a decomposition algorithm which includes a sub-algorithm that solves the subproblems was presented.

The proposed algorithm was compared with existing well-known algorithms: the FD algorithm and the D–S algorithm with respect to the storage and computation time requirements. It was proven that the proposed algorithm and the FD algorithm require the amount of storage of $O(n)$, whereas the D–S algorithm requires the amount of storage of $O(n \log(n))$. On the other hand, the proposed algorithm and the D–S algorithm have shown nearly the same convergence speed but the FD algorithm has shown much slower convergence than the other two algorithms.

Analytical and experimental results clearly indicated the outperformance of this algorithm over existing methods such as the FD algorithm and the D–S algorithm.

We also have studied the effects of link communication time and the node processing time on system performance measures in the optimal load balancing by parametric analysis. The following clear and interesting properties were derived for a tree hierarchy network.

Supposing that there are m $(m = |Nu|)$ nodes that have no link flow rates in the optimal solution of a certain network, we divide the tree hierarchy network into $m + 1$ sub-tree networks, each root of which has no link flow rate. The communication time of a link and the processing time of a node have the effects only on the link flow rates and the loads of nodes that are in the same sub-tree network.

An increase in the communication time of an arbitrary link causes a decrease in the link flow rates of its substantial direct line nodes, but causes an increase in the link flow rates of its substantial collateral nodes. Further, it also causes an increase in the loads on its substantial descendant nodes and itself, but causes a decrease in the loads of other nodes in the same sub-tree network.

Moreover, an increase in the processing time of an arbitrary node causes an increase in the link flow rates of its substantial ancestor nodes and itself, but causes a decrease in the link flow rates of its substantial descendant nodes and its substantial collateral nodes. Further, it causes a decrease in the load of itself, but causes an increase in the loads of other nodes in the same sub-tree network.

In gereral, either an increase in the communication time of a link or an increase in the processing time of a node will cause the increase in the mean response time.

Chapter 4

Load Balancing of Star Network Configurations with Two-way Traffic

4.1   Introduction

Tantawi and Towsley [TT84] have considered the optimal static load balancing in a general star network configuration. In Tantawi and Towsley model, however, there is only one-way traffic from satellite nodes to the central node in the sense that jobs can be forwarded for remote processing only from satellites to the central node.

Usually, satellite hosts send some of jobs arriving at them to a very powerful central host. However, we can think of the following case: when the central host has a job arrival rate more than its processing capacity, it would like to forward a part of these jobs to rather vacant satellite hosts. Therefore, in some cases, it is reasonable to consider the case where there are two ways between two nodes in a computer network in the sense that jobs can be forwarded from one node to another node or from the latter to the former for remote processing. In a star network with two-way traffic we can take the central node as a switch. A job arriving at a node can be sent to any nodes for processing. In the past few years, several star networks with two-way traffic have been developed [LB83, Alb83, RM78, Sch83]. Due to the development of fiber-optic technique, there are growing interests in the star network configurations with two-way traffic [AK91, Kam87, Kum89, MAHE88].

In this chapter, we study the static load balancing in a star network configuration with two-way traffic between the central host computer node and any satellite computer node. For the sake of convenience, we call the star computer network with two-way traffic the *star network*. Jobs arrive at each node according to a time-invariant Poisson process. Each node determines whether a job should be processed locally or scheduled to another

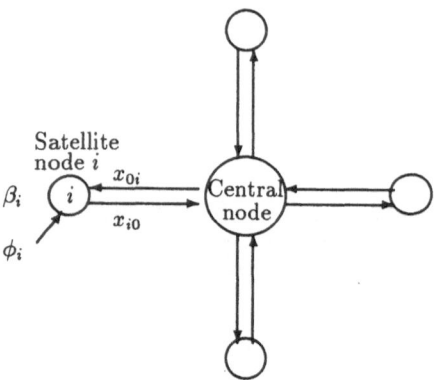

Figure 4.1: A star network with two-way traffic

node for remote processing. In the latter case, there is a delay incurred as a result of forwarding the job and sending its response back over the network. On the basis of these assumptions, we formulate the static load balancing problem as a nonlinear optimization problem which minimizes the mean response time of a job in a star network. The necessary and sufficient conditions for optimal solution are studied. An effective $O(n)$ algorithm is proposed to solve the optimization problem for an n-satellite system.

Note that the link communication time, i.e., the traffic–independent mean service time of each communication channel, is a critical system parameter that affects the network system performance. The effects of link communication time on the optimal static load balancing in a star network are also studied by parametric analysis. By using the proposed algorithm, we show that the effects of the link communication time in a star network are large in numerical experiments. The results of this chapter might be helpful in designing a network or making parametric adjustments to improve the system performance.

4.2   Model Description and Problem Formulation

A *star network* is shown in Figure 4.1. There are a *central node* and many *satellite nodes* in the network. Let N denote the set of nodes in a star network. Refer to node 0 as the central node in a star network. Let L be the set of satellite nodes in a star network $(L = N - \{0\})$. There are links from satellite nodes to the central node as well as links from the central node to satellite nodes.

The nodes in a star network may be heterogeneous computer systems; that is, they may have different configurations, number of resources, and speed characteristics. Jobs arrive at node i ($i \in N$) according to a time–invariant Poisson process. A job arriving from outside at a satellite is either processed locally or transferred to another node (which may be the central node or another satellite node) for remote processing. In an analogous way, a job arriving at the central node may either be processed locally or be sent to a satellite for remote processing.

For the description of the model, we introduce the following notation and assumptions.

- β_i: the rate at which jobs are processed at node i, $i \in N$, also referred to as *load*. That is, the average number of jobs processed at node i per unit interval of time.

- x_{ij}: the job flow rate from node i to node j, also called the *traffic*, on link (i,j). That is, the average number of jobs sent from node i through link (i,j) per unit interval of time. Note that there are two links $(i,0)$ and $(0,i)$ between a satellite node i and the central node 0.

- $\boldsymbol{\beta}$: $[\beta_i]$.

- $\boldsymbol{x}$: $[x_{i0}, x_{0i}]$.

- $\boldsymbol{\alpha}$: $[\alpha_i]$, where α_i is a Lagrange multiplier, $i \in N$.

- ϕ_i: the external job arrival rate at node i, $i \in N$.

- $F_i(\beta_i)$: the mean node delay (queuing and processing delays) of a job processed at node i, $i \in N$.

- $G_{ij}(\boldsymbol{x})$: the mean communication delay of sending a job from node i to node j and sending the response back from node j to node i. Assume that the mean communication delay $G_{ij}(\boldsymbol{x})$ depends on job flow rates x_{ij} on link (i,j) and x_{ji} on link (j,i). Then, we use $G_{ij}(x_{ij}, x_{ji})$ instead of $G_{ij}(\boldsymbol{x})$.

- $D(\boldsymbol{\beta}, \boldsymbol{x})$: the mean response time that a job spends in the network from the time of its arrival until the time of its departure.

Here we assume that the node delay function $F_i(\beta_i)$ is a differentiable, increasing, and convex function, and the communication delay function $G_{ij}(x_{ij}, x_{ji})$ is a positive, differentiable, nondecreasing, and convex function with respect to x_{ij} and x_{ji}. We note that the class of delay functions satisfying above assumption is large, and it contains the M/M/c model and the central server model [Kle76], etc. We express $D(\boldsymbol{\beta}, \boldsymbol{x})$ as the sum of the mean node delay at all nodes and the mean communication delay on all links by using the Little's result as follows,

$$D(\boldsymbol{\beta}, \boldsymbol{x}) = \frac{1}{\phi}\{\sum_{i \in N} \beta_i F_i(\beta_i) + \sum_{i \in L} x_{i0} G_{i0}(x_{i0}, x_{0i}) + \sum_{i \in L} x_{0i} G_{0i}(x_{0i,i0})\} \qquad (4.1)$$

where $\phi = \sum_{i \in N} \phi_i$.

The goal of load balancing is to minimize the mean response time with respect to β and x. We have the optimization problem of minimizing the mean response time of a job which we can write by considering the structure of star network configurations with two-way traffic as follows:

$$\text{Minimize} \quad D(\beta, x) = \frac{1}{\phi} \{ \sum_{i \in N} \beta_i F_i(\beta_i) + \sum_{i \in L} x_{i0} G_{i0}(x_{i0}, x_{0i}) + \sum_{i \in L} x_{0i} G_{0i}(x_{0i}, x_{i0}) \}, \quad (4.2)$$

with respect to the variables β_i, $i \in N$, x_{i0}, $i \in L$, and x_{0i}, $i \in L$, subject to

$$\beta_0 = \phi_0 + \sum_{i \in L} x_{i0} - \sum_{i \in L} x_{0i}, \quad (4.3a)$$

$$\beta_i = \phi_i + x_{0i} - x_{i0}, \quad (4.3b)$$

$$\beta_i \geq 0, \quad i \in N, \quad (4.3c)$$

$$x_{i0} \geq 0, \ x_{0i} \geq 0, \quad i \in L. \quad (4.3d)$$

The first constraint (4.3a) and the second constraint (4.3b) are the flow balance constraints which equate the traffic flows in and out of the central node and satellite node i, respectively. Note that the mean node delay $F_i(\beta_i)$ is an increasing and convex function, and the mean communication delay functions $G_{i0}(x_{i0}, x_{0i})$ and $G_{0i}(x_{0i}, x_{i0})$ are nondecreasing and convex functions. According to eq.(4.1), the mean response time $D(\beta, x)$ also is a convex function.

4.3   Necessary and Sufficient Conditions for Static Load Balancing

The load balancing strategy which minimizes the mean response time is obtained by solving the optimization problem (4.2).

We introduce the incremental delay functions as follows,

$$f_i(\beta_i) \quad = \quad \frac{d}{d\beta_i} \beta_i F_i(\beta_i), \quad (4.4a)$$

$$g_{i0}(x_{i0}, x_{0i}) \quad = \quad \frac{\partial}{\partial x_{i0}} x_{i0} G_{i0}(x_{i0}, x_{0i}) + x_{0i} \frac{\partial G_{0i}(x_{0i}, x_{i0})}{\partial x_{i0}}, \quad (4.4b)$$

$$g_{0i}(x_{0i}, x_{i0}) \quad = \quad \frac{\partial}{\partial x_{0i}} x_{0i} G_{0i}(x_{0i}, x_{i0}) + x_{i0} \frac{\partial G_{i0}(x_{i0}, x_{0i})}{\partial x_{0i}}. \quad (4.4c)$$

Due to our assumption on $F_i(\beta_i)$, note that $f_i(\beta_i)$ is an increasing function. We define the inverse of the incremental node delay function f_i^{-1} by

$$f_i^{-1}(x) = \begin{cases} a, & \text{if } f_i(a) = x, \text{ and } a > 0, \\ 0, & \text{if } f_i(0) \geq x. \end{cases}$$

Theorem 4.1 *The set of values of β and x is an optimal solution to problem (4.2) if and only if the following set of relations holds.*

$$\alpha_i - g_{i0}(x_{i0}, x_{0i}) = \alpha_0, \quad x_{i0} > 0 \quad (i \in L), \tag{4.5a}$$

$$\alpha_i - g_{i0}(x_{i0}, x_{0i}) \le \alpha_0, \quad x_{i0} = 0 \quad (i \in L), \tag{4.5b}$$

$$\alpha_i + g_{0i}(x_{0i,x_{i0}}) = \alpha_0, \quad x_{0i} > 0 \quad (i \in L), \tag{4.5c}$$

$$\alpha_i + g_{0i}(x_{0i}, x_{i0}) \ge \alpha_0, \quad x_{0i} = 0 \quad (i \in L), \tag{4.5d}$$

$$f_i^{-1}(\alpha_i) = \beta_i, \quad (i \in N), \tag{4.6}$$

subject to

$$\phi_0 + \sum_{i \in L} x_{i0} = \sum_{i \in L} x_{0i} + \beta_0, \tag{4.7a}$$

$$\phi_i + x_{0i} = x_{i0} + \beta_i, \quad (i \in L), \tag{4.7b}$$

where α_i $(i \in N)$ is the Lagrange multiplier.

PROOF. To obtain the optimal solution, we form the Lagrangian function as follows,

$$
\begin{aligned}
H(\boldsymbol{\beta}, \boldsymbol{x}, \boldsymbol{\alpha}) = \; & \phi D(\boldsymbol{\beta}, \boldsymbol{x}) + \alpha_0 \Big(\phi_0 + \sum_{i \in L} x_{i0} - \sum_{i \in L} x_{0i} - \beta_0\Big) \\
& + \sum_{i \in L} \alpha_i (\phi_i + x_{0i} - x_{i0} - \beta_i).
\end{aligned}
\tag{4.8}
$$

Since the objective function is convex by assumption and the feasible set is a convex set, a set of value of β and x, (β, x), is an optimal solution to the problem (4.2) if and only if it satisfies the following Kuhn-Tucker conditions (e.g., [Int71]),

$$\frac{\partial H}{\partial \beta_i} = f_i(\beta_i) - \alpha_i \ge 0, \tag{4.9a}$$

$$\beta_i \frac{\partial H}{\partial \beta_i} = \beta_i(f_i(\beta_i) - \alpha_i) = 0, \tag{4.9b}$$

$$\beta_i \ge 0, \tag{4.9c}$$

$$i \in N.$$

$$\frac{\partial H}{\partial x_{i0}} = g_{i0}(x_{i0}, x_{0i}) + \alpha_0 - \alpha_i \ge 0, \tag{4.10a}$$

$$x_{i0} \frac{\partial H}{\partial x_{i0}} = x_{i0}(g_{i0}(x_{i0}, x_{0i}) + \alpha_0 - \alpha_i) = 0, \tag{4.10b}$$

$$x_{i0} \ge 0, \tag{4.10c}$$

$$i \in L.$$

$$\frac{\partial H}{\partial x_{0i}} = g_{0i}(x_{0i}, x_{i0}) - \alpha_0 + \alpha_i \geq 0, \tag{4.11a}$$

$$x_{0i}\frac{\partial H}{\partial x_{0i}} = x_{0i}(g_{0i}(x_{0i}, x_{i0}) - \alpha_0 + \alpha_i) = 0, \tag{4.11b}$$

$$x_{0i} \geq 0, \tag{4.11c}$$

$$i \in L.$$

$$\phi_0 + \sum_{i \in L} x_{i0} - \sum_{i \in L} x_{0i} - \beta_0 = 0, \tag{4.12a}$$

$$\phi_i + x_{0i} - x_{i0} - \beta_i = 0, \quad (i \in L). \tag{4.12b}$$

The above set of conditions (4.9), (4.10) and (4.11) is identical with the following set of relations,

$$f_i(\beta_i) = \alpha_i, \quad \beta_i > 0, \tag{4.13a}$$

$$f_i(\beta_i) \geq \alpha_i, \quad \beta_i = 0, \tag{4.13b}$$

$$i \in N.$$

$$\alpha_i - g_{i0}(x_{i0}, x_{0i}) = \alpha_0, \quad x_{i0} > 0, \tag{4.14a}$$

$$\alpha_i - g_{i0}(x_{i0}, x_{0i}) \leq \alpha_0, \quad x_{i0} = 0, \tag{4.14b}$$

$$i \in L.$$

$$\alpha_i + g_{0i}(x_{0i}, x_{i0}) = \alpha_0, \quad x_{0i} > 0, \tag{4.15a}$$

$$\alpha_i + g_{0i}(x_{0i}, x_{i0}) \geq \alpha_0, \quad x_{0i} = 0, \tag{4.15b}$$

$$i \in L.$$

We can easily see the above set of relations is equivalent to the set in Theorem 4.1. Q.E.D.

Remark: The above theorem shows that the values of β and x can be determined by using relations (4.5) and (4.6) if the Lagrange multipliers can be determined. A limitation of the usefulness of the theorem may arise from the fact that it is not easy to determine all multipliers α_i $(i \in N)$ by directly using Theorem 4.1. We proceed to find more properties about the optimal solution to problem (4.2).

Theorem 4.2 *In the optimal solution to problem (4.2), the following relations hold true,*

$$x_{i0} = 0, \quad \text{if } x_{0i} > 0, \text{ and} \tag{4.16a}$$

$$x_{0i} = 0, \quad \text{if } x_{i0} > 0, \quad i \in L. \tag{4.16b}$$

PROOF. It is proved by contradiction. Suppose relation (4.16) is not true and there exists node i ($i \in L$) such that $x_{i0} > 0$ and $x_{0i} > 0$. According to Theorem 4.1, we have

$$\alpha_i - g_{i0}(x_{i0}, x_{0i}) = \alpha_0, \tag{4.17a}$$

$$\alpha_i + g_{0i}(x_{0i,x_{i0}}) = \alpha_0. \tag{4.17b}$$

Then
$$-g_{i0}(x_{i0}, x_{0i}) = g_{0i}(x_{0i}, x_{i0}), \tag{4.18}$$

which is a contradiction to the fact that $g_{i0}(x_{i0}, x_{0i})$ and $g_{0i}(x_{0i}, x_{i0})$ are positive if $x_{i0} > 0$ and $x_{0i} > 0$. Q.E.D.

Remark: Theorem 4.2 shows an important property in the optimal solution to (4.2) that the traffic from satellite node i to the central node, x_{i0}, and the traffic from the central node to satellite node i, x_{0i}, ($i \in L$) cannot be positive both at the same time.

Note that $g_{ij}(x_{ij}, x_{ji})$ is either $g_{ij}(x_{ij}, 0)$ or $g_{ij}(0, x_{ji})$ since $x_{ij}x_{ji} = 0$ in the optimal solution. Hereafter we denote $g_{ij}(x_{ij}, 0)$ by $g_{ij}(x_{ij})$ in the optimal solution.

Furthermore, we have the following definitions. Satellite node i is said to be a *source* if $x_{i0} > 0$, i.e., it sends all or part of jobs that arrive at it externally. Satellite node i is an *idle source* if $x_{i0} = \phi_i$, and an *active source* if $0 < x_{i0} < \phi_i$. If satellite node i only processes all the jobs that arrive at it externally ($\beta_i = \phi_i$, i.e., $x_{i0} = 0$ and $x_{0i} = 0$), it is a *neutral node*. If satellite node i does not only process all jobs that arrive at it externally but also processes some jobs that are forwarded from the central node, it is a *sink node*. We denote the sets of idle source nodes, active source nodes, neutral nodes and sink nodes by Rd, Ra, Nu, and S respectively. From Theorem 4.2, we have the following corollary.

Corollary 4.3 *The sets of Rd, Ra, Nu, and S are disjoint and*

$$L = Rd \cup Ra \cup Nu \cup S. \tag{4.19}$$

PROOF. It is the direct result from Theorem 4.2. Q.E.D.

Now we proceed to give the following theorem based on which we will propose a simple algorithm.

Define the following functions,

$$p_{i0}(x_{i0}) = f_i(\phi_i - x_{i0}) - g_{i0}(x_{i0}), \tag{4.20a}$$

$$q_{0i}(x_{0i}) = f_i(\phi_i + x_{0i}) + g_{0i}(x_{0i}). \tag{4.20b}$$

Note that $p_{i0}(x_{i0})$ is a monotonically decreasing function and $q_{0i}(x_{0i})$ is a monotonically increasing function. The two functions $p_{i0}(x_{i0})$ and $q_{0i}(x_{0i})$ have their inverse functions,

p_{io}^{-1} and q_{0i}^{-1}, which are defined similarly as the inverse function of $f_i(\beta_i)$. According to the above definitions, we also have the following relations,

$$q_{0i}(x_{0i}) > q_{0i}(0) \geq p_{io}(0) > p_{io}(\phi_i), \quad x_{0i} > 0, \; \forall i \in L. \tag{4.21}$$

Theorem 4.4 *The set of values of β and x is an optimal solution to problem (4.2) if and only if the following set of relations holds.*

$$p_{io}(x_{io}) = \alpha, \qquad if \; 0 < x_{io} < \phi_i, \; (i \in Ra), \tag{4.22a}$$

$$p_{io}(x_{io}) \geq \alpha, \qquad if \; x_{io} = \phi_i, \; (i \in Rd), \tag{4.22b}$$

$$q_{0i}(x_{0i}) = \alpha, \qquad if \; x_{0i} > 0, \; (i \in S), \tag{4.22c}$$

$$q_{0i}(x_{0i}) \geq \alpha \geq p_{io}(x_{io}), \quad if \; x_{io} = x_{0i} = 0, \; (i \in Nu), \tag{4.22d}$$

and
$$\beta_0 = f_0^{-1}(\alpha), \tag{4.23}$$

subject to

$$\sum_{i \in Ra} p_{io}^{-1}(\alpha) + \sum_{i \in Rd} \phi_i = f_0^{-1}(\alpha) - \phi_0 + \sum_{i \in S} q_{0i}^{-1}(\alpha). \tag{4.24}$$

PROOF [NECESSITY]. By noting $\phi_i + x_{0i} = x_{io} + \beta_i \; (i \in L)$, relations (4.22) and (4.23) can be derived directly by relations (4.5) and (4.6) in Theorem 4.1. Furthermore, according to eq.(7a), using Corollary 4.3, we have eq.(4.24).

[SUFFICIENCY]: Note that the set of values of β and x is determined uniquely by a given α according to relations (4.22) and (4.23). We prove that constraint (4.24) determines the value of α, denoted by α^*, uniquely, and that the set of values of β and x, (β^*, x^*), is the optimal solution to problem (4.2) where (β^*, x^*) is determined uniquely by α^*.

From relations (4.22), we see that the sets of Ra, Rd, S and Nu are determined uniquely by a given α. For an arbitrary α, define

$$Ra(\alpha) = \{i \mid p_{io}(x_{io}) = \alpha, \; 0 < x_{io} < \phi_i, \; i \in L\}, \tag{4.25a}$$

$$Rd(\alpha) = \{i \mid p_{io}(\phi_i) \geq \alpha, \; i \in L\}, \tag{4.25b}$$

$$S(\alpha) = \{i \mid q_{0i}(x_{0i}) = \alpha, \; x_{0i} > 0, \; i \in L\}, \tag{4.25c}$$

$$Nu(\alpha) = \{i \mid q_{0i}(0) \geq \alpha \geq p_{io}(0), \; i \in L\}, \tag{4.25d}$$

$$\lambda_R(\alpha) = \sum_{i \in Ra(\alpha)} p_{io}^{-1}(\alpha) + \sum_{i \in Rd(\alpha)} \phi_i, \tag{4.25e}$$

$$\lambda_S(\alpha) = f_0^{-1}(\alpha) - \phi_0 + \sum_{i \in S(\alpha)} q_{0i}^{-1}(\alpha). \tag{4.25f}$$

Note that $p_{io}(x_{io})$ is a monotonically decreasing function, $f_0(\beta_0)$ and $q_{0i}(x_{0i})$ are monotonically increasing functions. We have the following. As α increases, $\lambda_R(\alpha)$ first stays to be $\sum_{i \in L} \phi_i$ while $\min_{i \in L} p_{i0}(\phi_i) \geq \alpha$, and then decreases monotonically to be zero while $\alpha \geq \max_{i \in L} p_{i0}(0)$. On the other hand, as α increases, $\lambda_S(\alpha)$ first stays to be $-\phi_0$ while $f_0(0) \geq \alpha$ and $\min_{i \in L} q_{0i}(0) \geq \alpha$, and then increases monotonically. Noting (4.21), we see that α is determined uniquely by constraint (4.24), i.e.,

$$\lambda_R(\alpha^*) = \lambda_S(\alpha^*), \tag{4.26}$$

where α^* is the value of α.

Furthermore, α^* determines the set of values of β and x, (β^*, x^*), uniquely, according to relations (4.22) and (4.23).

Now we prove that (β^*, x^*) is the optimal solution to problem (4.2). Assume that $(\tilde{\beta}, \tilde{x})$ is the optimal solution to problem (4.2) and $(\tilde{\beta}, \tilde{x}) \neq (\beta^*, x^*)$. Then there should be a value of α, $\tilde{\alpha}$, which determines $(\tilde{\beta}, \tilde{x})$ uniquely. We note that $\tilde{\alpha}$ satisfies constraint (4.24), i.e., $\lambda_R(\tilde{\alpha}) = \lambda_S(\tilde{\alpha})$. Since constraint (4.24) determines the value of α uniquely, we have $\tilde{\alpha} = \alpha^*$. Then we have $(\tilde{\beta}, \tilde{x}) = (\beta^*, x^*)$. It is a contradiction to the assumption that $(\tilde{\beta}, \tilde{x}) \neq (\beta^*, x^*)$. Q.E.D.

Remark: We may have a simple interpretation of Theorem 4.4 as follows. If the value of α is determined, in the optimal solution, a satellite node may be an active source node, or an idle node, or a sink node, or a neutral node, according to the set of relations (4.22). The value of α can be evaluated directly by equation (4.24).

4.4 Proposed Algorithm

According to Theorem 4.4, we can provide an algorithm to solve the optimization problem (4.2). The main idea of the proposed algorithm is to determine the value of α in Theorem 4.4 by using a simple one-dimensional search method, then to determine the node partition (active source, idle source, neutral and sink) and the optimal load in the star network according to relations (4.22) and (4.23).

- Proposed algorithm

1. Compare. $O(n)$ $(n = |L|)$

   If $(\forall i \in L)$ $q_{0i}(0) \geq f_0(\phi_0) \geq p_{i0}(0)$, then no load balancing is required. In this case, let

   $\beta_i = \phi_i$, $\forall i \in N$, $x_{i0} = x_{0i} = 0$, $\forall i \in L$.

   Otherwise, proceed to step 2.

2. Determine α and the node partition. $O(n)$

Find such α that satisfies

$$\lambda_S(\alpha) = \lambda_R(\alpha), \quad (\text{see eq.}(4.26))$$

by applying a golden section search, where

$$
\begin{aligned}
Ra(\alpha) &= \{i|\; p_{i0}(x_{i0}) = \alpha,\; 0 < x_{i0} < \phi_i,\; i \in L\}, \\
Rd(\alpha) &= \{i|\; p_{i0}(\phi_i) \geq \alpha,\; i \in L\}, \\
S(\alpha) &= \{i|\; q_{0i}(x_{0i}) = \alpha,\; x_{0i} > 0,\; i \in L\}, \\
Nu(\alpha) &= \{i|\; q_{0i}(0) \geq \alpha \geq p_{i0}(0),\; i \in L\}, \\
\lambda_R(\alpha) &= \sum_{i \in Ra(\alpha)} p_{i0}^{-1}(\alpha) + \sum_{i \in Rd(\alpha)} \phi_i, \\
\lambda_S(\alpha) &= f_0^{-1}(\alpha) - \phi_0 + \sum_{i \in S(\alpha)} q_{0i}^{-1}(\alpha).
\end{aligned}
$$

The range of the value of α can be given according to the proof of sufficiency of Theorem 4.4.

Proceed to step 3.

3. Determine the optimal load. $O(n)$

$x_{i0} = \phi_i,\; x_{0i} = 0,\; \beta_i = 0,\; \forall i \in Rd(\alpha),$

$x_{i0} = p_{i0}^{-1}(\alpha),\; x_{0i} = 0,\; \beta_i = \phi_i - x_{i0},\; \forall i \in Ra(\alpha),$

$x_{i0} = 0,\; x_{0i} = q_{0i}^{-1}(\alpha),\; \beta_i = \phi_i + x_{0i},\; \forall i \in S(\alpha),$

$x_{i0} = 0,\; x_{0i} = 0,\; \beta_i = \phi_i,\; \forall i \in Nu(\alpha).$

4.5   Parametric Analysis

We have obtained the necessary and sufficient condition for the optimal solution to the load-balancing problem (4.2). In this section, we study the effects of a communications network parameter on the optimal load balancing by parametric analysis. Specifically, we consider the parameter t_{ij} (referred to as the *communication time*), which is the traffic-independent mean service time of the communication channel, on link (i, j). We use the notation $G_{i0}(x_{i0}, x_{0i}, t_{i0})$ and $G_{0i}(x_{0i}, x_{i0}, t_{0i})$ for the expected communication delays in order to make the parameters t_{i0} and t_{0i} explicit. Note that $x_{i0}x_{0i} = 0$, $x_{i0} \geq 0$ and $x_{0i} \geq 0$ in optimal solution. We denote $G_{i0}(x_{i0}, 0, t_{i0})$ and $G_{0i}(x_{0i}, 0, t_{0i})$ by $G_{i0}(x_{i0}, t_{i0})$ and $G_{0i}(x_{0i}, t_{0i})$ in optimal solution, respectively. Similarly, we denote by $g_{i0}(x_{i0}, t_{i0})$ and $g_{0i}(x_{0i}, t_{0i})$ in optimal solution.

Note that the optimal solution is characterized by the partition of nodes into different types (active source, idle source, sink, and neutral). Using Theorem 4.4, we can study the effects of the link communication time on the optimal solution. Here we consider an interval of the link communication time during which the node partition remains the same. For the sake of simplicity, we denote t_q as the communication time in link q, where q may be either $(q, 0)$ or $(0, q)$. For $q \in Ra$ (or $q \in S$), we mean that node q is an active source node (or a sink node).

As the communication time of a certain link q, t_q, increases, it would cause some nodes to process more jobs; on the other hand, it would also cause some other nodes to send more jobs to their parent nodes in a star network. This means that the increase in t_q has the effects on the node loads and the link flow rates, and then on the mean response time in the network. In the limit, large t_q may lead to an infinitely long response time if the loads on some nodes exceed the processing capacities of these nodes. The results of parametric analysis is important in designing a network or making parametric adjustments to improve the system performance.

Assume the incremental communication delays on links $(i, 0)$ and $(0, i)$ increase with t_{i0} and t_{0i} respectively, and

$$\begin{aligned}
\frac{\partial g_{i0}(x_{i0}, t_{i0})}{\partial t_q} &> 0, \quad \text{if } q = (i, 0), \\
&= 0, \quad \text{if } q \neq (i, 0), \\
\frac{\partial g_{0i}(x_{0i}, t_{0i})}{\partial t_q} &> 0, \quad \text{if } q = (0, i), \\
&= 0, \quad \text{if } q \neq (0, i).
\end{aligned}$$

Let t denote $[t_q]$. Link traffic x_{i0}, x_{0i}, node load β_i, and the Lagrange multiplier α in the optimal solution are determined by given t, and are written as $x_{i0}(t)$, $x_{0i}(t)$, $\beta_i(t)$, and $\alpha(t)$.

The effects of the link communication time on the link traffic in a star network are specified in Theorem 4.5.

Theorem 4.5 *In optimal solution to problem (4.2), the first derivates of the traffic on link $(i, 0)$ x_{i0} and on link $(0, i)$ x_{0i} $(i \in L)$ satisfy the following relations,*

$$\begin{aligned}
\frac{\partial x_{i0}}{\partial t_q} &> 0, \quad \text{if } i \in Ra, \ q \in Ra \text{ and } q \neq (i, 0), \\
&< 0, \quad \text{if } i \in Ra, \text{ and } q = (i, 0), \\
&< 0, \quad \text{if } i \in Ra, \text{ and } q \in S, \\
&= 0, \quad \text{else},
\end{aligned} \tag{4.27}$$

$$\begin{aligned}
\frac{\partial x_{0i}}{\partial t_q} &< 0, \quad \text{if } i \in S, \text{ and } q \in Ra,
\end{aligned}$$

$$> \dot{0}, \quad \text{if } i \in S, \, q \in S, \text{ and } q \neq (0, i),$$
$$< 0, \quad \text{if } i \in S, \text{ and } q = (0, i), \qquad (4.28)$$
$$= 0, \quad \text{else.}$$

PROOF. Let us consider the following cases.

Case 1. $i \in Ra$.

From eq.(4.22a) in Theorem 4.4, we have,

$$f_i(\beta_i(t)) - g_{i0}(x_{i0}(t), t_{i0}) = \alpha. \qquad (4.29)$$

Noting that $\beta_i = \phi_i - x_{i0}$ $(i \in Ra)$, we have

$$\begin{aligned}
\frac{\partial \alpha}{\partial t_q} &= \frac{\partial f_i}{\partial \beta_i} \frac{\partial \beta_i}{\partial t_q} - \frac{\partial g_{i0}}{\partial x_{i0}} \frac{\partial x_{i0}}{\partial t_q} - \frac{\partial g_{i0}}{\partial t_q} \\
&= -(\frac{\partial f_i}{\partial \beta_i} + \frac{\partial g_{i0}}{\partial x_{i0}}) \frac{\partial x_{i0}}{\partial t_q} - \frac{\partial g_{i0}}{\partial t_q}. \qquad (4.30)
\end{aligned}$$

Rearranging (4.30), we have

$$\frac{\partial x_{i0}}{\partial t_q} = -(\frac{\partial \alpha}{\partial t_q} + \frac{\partial g_{i0}}{\partial t_q})/A_{i0} \qquad (4.31)$$

where

$$A_{i0} = \frac{\partial f_i}{\partial \beta_i} + \frac{\partial g_{i0}}{\partial x_{i0}}.$$

Since $x_{i0} > 0$ $(i \in Ra)$, according to Theorem 4.2, we also have

$$\frac{\partial x_{0i}}{\partial t_q} = 0. \qquad (4.32)$$

Case 2. $i \in S$.

From eq.(4.22c) in Theorem 4.4, we have,

$$f_i(\beta_i(t)) + g_{0i}(x_{0i}(t), t_{0i}) = \alpha. \qquad (4.33)$$

Noting that $\beta_i = \phi_i + x_{0i}$ $(i \in S)$, we have

$$\begin{aligned}
\frac{\partial \alpha}{\partial t_q} &= \frac{\partial f_i}{\partial \beta_i} \frac{\partial \beta_i}{\partial t_q} + \frac{\partial g_{0i}}{\partial x_{0i}} \frac{\partial x_{0i}}{\partial t_q} + \frac{\partial g_{0i}}{\partial t_q} \\
&= (\frac{\partial f_i}{\partial \beta_i} + \frac{\partial g_{0i}}{\partial x_{0i}}) \frac{\partial x_{0i}}{\partial t_q} + \frac{\partial g_{0i}}{\partial t_q}. \qquad (4.34)
\end{aligned}$$

Rearranging (4.34), we have,

$$\frac{\partial x_{0i}}{\partial t_q} = (\frac{\partial \alpha}{\partial t_q} - \frac{\partial g_{0i}}{\partial t_q})/B_{0i}, \qquad (4.35)$$

where

$$B_{0i} = \frac{\partial f_i}{\partial \beta_i} + \frac{\partial g_{0i}}{\partial x_{0i}}.$$

Since $x_{0i} > 0$, $(i \in S)$, according to Theorem 4.2, we also have

$$\frac{\partial x_{i0}}{\partial t_q} = 0. \tag{4.36}$$

Case 3. $i \in Nu$.

It is easy to see that

$$\frac{\partial x_{i0}}{\partial t_q} = \frac{\partial x_{0i}}{\partial t_q} = 0. \tag{4.37}$$

Case 4. $i \in Rd$.

It is easy to see that

$$\frac{\partial x_{i0}}{\partial t_q} = \frac{\partial x_{0i}}{\partial t_q} = 0. \tag{4.38}$$

By noting $\beta_0 = \phi_0 + \sum_{i \in Ra} x_{i0} + \sum_{i \in Rd} \phi_i - \sum_{i \in S} x_{0i}$, using (4.31) and (4.35), we have

$$
\begin{aligned}
\frac{\partial \alpha}{\partial t_q} &= \frac{\partial f_0}{\partial \beta_0} \frac{\partial \beta_0}{\partial t_q} \\
&= \frac{\partial f_0}{\partial \beta_0} \left(\sum_{i \in Ra} \frac{\partial x_{i0}}{\partial t_q} - \sum_{i \in S} \frac{\partial x_{0i}}{\partial t_q} \right) \\
&= \frac{\partial f_0}{\partial \beta_0} \left(- \sum_{i \in Ra} \frac{\partial g_{i0}}{\partial t_q} / A_{i0} + \sum_{i \in S} \frac{\partial g_{0i}}{\partial t_q} / B_{0i} \right) / Q,
\end{aligned}
\tag{4.39}
$$

where

$$Q = 1 + \frac{\partial f_0}{\partial \beta_0} \left(\sum_{i \in Ra} 1/A_{i0} + \sum_{i \in S} 1/B_{0i} \right).$$

Note that $f_i(\beta_i)$ is the increasing function with respect to β_i, and $g_{i0}(x_{i0})$, $g_{0i}(x_{0i})$ are the nondecreasing functions with respect to x_{i0} and x_{0i} respectively. Then

$$A_{i0} > 0, \quad B_{0i} > 0, \quad Q > 0.$$

Substituting (4.39) into (4.31), we have the following relations for $i \in Ra$,

$$\frac{\partial x_{i0}}{\partial t_q} = -\left(\frac{\partial \alpha}{\partial t_q} + \frac{\partial g_{i0}}{\partial t_q} \right) / A_{i0}$$

$$= \begin{cases} \frac{\partial f_0}{\partial \beta_0}\frac{\partial g_q}{\partial t_q}/(A_q * A_{i0} * Q) > 0, & \text{if } q \in Ra \text{ and } q \neq (i,0), \\[2ex] -[1 + \frac{\partial f_0}{\partial \beta_0}(\sum_{l \in Ra, l \neq i} 1/A_{l0} \\ \quad + \sum_{l \in S} 1/B_{0l}\frac{\partial g_{i0}}{\partial t_q}]/(A_{i0} * Q) < 0, & \text{if } q = (i,0), \\[2ex] -\frac{\partial f_0}{\partial \beta_0}\frac{\partial g_q}{\partial t_q}/(B_q * A_{i0} * Q) < 0, & \text{if } q \in S, \\[2ex] 0, & \text{else.} \end{cases} \qquad (4.40)$$

Substituting (4.39) into (4.35), we have the following relations for $i \in S$,

$$\frac{\partial x_{0i}}{\partial t_q} = (\frac{\partial \alpha}{\partial t_q} - \frac{\partial g_{0i}}{\partial t_q})/B_{0i}$$

$$= \begin{cases} -\frac{\partial f_0}{\partial \beta_0}\frac{\partial g_q}{\partial t_q}/(A_q * B_{0i} * Q) < 0, & \text{if } q \in Ra, \\[2ex] \frac{\partial f_0}{\partial \beta_0}\frac{\partial g_q}{\partial t_q}/(B_q * B_{0i} * Q) > 0, & \text{if } q \in S \text{ and } q \neq (0,i), \\[2ex] -[1 + \frac{\partial f_0}{\partial \beta_0}\frac{\partial g_q}{\partial t_q}(\sum_{l \in Ra} 1/A_{l0} \\ \quad + \sum_{l \in S, l \neq i} 1/B_{0l})]/(B_{0i} * Q) < 0, & \text{if } q \in S, \text{ and } q = (0,i), \\[2ex] 0, & \text{else.} \end{cases} \qquad (4.41)$$

From relations (4.32), (4.36), (4.37), (4.38), (4.40) and (4.41), we have the theorem. Q.E.D.

Remark: The relation (4.27) may be interpreted as follows. The increase in the communication time of a link from an active source node to the central node causes the increases in the link traffic from other active source nodes to the central node, and causes the decrease in the link traffic from this active source node itself to the central node. The increase in the communication time of a link from the central node to a sink node causes the decreases in the link traffic from a active source node to the central node. According to relation (4.28), we can have the following. The increase in the communication time of a link from an active source node to the central node causes the decrease in the link traffic from the central node to a sink node. The increase in the communication time of a link from the central node to a sink node causes the increases in link traffic from the central node to other sink nodes, and causes the decrease in the link traffic from the central node to this sink node itself. Otherwise, the communication time of a link has no effects on the link traffic in a network.

The effects of the link communication time on load of an arbitrary node are specified in Theorem 4.6.

Theorem 4.6 *In an optimal solution, the following relations hold true for the processing capacity in node* i $(i \in N)$, β_i,

$$
\begin{aligned}
\frac{\partial \beta_i}{\partial t_q} \quad &< \quad 0, \quad \text{if } i \in Ra, \ q \in Ra \text{ and } q \neq (i,0), \\
&> \quad 0, \quad \text{if } i \in Ra, \text{ and } q = (i,0), \\
&> \quad 0, \quad \text{if } i \in Ra, \text{ and } q \in S, \\
&< \quad 0, \quad \text{if } i \in S, \text{ and } q \in Ra, \\
&> \quad 0, \quad \text{if } i \in S, \ q \in S, \text{ and } q \neq (0,i), \\
&< \quad 0, \quad \text{if } i \in S, \text{ and } q = (0,i), \\
&= \quad 0, \quad else.
\end{aligned}
\tag{4.42}
$$

PROOF.

For $i \in Ra$, noting that $\beta_i = \phi_i - x_{i0}$, we have

$$
\frac{\partial \beta_i}{\partial t_q} = -\frac{\partial x_{i0}}{\partial t_q}.
\tag{4.43}
$$

For $i \in S$, noting that $\beta_i = \phi_i + x_{0i}$, we have

$$
\frac{\partial \beta_i}{\partial t_q} = \frac{\partial x_{0i}}{\partial t_q}.
\tag{4.44}
$$

Also note that

$$
\frac{\partial \beta_i}{\partial t_q} = 0. \quad \text{if } i \in Nu \cup Rd.
\tag{4.45}
$$

According to Theorem 4.5, we have the results. Q.E.D.

Remark: Theorem 4.6 may be interpreted as follows. The increase in the communication time of a link from an active source node to the central node causes the decreases in the loads of other active source nodes and sink nodes, but causes the increase in the load of this node itself. The increase in the communication time of a link from the central node to a sink node causes the increases in the loads of active source nodes and other sink node, and causes the decrease in the load of this sink node itself. Otherwise, the communication time of a link has no effects on the node loads in a network.

We also note that the increase in the communication time of a link in a network will generally cause the increase of the mean system response time. Here we do not discuss the effects of the communication time on the mean system response time in detail.

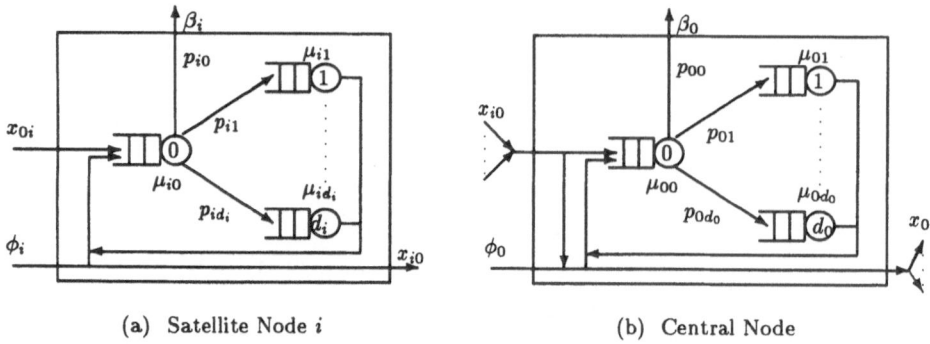

(a) Satellite Node i (b) Central Node

Figure 4.2: Node model

4.6 A Numerical Example

Here we consider a star computer network that consists of five nodes (host computers). Let us number the nodes in a star network as nodes 0, 1, 2, 3, and 4 where node 0 is the central node. Each node is modeled as a central–server model as illustrated in Figure 4.2. In satellite node i model (Figure4.2(a)), exponential server 0 is a CPU that processes jobs according to the processor sharing discipline, exponential servers 1, 2, ..., d_i are I/O devices which process jobs according to FCFS. Let p_{i0} and p_{ij}, $j = 1, 2, ..., d_i$, be the probabilities that, after departing from the CPU, a job leaves node i or requests I/O service at device j, $j = 1, 2, ..., d_i$, respectively. The central node model (Figure4.2(b)) is almost the same as that of a satellite node except that there are much link traffic from satellite nodes to the central node and link traffic from the central node to satellite nodes. The expected node delay of a job in such node model is given as

$$F_i(\beta_i) = \sum_{j=0}^{d_i} \frac{q_{ij}}{\mu_{ij} - q_{ij}\beta_i}, \tag{4.46}$$

where $q_{i0} = 1/p_{i0}$ and $q_{ij} = p_{ij}/p_{i0}$, and μ_{ij} is the service rate of node i at server j, $j = 1, 2, ..., d_i$, $i = 0, 1, ..., 4$. Substituting the above equation into eq.(4.4a) yields the incremental node delay function

$$f_i(\beta_i) = \sum_{j=0}^{d_i} \frac{q_{ij}\mu_{ij}}{(\mu_{ij} - q_{ij}\beta_i)^2}, \quad \beta_i < \frac{\mu_{ij}}{q_{ij}}, \quad j = 0, 1, ..., d_i \tag{4.47}$$

In this case, the inverse function may not have a closed-form expression; generally, it can be easily obtained numerically by solving a nonlinear equation of a single variable.

Table 4.1: Parameters of node models

Node	Processing rates of servers(jobs/sec.)					Probabilities of a job leaving CPU					Job arrival rate (jobs/min)
	μ_{i0}	μ_{i1}	μ_{i2}	μ_{i3}	μ_{i4}	p_{i0}	p_{i1}	p_{i2}	p_{i3}	p_{i4}	
0	1000	200	200	200	300	0.1	0.2	0.2	0.2	0.3	95
1	500	200	250	-	-	0.1	0.4	0.5	-	-	40
2	100	30	30	30	-	0.1	0.3	0.3	0.3	-	9
3	1000	500	400	-	-	0.1	0.5	0.4	-	-	70
4	200	40	40	40	60	0.1	0.2	0.2	0.2	0.3	19

For the link communication model, we have the following consideration. Denote the mean communication time for sending a packet from node i to node j on link (i, j) by t_{ij} (min.). Assume that on average a packets are required in sending the program and the data of a job, and b packets are used in sending back a response for the job. Note therefore that $ax_{ij} + bx_{ji}$ packets per minute are sent from node i to node j, and $bx_{ij} + ax_{ji}$ packets per minute are sent from node j to node i. We assume that the mean communication delay, $G_{i0}(x_{i0}, x_{0i})$, from node i to node 0 on link $(i, 0)$ and the mean communication delay, $G_{0i}(x_{0i}, x_{i0})$, have the following forms:

$$G_{i0}(x_{i0}, x_{0i}) = \frac{at_{i0}}{1 - t_{i0}(ax_{i0} + bx_{0i})} + \frac{bt_{0i}}{1 - t_{0i}(bx_{i0} + ax_{0i})}, \tag{4.48}$$

$$G_{0i}(x_{0i}, x_{i0}) = \frac{at_{0i}}{1 - t_{0i}(ax_{0i} + bx_{i0})} + \frac{bt_{i0}}{1 - t_{i0}(bx_{0i} + ax_{i0})}, \tag{4.49}$$
$$1 > t_{i0}(ax_{i0} + bx_{0i}) \text{ and } 1 > t_{0i}(bx_{i0} + ax_{0i}).$$

Table 4.1 gives the set of parameters of node models in the example. In this example, we examine the case where $t_{i0} = t_{0i} = 0.001$ minute $(i = 1, 2, 3, 4)$, and where $a = 80$ and $b = 20$. All parameters in the example are fixed except that the communication time on link (0,3) increases from 0.0 second.

The effects of the link communication time on the optimal load balancing strategy are studied by using proposed algorithm. Figure 4.3 shows that the optimal mean response time increases as the communication time on link (0,3) increases. We note that, without load balancing (i.e., each node processes its own local stream of jobs), the mean response time is 1.051 min., which is much higher than the mean response time with load balancing shown in Figure 4.3. In the optimal solution, nodes 1 and 3 are the sink nodes, nodes 2 and 4 are the active nodes. As the communication time on link (0,3) increases from 0.0

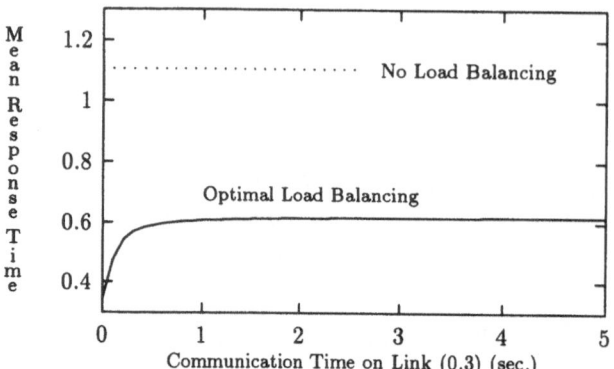

Figure 4.3: The effect on mean response time

sec. to 5 sec., we note that the node partitions remain the same and the link traffic on links (1,0), (0,2), (3,0) and (0,4) are zero. The link traffic of links (2,0), (0,3) and (4,0) decrease and the link traffic of link (0,1) increases, as shown in Figure 4.4.

4.7   Discussion

The results in our above numerical experiment agree with that in the theorems 4.5 and 4.6. Some other numerical experiments wherein we examined different system models were also conducted. They showed results similar to the above example. We note that, by using numerical experiments, we can study the behavior of system performance quantitatively during an interval of the communication time regardless of whether the node partition remains the same or not. Note that it is necessary to make numerical experiments for analyzing the behavior of system performance quantitatively in designing a network. Through the numerical experiments, we observe that the effects of the communication time in star networks are large.

To study the behavior of system performance numerically, the proposed algorithm is quite efficient. We program the proposed algorithm in FORTRAN and run it in a SPARC workstation. In our numerical experiments, it always takes less than 0.5 second to obtain the optimal solution to problem (4.2) for a given parameter set. Notice that the main process of the proposed algorithm is determining a Lagrange multiplier by applying a golden section search (in step 2). This algorithm also makes use of the important property in the optimal solution that the job flow rates x_{i0} and x_{0i} cannot be positive

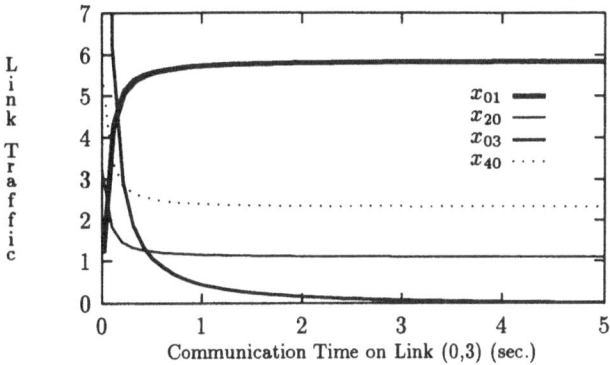

Figure 4.4: The effects on the link traffic

both in the same time. The above results are expected because the proposed algorithm is straightforward and makes use of the special structure of the problem.

4.8 Conclusion

We have studied the optimal static load balancing problem in a star network with two-way traffic. The static load balancing problem was formulated as a nonlinear optimization problem which minimizes the mean response time.

It was proven that in the optimal solution the satellite nodes in the star network are divided into following types, the idle source nodes, the active source nodes, the neutral nodes, and the sink nodes.

The necessary and sufficient conditions for optimal solution were studied and an efficient algorithm with $O(n)$ complexity was proposed for the optimal load balancing of an n-satellite system.

The effects of link communication time on optimal load balancing in a star network were studied by parametric analysis. A significant system performance improvement achieved by employing the proposed algorithm, over that without load balancing was illustrated in numerical examples. The numerical examples also showed that the effects of the link communication time in a star network are significant.

Chapter 5

Load Balancing in Tree Network Configurations with Two-way Traffic

5.1 Introduction

In previous chapter, we have studied the optimal static load balancing in star network configurations with two-way traffic. By using access couplers, Tamura *et al.* [Tam84] propose the cascade star network configurations and Ota [Ota92] proposes the coupled star network configurations for optical local area networks. Note that in cascade star networks and coupled star network, there is two-way traffic in any links. All star network configurations with two-way, cascade star network configurations and coupled star network configurations are examples of tree network configurations with two-way traffic.

In the tree networks with two-way traffic there are not hierarchy structures. Every node is connected to one of its neighbor nodes by one and only one communication line. Since there is two-way traffic between any two nodes (host computers) in this configuration, a job arriving at a node can be processed locally or be forwarded to any other nodes in the network for processing. One also may note that the X-tree multi-processor computer network [DP78] is an example of the tree-configured networks. Due to the development of fiber-optic technique, there are also growing interests in tree network with two-way traffic [SM89, GF88, Yem83].

In this chapter, we focus on the static load balancing in a tree computer network with two-way traffic. For the sake of convenience, we also call the tree computer network with two-way traffic the *tree network*. Consider a set of heterogeneous host computers (nodes) interconnected by a tree network. Jobs arrive at each node according to a time-invariant Poisson process. Each node determines whether a job should be processed locally or scheduled to another node for remote processing. In the latter case, there is a delay

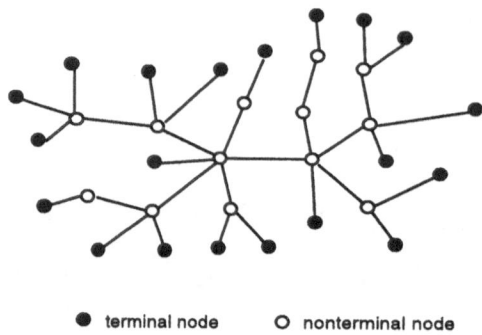

● terminal node   ○ nonterminal node

Figure 5.1: A tree network with two-way traffic

incurred as a result of forwarding the job and sending its response back over the network. Based on these assumptions, we formulate the static load balancing problem as a nonlinear optimization problem which minimizes the mean response time of a job in a tree network. The optimal solution to this problem is referred to as the optimal load balancing strategy. We demonstrate that the optimization problem of the particular structure can be solved by a decomposition technique. An algorithm is developed to obtain the optimal solution.

The proposed algorithm is compared with existing well-known algorithms: the FD (Flow Deviation) algorithm [FGK73] and the Dafermos algorithm [Daf72] with respect to the storage and computation time requirements. It is demonstrated that for an n-node system, the proposed algorithm and the FD algorithm require only the amount of storage of $O(n)$, whereas the Dafermos algorithm requires the amount of storage of $O(n^2)$. It is also shown that the proposed algorithm converges to the optimal solution most quickly, whereas the FD algorithm converges to the optimal solution most slowly.

5.2   Model Description and Problem Formulation

We call a *network* G where G is a pair (N, E). Here N denotes a set whose elements are called *nodes*, E is a set whose elements are *unordered pairs* of distinct elements of N. Each unordered pair $e = < i, j >$ in E is called *edge* .

A *cycle* is a sequence of edges $< e_1, e_2 >$, $< e_2, e_3 >$,..., $< e_{n-1}, e_n >$ where e_1, e_2,..., e_{n-1} are distinct nodes and the first node e_1 is the same as the last node e_n, $e_1 = e_n$. A *tree network* is the network that contains no cycles. Figure 5.1 shows a tree network.

A *neighbor node* to node i is the node which is connected with node i by an edge. In a tree network, the node which has only one neighbor node is called a *terminal node*, and the node which has at least two neighbor nodes is called a *nonterminal node*.

For each edge $< i, j >$, we define two ordered pairs (i, j) and (j, i) which are called *links*. Denote the set of links by L. By a *path* connecting an ordered pair (x, y), we mean a sequence of links $(p_1, p_2), (p_2, p_3), ..., (p_{n-1}, p_n)$ where $p_1, p_2, ..., p_n$ are distinct nodes, $p_1 = x$, and $p_n = y$. In particular, each link is a path.

The nodes in a tree network may be heterogeneous computer systems; that is, they may have different configurations, number of resources, and speed characteristics. Each node contains one or more resources (such as CPU and I/O devices) contended for by the jobs processed at the node. Jobs arrive at node i ($i \in N$) according to a time–invariant Poisson process. Since there are two links between two nodes in each edge, a job arriving at node i can either be processed locally or be transferred to another node through a path for remote processing.

For our analyses, notation and assumptions are introduced as follows.

- β_i: the rate at which jobs are processed at node i, $i \in N$, also referred to as *load*. That is, the average number of jobs processed at node i per unit interval of time.

- x_{ij}: the job flow rate from node i to node j, also called the *traffic*, on link (i, j), $(i, j) \in L$. That is, the average number of jobs sent from node i through link (i, j) per unit interval of time. Note that on each edge in a tree network there is two-way traffic which includes both the possibilities of traffic from node i to node j as well as that from node j to node i.

- y_{sd}: the rate of jobs that arrive at node s and are transferred to node d for processing. That is, the average number of jobs arriving at node s and be sent to node d per unit interval of time. Note that if $s = d$, y_{sd} means the rate of jobs that arrive at node s and are processed locally.

- B: the set of nonterminal nodes.

- T: the set of terminal nodes.

- V_j: the set of neighbor nodes to node j, i.e.,
  $$V_j = \{i \mid (i, j) \in L\}.$$

- α: $[\alpha_i]$, where α_i is the Lagrange multiplier, $i \in N$.

- β: $[\beta_i]$.

- x: $[x_{ij}]$.

- ϕ_i: the external job arrival rate at node i, $i \in N$.

- $F_i(\beta_i)$: the mean node delay (queuing and processing delays) of a job processed at node i, $i \in N$.

- $G_{ij}(\boldsymbol{x})$: the mean communication delay of sending a job from node i to node j and sending the response back from node j to node i on link (i,j) $((i,j) \in L)$. We assume that the mean communication delay $G_{ij}(\boldsymbol{x})$ depends upon job flow rates x_{ij} on link (i,j) and x_{ji} on link (j,i). Then, we use $G_{ij}(x_{ij}, x_{ji})$ instead of $G_{ij}(\boldsymbol{x})$.

- $D(\boldsymbol{\beta}, \boldsymbol{x})$: mean response time that a job spends in a network from the time of its arrival until the time of its departure.

Note that there is only one path from a node to another node in the tree network. We have the following relations.

$$x_{ij} = \sum_{s,d \in N} \delta_{ij}^{sd} y_{sd}, \quad y_{sd} \geq 0,$$

where

$$\delta_{ij}^{sd} = \begin{cases} 1 & \text{if link } (i,j) \text{ is contained in path } (s,d), \\ 0 & \text{else.} \end{cases}$$

We can obtain the set of values of $\{y_{sd}\}$ if we have the set of values of $\boldsymbol{x}$ even though $\{y_{sd}\}$ may not be unique.

Assume that the node delay function $F_i(\beta_i)$ $(i \in N)$ is a differentiable, increasing, and convex function, and the communication delay function $G_{ij}(x_{ij}, x_{ji})$ $((i,j) \in L)$ is a positive, differentiable, nondecreasing, and convex function with respect to x_{ij} and x_{ji}. We note that the class of delay functions satisfying the above assumptions is large, and it contains those of the M/M/c model and the central server model [Kle76], etc. We express $D(\boldsymbol{\beta}, \boldsymbol{x})$ as the sum of the mean node delay at all nodes and the mean communication delay on all links by using the Little's result as follows,

$$D(\boldsymbol{\beta}, \boldsymbol{x}) = \frac{1}{\Phi} \{ \sum_{i \in N} \beta_i F_i(\beta_i) + \sum_{(i,j) \in L} x_{ij} G_{ij}(x_{ij}, x_{ji}) \} \qquad (5.1)$$

where $\Phi = \sum_{i \in N} \phi_i$.

The goal of load balancing is to minimize the mean response time with respect to $\boldsymbol{\beta}$ and $\boldsymbol{x}$. The optimization problem of minimizing the mean response time of a job can be written by considering the structure of tree networks as follows:

$$\text{Minimize} \quad D(\boldsymbol{\beta}, \boldsymbol{x}) = \frac{1}{\Phi} \{ \sum_{i \in N} \beta_i F_i(\beta_i) + \sum_{(i,j) \in L} x_{ij} G_{ij}(x_{ij}, x_{ji}) \} \qquad (5.2)$$

with respect to the variables β_i, $i \in N$, x_{ij}, $(i,j) \in L$,

subject to

$$\phi_i + \sum_{l \in V_i} x_{li} = \beta_i + \sum_{l \in V_i} x_{il}, \quad i \in N, \tag{5.3a}$$

$$\beta_i \geq 0, \quad i \in N, \tag{5.3b}$$

$$x_{ij} \geq 0, \quad (i,j) \in L. \tag{5.3c}$$

The first constraint (5.3a) is the flow balance constraint which equates the job flow rates in and out of node i. Note that the mean node delay $F_i(\beta_i)$ $(i \in N)$ is an increasing and convex function, and the mean communication delay function $G_{ij}(x_{ij}, x_{ji})$ $((i,j) \in L)$ is a nondecreasing, and convex function. According to eq.(5.1), the mean response time also is a convex function.

We define the set of feasible solutions of the above optimization problem FS as follows, $FS = \{(\boldsymbol{\beta}, \boldsymbol{x}) \mid (\boldsymbol{\beta}, \boldsymbol{x}) \ satisfies \ the \ constraints \ (5.3)\}$. We note that the set FS is a convex polyhedron. We also note that the set FS is closed according to the constraints (5.3) which the job flow pattern $(\boldsymbol{\beta}, \boldsymbol{x})$ in the set FS must satisfy.

5.3 Necessary and Sufficient Conditions for Static Load Balancing

We introduce the *incremental node delay function* $f_i(\beta_i)$ at node i, and the *incremental communication delay function* $g_{ij}(x_{ij}, x_{ji})$ on link (i,j) as follows,

$$f_i(\beta_i) = \frac{d}{d\beta_i}\beta_i F_i(\beta_i), \tag{5.4a}$$

$$g_{ij}(x_{ij}, x_{ji}) = \frac{\partial}{\partial x_{ij}}x_{ij}G_{ij}(x_{ij}, x_{ji}) + x_{ji}\frac{\partial G_{ji}(x_{ji}, x_{ji})}{\partial x_{ij}}. \tag{5.4b}$$

Due to our assumptions about $F_i(\beta_i)$ and $G_{ij}(x_{ij}, x_{ji})$, we note that $f_i(\beta_i)$ is an increasing function, $g_{ij}(x_{ij}, x_{ji})$ is a nondecreasing function with respect to x_{ij} and x_{ji}. Also note that $g_{ij}(x_{ij}, x_{ji})$ is positive if $x_{ij} > 0$. We define the inverse of an increasing function $f_i(\bullet)$ as follows:

$$f_i^{-1}(x) = \begin{cases} a, & \text{if } f_i(a) = x, \text{ and } a > 0, \\ 0, & \text{if } f_i(0) \geq x. \end{cases}$$

Theorem 5.1 *The set of values of β and x is an optimal solution to problem (5.2) if and only if the following set of relations hold.*

$$\alpha_i - g_{ij}(x_{ij}, x_{ji}) = \alpha_j, \quad if \ x_{ij} > 0 \ ((i,j) \in L), \tag{5.5a}$$

$$\alpha_i - g_{ij}(x_{ij}, x_{ji}) \leq \alpha_j, \quad if \ x_{ij} = 0 \ ((i,j) \in L), \tag{5.5b}$$

$$f_i^{-1}(\alpha_i) = \beta_i, \quad (i \in N), \tag{5.6}$$

subject to

$$\phi_i + \sum_{l \in V_i} x_{li} = \sum_{l \in V_i} x_{il} + \beta_i, \quad i \in N, \tag{5.7}$$

where α_i ($i \in N$) is the Lagrange multiplier.

PROOF. To obtain the optimal solution to problem (5.2), we form the Lagrangian function as follows,

$$H(\boldsymbol{\beta}, \boldsymbol{x}, \boldsymbol{\alpha}) = \boldsymbol{\Phi} D(\boldsymbol{\beta}, \boldsymbol{x}) + \sum_{j \in N} \alpha_j (\phi_j + \sum_{i \in V_j} x_{ij} - \sum_{i \in V_j} x_{ji} - \beta_j). \tag{5.8}$$

Since the objective function is convex by assumption and the feasible set FS is a convex set, a set of values of $\boldsymbol{\beta}$ and $\boldsymbol{x}$ is an optimal solution to problem (5.2) if and only if it satisfies the following Kuhn-Tucker conditions (e.g., [Int71]),

$$\frac{\partial H}{\partial \beta_i} = f_i(\beta_i) - \alpha_i \geq 0, \tag{5.9a}$$

$$\beta_i \frac{\partial H}{\partial \beta_i} = \beta_i (f_i(\beta_i) - \alpha_i) = 0, \tag{5.9b}$$

$$\beta_i \geq 0, \tag{5.9c}$$

$$i \in N.$$

$$\frac{\partial H}{\partial x_{ij}} = g_{ij}(x_{ij}, x_{ji}) + \alpha_j - \alpha_i \geq 0, \tag{5.10a}$$

$$x_{ij} \frac{\partial H}{\partial x_{ij}} = x_{ij}(g_{ij}(x_{ij}, x_{ji}) + \alpha_j - \alpha_i) = 0, \tag{5.10b}$$

$$x_{ij} \geq 0, \tag{5.10c}$$

$$(i, j) \in L.$$

$$\phi_i + \sum_{j \in V_i} x_{ji} - \sum_{j \in V_i} x_{ij} - \beta_i = 0, \quad i \in N. \tag{5.11}$$

The above set of conditions (5.9) and (5.10) are identical with the following set of relations,

$$f_i(\beta_i) = \alpha_i, \quad \beta_i > 0, \tag{5.12a}$$

$$f_i(\beta_i) \geq \alpha_i, \quad \beta_i = 0, \tag{5.12b}$$

$$i \in N.$$

$$\alpha_i - g_{ij}(x_{ij}, x_{ji}) = \alpha_j, \quad x_{ij} > 0, \tag{5.13a}$$

$$\alpha_i - g_{ij}(x_{ij}, x_{ji}) \leq \alpha_j, \quad x_{ij} = 0, \tag{5.13b}$$

$$(i, j) \in L.$$

We can easily see the above set of relations are equivalent to the set in Theorem 1. Q.E.D.

Let us proceed to find more properties of the optimal solution to problem (5.2).

Theorem 5.2 *In the optimal solution to problem (5.2), the following relations hold true,*

$$x_{ij} = 0, \quad if \ x_{ji} > 0, \ < i, j > \in E. \tag{5.14}$$

PROOF. It is proved by contradiction. Suppose the relation (5.14) is not true and there exists link (i, j) $((i, j) \in L)$ such that $x_{ij} > 0$ and $x_{ji} > 0$. According to Theorem 1, we have

$$\alpha_i - g_{ij}(x_{ij}, x_{ji}) = \alpha_j, \quad ((i, j) \in L) \tag{5.15a}$$
$$\alpha_j - g_{ji}(x_{ji}, x_{ij}) = \alpha_i, \quad ((j, i) \in L). \tag{5.15b}$$

Then
$$g_{ij}(x_{ij}, x_{ji}) + g_{ji}(x_{ji}, x_{ij}) = 0. \tag{5.16}$$

which is a contradiction to the fact that $g_{ij}(x_{ij}, x_{ji})$ and $g_{ji}(x_{ji}, x_{ij})$ are both positive functions. Q.E.D.

Theorem 5.2 shows an important property in the optimal solution that the job flow rate from the node i to node j, x_{ij}, and the job flow rate from the node j to node i, x_{ji}, on edge $< i, j >$ cannot be positive both at the same time.

Note that $g_{ij}(x_{ij}, x_{ji})$ is either $g_{ij}(x_{ij}, 0)$ or $g_{ij}(0, x_{ji})$ since $x_{ij}x_{ji} = 0$ in the optimal solution. Hereafter we denote $g_{ij}(x_{ij}, 0)$ by $g_{ij}(x_{ij})$ in the optimal solution.

Define ϕ_{ij} as follows.
$$\phi_{ij} = \begin{cases} \phi_i + \sum_{l \in V_i - \{j\}} x_{li} - \sum_{l \in V_i - \{j\}} x_{il}, & i \in B, \\ \phi_i, & i \in T. \end{cases}$$
Note that $0 \le x_{ij} \le \phi_{ij}$. We have the following definitions further. A node i in the set V_j is said to be a *source* to node j if $x_{ij} > 0$, i.e., it sends some jobs to node j. Node i is an *idle source* if $x_{ij} = \phi_{ij}$, or an *active source* if $0 < x_{ij} < \phi_{ij}$. If node i does not send (or receive) any jobs to (or from) node j (i.e., $x_{ij} = 0$ and $x_{ji} = 0$), it is a *neutral node*. If node i receives some jobs from node j (i.e., $x_{ji} > 0$), it is a *sink node*. Denote the sets of idle source nodes, active source nodes, neutral nodes and sink nodes in V_j by Rd_j, Ra_j, Nu_j, and S_j, respectively. We have the following corollary and theorem.

Corollary 5.3 *The sets of Rd_j, Ra_j, Nu_j, and S_j are mutually disjoint and*

$$V_j = Rd_j \cup Ra_j \cup Nu_j \cup S_j. \tag{5.17}$$

PROOF. It is a direct result from Theorem 2. Q.E.D.

Theorem 5.4 *The set of values of β and x is an optimal solution to problem (5.2) if and only if the following set of relations hold.*

$$f_i(\beta_i) = \alpha_i = \alpha_j + g_{ij}(x_{ij}), \qquad if\ 0 < x_{ij} < \phi_{ij},\ (i \in Ra_j,\ j \in B), \qquad (5.18a)$$

$$f_i(\beta_i) \geq \alpha_i = \alpha_j + g_{ij}(x_{ij}), \qquad if\ x_{ij} = \phi_{ij},\ (i \in Rd_j,\ j \in B), \qquad (5.18b)$$

$$f_i(\beta_i) = \alpha_i = \alpha_j - g_{ji}(x_{ji}), \qquad if\ x_{ji} > 0,\ (i \in S_j,\ j \in B), \qquad (5.18c)$$

$$g_{ji}(x_{ji}) + \alpha_i \geq \alpha_j \geq \alpha_i - g_{ij}(x_{ij}), \quad if\ x_{ij} = x_{ji} = 0,\ (i \in Nu_j,\ j \in B), \qquad (5.18d)$$

and
$$\beta_i = f_i^{-1}(\alpha_i),\ \ i \in N, \qquad (5.19)$$

subject to

$$\phi_i + \sum_{l \in V_i} x_{li} = \sum_{l \in V_i} x_{il} + \beta_i, \ \ i \in N, \qquad (5.20)$$

PROOF. This theorem can be directly derived by using Theorem 1, Theorem 2 and Corollary 3. Q.E.D.

5.4 Decomposition

Let us examine the structure of problem (5.2). Consider a particular nonterminal node j $(j \in B)$ as the central node and the nodes in set V_j as the satellite nodes in a star network with two-way traffic [LK94b]. We note that the values of variables β_i, x_{ij}, x_{ji} $(i \in V_j)$ and β_j can be varied while keeping the values of other variables in β and x constant without violating the constraints (5.3). If we keep such other variables constant, the minimization problem (5.2) reduces to the following star subproblem,

$$\text{Minimize } \frac{1}{\Phi}\{\beta_j F_j(\beta_j) + \sum_{i \in V_j}[\beta_i F_i(\beta_i) + x_{ij}G_{ij}(x_{ij}, x_{ji}) + x_{ji}G_{ji}(x_{ji}, x_{ij})] + C\}, \quad (5.21)$$

with respect to the variables β_i, x_{ij}, x_{ji}, $(i \in V_j)$, and β_j,
subject to

$$\phi_j + \sum_{i \in V_j} x_{ij} = \sum_{i \in V_j} x_{ji} + \beta_j, \qquad (5.22a)$$

$$\phi_{ij} = \beta_i + x_{ij} - x_{ji},\ \ i \in V_j, \qquad (5.22b)$$

$$(i.e.,\ \phi_i + \sum_{l \in V_i - \{j\}} (x_{li} - x_{il}) = \beta_i + x_{ij} - x_{ji})$$

$$\beta_j \geq 0, \qquad (5.22c)$$

$$\beta_i \geq 0,\ x_{ij} \geq 0,\ x_{ji} \geq 0,\ \ i \in V_j. \qquad (5.22d)$$

(C in (5.21) denotes a variable that is independent of β_i, x_{ij}, x_{ji}, $i \in V_j$, and β_j.)

The observation would lead us to an interesting decomposition technique to solve problem (5.2). In relation to this, we have the following theorems.

Theorem 5.5 *The set of values of β_i, x_{ij}, x_{ji} ($i \in V_j$) and β_j is an optimal solution to subproblem (5.21) if and only if the following set of relations hold true for all $i \in V_j$,*

$$f_i(\beta_i) = \alpha_i = \alpha_j + g_{ij}(x_{ij}), \quad if \ 0 < x_{ij} < \phi_{ij}, \ (i \in Ra_j), \tag{5.23a}$$
$$f_i(\beta_i) \geq \alpha_i = \alpha_j + g_{ij}(x_{ij}), \quad if \ x_{ij} = \phi_{ij}, \ (i \in Rd_j), \tag{5.23b}$$
$$f_i(\beta_i) = \alpha_i = \alpha_j - g_{ji}(x_{ji}), \quad if \ x_{ji} > 0, \ (i \in S_j), \tag{5.23c}$$
$$g_{ji}(x_{ji}) + \alpha_i \geq \alpha_j \geq \alpha_i - g_{ij}(x_{ij}), \quad if \ x_{ij} = x_{ji} = 0, \ (i \in Nu_j), \tag{5.23d}$$

$$\beta_i = f_i^{-1}(\alpha_i), \ i \in V_j, \tag{5.24a}$$
$$\beta_j = f_j^{-1}(\alpha_j), \tag{5.24b}$$

subject to

$$\phi_j + \sum_{i\in V_j} x_{ij} = \sum_{i\in V_j} x_{ji} + \beta_j, \tag{5.25a}$$
$$\phi_{ij} = \beta_i + x_{ij} - x_{ji} \ i \in V_j, \tag{5.25b}$$

where α_i ($i \in V_j$) and α_j are the Lagrange multipliers.

PROOF. The proof of the above theorem is similar to that of Theorem 4, by regarding node j as the only one nonterminal node in a tree network. Q.E.D.

The above theorem shows that the values of β_i, x_{ij}, x_{ji} ($i \in V_j$) and β_j can be determined by using (5.23) and (5.24) if the Lagrange multipliers are given. A limitation of the usefulness of this theorem may arise from the fact that it is not easy to determine all the Lagrange multipliers directly by using this theorem. We proceed to provide another equivalent theorem showing the necessary and sufficient conditions of the solution to subproblem (5.21) in chapter 4 of this book.

For the sake of convenience, we introduce the following notations,

$$p_i(x_{ij}) = f_i(\phi_{ij} - x_{ij}) - g_{ij}(x_{ij}), \ i \in V_j, \ j \in B; \tag{5.26a}$$
$$q_i(x_{ji}) = f_i(\phi_{ij} + x_{ji}) + g_{ji}(x_{ji}), \ i \in V_j, \ j \in B. \tag{5.26b}$$

Note that $p_i(x_{ij})$ is a monotonically decreasing function and that $q_i(x_{ji})$ is a monotonically increasing function. The two functions $p_i(x_{ij})$ and $q_i(x_{ji})$ have their inverse functions, p_i^{-1} and q_i^{-1}, which are defined similarly as the inverse function of $f_i(\beta_i)$. According to the above definitions, we have the following relations,

$$q_i(x_{ji}) > q_i(0) \geq p_i(0) > p_i(\phi_{ij}), \ x_{ji} > 0, \ \forall i \in V_j, \ j \in B. \tag{5.27}$$

Theorem 5.6 *The set of values of β_i, x_{ij}, x_{ji} ($i \in V_j$) and β_j is an optimal solution to subproblem (5.21) if and only if the following set of relations hold true for all $i \in V_j$,*

$$p_i(x_{ij}) = \alpha_j, \quad \text{if } 0 < x_{ij} < \phi_{ij}, \ (i \in Ra_j), \tag{5.28a}$$

$$p_i(x_{ij}) \geq \alpha_j, \quad \text{if } x_{ij} = \phi_{ij}, \ (i \in Rd_j), \tag{5.28b}$$

$$q_i(x_{ji}) = \alpha_j, \quad \text{if } x_{ji} > 0, \ (i \in S_j), \tag{5.28c}$$

$$q_i(x_{ji}) \geq \alpha_j \geq p_i(x_{ij}), \quad \text{if } x_{ij} = x_{ji} = 0, \ (i \in Nu_j), \tag{5.28d}$$

$$\beta_j = f_j^{-1}(\alpha_j), \tag{5.29}$$

subject to

$$\phi_j + \sum_{i \in Ra_j} p_i^{-1}(\alpha_j) + \sum_{i \in Rd_j} \phi_{ij} = f_j^{-1}(\alpha_j) + \sum_{i \in S_j} q_i^{-1}(\alpha_j). \tag{5.30}$$

PROOF [NECESSITY]. By noting $\phi_{ij} + x_{ji} = x_{ij} + \beta_i$ ($i \in V_j$), relations (5.28) and (5.29) can be derived directly by relations (5.23) and (5.24) in Theorem 5. Furthermore, according to eq.(5.25a), using Corollary 3, we have eq.(5.30).

[SUFFICIENCY]: Note that the set of values of β_i, x_{ij}, x_{ji} ($i \in V_j$) and β_j is uniquely determined by a given α_j according to relations (5.28) and (5.29). We prove that constraint (5.30) determines the value of α_j, denoted by α_j^*, uniquely, and that the set of values of β_i, x_{ij}, x_{ji} ($i \in V_j$) and β_j, $(\boldsymbol{\beta}_j^*, \boldsymbol{x}_j^*)$, is the optimal solution to subproblem (5.21) where $(\boldsymbol{\beta}_j^*, \boldsymbol{x}_j^*)$ is determined uniquely by α_j^*.

From relations (5.28), we see that the sets of Ra_j, Rd_j, S_j and Nu_j are determined uniquely by a given α_j. For an arbitrary α_j, we define

$$Ra_j(\alpha_j) = \{i|\, p_i(x_{ij}) = \alpha_j, \ 0 < x_{ij} < \phi_{ij}, \ i \in V_j\}, \tag{5.31a}$$

$$Rd_j(\alpha_j) = \{i|\, p_i(\phi_{ij}) \geq \alpha_j, \ i \in V_j\}, \tag{5.31b}$$

$$S_j(\alpha_j) = \{i|\, q_i(x_{ji}) = \alpha_j, \ x_{ji} > 0, \ i \in V_j\}, \tag{5.31c}$$

$$Nu_j(\alpha_j) = \{i|\, q_i(0) \geq \alpha \geq p_i(0), \ i \in V_j\}, \tag{5.31d}$$

$$\lambda_R(\alpha_j) = \phi_j + \sum_{i \in Ra_j(\alpha_j)} p_i^{-1}(\alpha_j) + \sum_{i \in Rd_j(\alpha_j)} \phi_{ij}, \tag{5.31e}$$

$$\lambda_S(\alpha_j) = f_j^{-1}(\alpha_j) + \sum_{i \in S_j(\alpha_j)} q_i^{-1}(\alpha_j). \tag{5.31f}$$

Note that $p_i(x_{ij})$ is a monotonically decreasing function, and that $f_j(\beta_j)$ and $q_i(x_{ji})$ are monotonically increasing functions. We have the following. As α_j increases, $\lambda_R(\alpha_j)$ first stays to be $\phi_j + \sum_{i \in V_j} \phi_{ij}$ while $\min_{i \in V_j} p_i(\phi_{ij}) \geq \alpha_j$, and then decreases monotonically to be ϕ_j while $\alpha_j \geq \max_{i \in V_j} p_i(0)$. On the other hand, as α_j increases, $\lambda_S(\alpha_j)$ first stays to be zero while $f_j(0) \geq \alpha_j$ and $\min_{i \in V_j} q_i(0) \geq \alpha_j$, and then increases monotonically. Noting (5.27), we can see that α_j is determined uniquely by constraint (5.30), i.e.,

$$\lambda_R(\alpha_j^*) = \lambda_S(\alpha_j^*), \tag{5.32}$$

where α_j^* is the value of α_j.

Furthermore, α_j^* uniquely determines the set of values of β_i, x_{ij}, x_{ji} ($i \in V_j$) and β_j, $(\boldsymbol{\beta}_j^*, \boldsymbol{x}_j^*)$, according to relations (5.28) and (5.29),

Now we prove that $(\boldsymbol{\beta}_j^*, \boldsymbol{x}_j^*)$ is the optimal solution to subproblem (5.21). Assume that $(\tilde{\boldsymbol{\beta}}_j, \tilde{\boldsymbol{x}}_j)$ is the optimal solution to subproblem (5.21) and $(\tilde{\boldsymbol{\beta}}_j, \tilde{\boldsymbol{x}}_j) \neq (\boldsymbol{\beta}_j^*, \boldsymbol{x}_j^*)$. Then there should be a value of α_j, $\tilde{\alpha}_j$, which determines $(\tilde{\boldsymbol{\beta}}_j, \tilde{\boldsymbol{x}}_j)$ uniquely. We note that $\tilde{\alpha}_j$ satisfies the constraint (5.30), i.e., $\lambda_R(\tilde{\alpha}_j) = \lambda_S(\tilde{\alpha}_j)$. Since constraint (5.30) determines the value of α_j uniquely, we have $\tilde{\alpha}_j = \alpha_j^*$. Then we have $(\tilde{\boldsymbol{\beta}}_j, \tilde{\boldsymbol{x}}_j) = (\boldsymbol{\beta}_j^*, \boldsymbol{x}_j^*)$. It is a contradiction to the assumption. Q.E.D.

Remark: We have a simple interpretation of the relations in Theorem 6 as follows. If the value of α_j is determined, in the optimal solution to star subproblem (5.21), the satellite node i ($i \in V_j$) may be an active source node, or an idle node, or a sink node, or a neutral node according to the set of relations (5.28). The value of α_j can be evaluated directly by equation (5.30).

Theorem 5.7 (Decomposition Theorem) *The set of values of β and x is an optimal solution to problem (5.2) if and only if the set of values of β_i, x_{ij}, $x_{ji}(i \in V_j)$ and β_j is an optimal solution to star subproblem (5.21) for all $j \in B$ at the same time.*

PROOF [NECESSITY]. It is clear since a set of values of $\boldsymbol{\beta}$, $\boldsymbol{x}$ and $\boldsymbol{\alpha}$ that satisfies the relations of Theorem 4 also satisfies the relations in Theorems 5 and 6 for all $j \in B$ at the same time.

[SUFFICIENCY]: Note that the set of values of β_i, x_{ij}, x_{ji} ($i \in V_j$, $j \in B$) and β_j is determined if α_j ($j \in B$) is given. The sufficiency can be proved if we show that the value of α_j is determined uniquely and consistently for all nonterminal node j, $j \in B$, by the relations given in Theorems 5 and 6.

We have shown that α_j ($\forall j \in B$) is determined uniquely by eq.(5.30) in Theorem 6. For node i in set V_j ($j \in B$), we note that it is not needed to determine the value of α_i if $i \in T$ (i.e., node i is a terminal node). If $i \in V_j \cap B$ (i.e., node i in set V_j also is a nonterminal node), it is easy to see that the values of α_i and α_j satisfy the relations (5.23) in Theorem 5. Then the values of α_i ($i \in V_j$) and α_j satisfy the relations (5.18) in Theorem 4. That is, the values of α_j ($\forall j \in B$) are also determined consistently by eq.(5.30) in Theorem 6. Q.E.D.

Remark: Theorem 7 shows that the optimization problem (5.2) can be solved by solving much simpler subproblem (5.21).

5.5   Proposed Algorithm

The optimal load balancing algorithm solving the optimization problem (5.2) is derived
by using Theorem 7 (Decomposition theorem) and Theorem 6 above. This algorithm
(called a single-point algorithm) can obtain the optimal solution to optimization problem
(5.2) for an arbitrary set of parameter values.

   o Single-point algorithm

1. Initialization.

   $r = 0$. (r : iteration number)

   Find $(\boldsymbol{x}^{(0)}, \boldsymbol{\beta}^{(0)})$ ($\in FS$) as an initial feasible solution to the problem (5.2).

2. Solve the star subproblem (5.21) iteratively to obtain $\boldsymbol{\beta}^{(r)}$, $\boldsymbol{x}^{(r)}$.

   $r = r + 1$.

   Let $\boldsymbol{x}^{(r)} = \boldsymbol{x}^{(r-1)}$, $\boldsymbol{\beta}^{(r)} = \boldsymbol{\beta}^{(r-1)}$. Execute the following procedure for every nonterminal node j ($j \in B$) in the tree network. Take node j as the central node and node
i ($i \in V_j$) as a satellite node in the star subproblem (5.21). Let $\beta_j^{(r)}$, $\beta_i^{(r)}$, $x_{ij}^{(r)}$, $x_{ji}^{(r)}$
and $\phi_{ij}^{(r)}$ ($i \in V_j$) be β_j, β_i, x_{ij}, x_{ji} and ϕ_{ij} in subproblem (5.21), respectively, with

$$\phi_{ij}^{(r)} \text{ being given, where } \phi_{ij}^{(r)} = \begin{cases} \phi_i + \sum_{l \in V_i - \{j\}} x_{li}^{(r)} - \sum_{l \in V_i - \{j\}} x_{il}^{(r)}, & i \in B, \\ \phi_i, & i \in T. \end{cases} \text{ Apply}$$

the following sub-algorithm to solve subproblem (5.21).

3. Stopping rule.

   If $|D(\boldsymbol{\beta}^{(r)}, \boldsymbol{x}^{(r)}) - D(\boldsymbol{\beta}^{(r-1)}, \boldsymbol{x}^{(r-1)})|/D(\boldsymbol{\beta}^{(r)}, \boldsymbol{x}^{(r)}) < \varepsilon$ then STOP, where ε is a proper
acceptance tolerance; otherwise, go to step 2.

   o Sub-algorithm that computes the values of β_i, x_{ij}, x_{ji} $i \in V_j$, and β_j by solving star
sub-problem (5.21).

1. Compare. $O(v)$ ($v = |V_j| + 1$)

   If ($\forall i \in V_j$) $q_i(0) \geq f_j(\phi_j) \geq p_i(0)$, then no load balancing is required. In this case,
let $\beta_i = \phi_i$, $x_{ij} = x_{ji} = 0$, $\forall i \in V_j$, and $\beta_j = \phi_j$, stop.

   Otherwise, proceed to step 2.

2. Determine α_j and the node partition. $O(v)$

   Find such α_j that satisfies

$$\lambda_S(\alpha_j) = \lambda_R(\alpha_j),$$

by applying a golden section search, where each value is calculated in the following order for a given α_j.

$$
\begin{aligned}
Ra_j(\alpha_j) &= \{i|\ p_i(x_{ij}) = \alpha_j,\ 0 < x_{ij} < \phi_{ij},\ i \in V_j\}, \\
Rd_j(\alpha_j) &= \{i|\ p_i(\phi_{ij}) \geq \alpha_j,\ i \in V_j\}, \\
S_j(\alpha_j) &= \{i|\ q_i(x_{ji}) = \alpha_j,\ x_{ji} > 0,\ i \in V_j\}, \\
Nu_j(\alpha_j) &= \{i|\ q_i(0) \geq \alpha \geq p_i(0),\ i \in V_j\}, \\
\lambda_R(\alpha_j) &= \phi_j + \sum_{i \in Ra_j(\alpha_j)} p_i^{-1}(\alpha_j) + \sum_{i \in Rd_j(\alpha_j)} \phi_{ij}, \\
\lambda_S(\alpha_j) &= f_j^{-1}(\alpha_j) + \sum_{i \in S_j(\alpha_j)} q_i^{-1}(\alpha_j).
\end{aligned}
$$

Proceed to step 3.

3. Determine the optimal load. $O(v)$

$$
\begin{aligned}
x_{ij} &= \phi_{ij},\ x_{ji} = 0,\ \beta_i = 0,\ \forall i \in Rd_j(\alpha_j), \\
x_{ij} &= p_i^{-1}(\alpha_j),\ x_{ji} = 0,\ \beta_i = \phi_{ij} - x_{ij},\ \forall i \in Ra_j(\alpha_j), \\
x_{ij} &= 0,\ x_{ji} = q_i^{-1}(\alpha_j),\ \beta_i = \phi_{ij} + x_{ji},\ \forall i \in S_j(\alpha_j), \\
x_{ij} &= 0,\ x_{ji} = 0,\ \beta_i = \phi_{ij},\ \forall i \in Nu_j(\alpha_j).
\end{aligned}
$$

The convergence of the proposed single-point algorithm can be shown in a similar way as given by Dafermos and Sparrow [DS69] as follows.

Let us call the sub-algorithm for nonterminal node j ($j \in B$) "equilibration operator" $E^{(j)} : FS \longrightarrow FS$, $j \in B$, where node j is taken as the central node in subproblem (5.21).

Note that the proposed algorithm applies sub-algorithm iteratively in step 2. Then we have another equilibration operator E as a collection of $E^{(j)}$, $E = E^{(b-1)} \circ E^{(b-2)} \circ \ldots \circ E^{(1)} \circ E^{(0)}$

where b is the number of elements of set B ($b = |B|$). It is clear that E is a map, $E : FS \longrightarrow FS$.

Note that $\boldsymbol{\beta}$ is determined if $\boldsymbol{x}$ is given. We can use $D(\boldsymbol{x})$ instead of $D(\boldsymbol{\beta}, \boldsymbol{x})$. Like Dafermos and Sparrow (Theorem (2.1) and Theorem (2.2) in [DS69]), we have the following lemma.

Lemma 5.8 *If the set of feasible solution to problem (5.2) FS is closed and convex, and the equilibration operator E has the following properties, then the proposed algorithm can have an optimal solution to problem (5.2).*

(1) $E\boldsymbol{x} = \boldsymbol{x}$ for some $\boldsymbol{x} \in FS$ implies that $\boldsymbol{x}$ satisfies the relations given in Theorem 5 for all $j \in B$.

(2) E is a continuous mapping from FS to FS.

(3) $D(\boldsymbol{x}) \geq D(E\boldsymbol{x})$ for all $\boldsymbol{x} \in FS$.

(4) $D(\boldsymbol{x}) = D(E\boldsymbol{x})$ for some $\boldsymbol{x} \in FS$ implies that $E\boldsymbol{x} = \boldsymbol{x}$.

Theorem 5.9 *The sequence $\{\boldsymbol{x}^{(r)}\}$ in the proposed algorithm converges to the optimal solution to the optimization problem (5.2).*

PROOF. If we can show that E satisfies the properties given in Lemma 5.8, the proof of the theorem can be completed. Now we prove that E has the four properties given in Lemma 5.8.

(1) $E\boldsymbol{x} = \boldsymbol{x}$ implies $E^{(j)}\boldsymbol{x} = \boldsymbol{x}$ for all $j \in B$. Then we have property (1).

(2) Note that $f_i(\beta_i)$ is continuous and increasing, $p_i(x_{ij})$ is continuous and decreasing, and $q_i(x_{ji})$ is continuous and increasing. It is easy to see that $E^{(j)}$ ($j \in B$) is a continuous mapping from FS to FS. Then E is also a continuous mapping from FS to FS by the definition.

(3) Consider an arbitrary $\boldsymbol{x} \in FS$. Let us define

$$\boldsymbol{x}^{(j)} = E^{(j)} \circ E^{(j-1)} \circ ... \circ E^{(0)}\boldsymbol{x}, \quad j = 0,1,2,...,b-1.$$

Then we have

$$\boldsymbol{x}^{(j)} = E^{(j)}\boldsymbol{x}^{(j-1)}, \quad j = 0,1,2,...,b-1,$$

where we let $\boldsymbol{x}^{(-1)} = \boldsymbol{x}$.

By examining the sub-algorithm, we easily see

$$D(\boldsymbol{x}^{(j-1)}) \geq D(E^{(j)}\boldsymbol{x}^{(j-1)}), \quad j = 0,1,2,...,b-1.$$

Therefore we have

$$D(\boldsymbol{x}) \geq D(\boldsymbol{x}^{(0)}) \geq \cdots \geq D(\boldsymbol{x}^{(b-1)}),$$

where $E\boldsymbol{x} = \boldsymbol{x}^{(b-1)}$.

Thus we have $D(\boldsymbol{x}) \geq D(E\boldsymbol{x})$.

(4) If $D(\boldsymbol{x}) = D(E\boldsymbol{x})$, from property (3) above,

$$D(\boldsymbol{x}) = D(\boldsymbol{x}^{(0)}) = \cdots = D(\boldsymbol{x}^{(b-1)}).$$

That is, $D(E^{(j)}\boldsymbol{x}^{(j)}) = D(\boldsymbol{x}^{(j)})$ for all $j \in B$.

Note that $\boldsymbol{x}$ consists of $\boldsymbol{x}_j$, $j = 0, 1, ..., b-1$, where $\boldsymbol{x}_j$ consists of x_{ij} and x_{ji}, $i \in V_j$. From above, we have $D(E^{(j)}\boldsymbol{x}_j^{(j)}) = D(\boldsymbol{x}_j^{(j)})$, which means that $\boldsymbol{x}_j^{(j)}$ satisfies the optimal conditions given in Theorem 5.5. Then $E^{(j)}\boldsymbol{x}_j^{(j)} = \boldsymbol{x}_j^{(j)}$. Therefore we have $E^{(j)}\boldsymbol{x}^{(j)} = \boldsymbol{x}^{(j)}$, and

$$E\boldsymbol{x} = E^{(b-1)} \circ E^{(b-2)} \circ ... E^{(1)} \circ E^{(0)}\boldsymbol{x} = \boldsymbol{x}. \quad \text{Q.E.D.}$$

5.6 Comparison of Algorithm Performance

Suppose there are n nodes ($n = |N|$) in a tree network G under consideration. Add to G another node $n+1$ and n links each of which connects a node in G to node $n+1$. Thus we obtain a new network G'. In G', let the traffic from node i to node $n+1$, x_{in+1}, be denoted by β_i. Also let the mean communication delay $G_{in+1}(x_{in+1})$ ($i \in N$) be equal to $F_i(\beta_i)$. Let us regard in such way that, in G', jobs that arrive at each node in G are routed through a sequence of links to node $n+1$. We can regard each node in G as the origin, and node $n+1$ as the destination for such jobs in G'. We refer to such an origin-destination pair as an *O/D pair*. By a *path* connecting an O/D pair (x, y), we mean a sequence of links (p_1, p_2), (p_2, p_3), ..., (p_{n-1}, p_n) where p_1, p_2, ..., p_n are distinct nodes, $p_1 = x$, and $p_n = y$. By the mean job flow time of G', we mean the overall average of time length that a job takes from the origin to the destination. Note that in G', we do not consider $F_i(\beta_i)$ to be a node delay because it is regarded as the communication delay of link $(i, n+1)$. By the job flow rate of a path for an O/D pair, we mean the rate of jobs that go through the path. We can think of such an optimization problem that gives a solution, i.e., the set of the job flow rates of all paths for all O/D pairs, that minimizes the mean job flow time. It is clear that such an optimization problem in G' is identical with our optimal load balancing problem in G. The FD (Flow Deviation) algorithm [FGK73] and the Dafermos algorithm [Daf71] both give such an optimal solution. Both algorithms repeat the following step until a reasonably optimal solution is obtained within an acceptance tolerance. In each such step, from a feasible solution they obtain another feasible solution that gives a shorter mean job flow time than the former.

The FD algorithm takes the set of the job flow rate of each link as the variables in the optimization problem. The job flow rate of each path can be easily obtained from the set of the job flow rate of each link. This algorithm takes the shortest route flow as the steepest descent direction. Note that the shortest route flow is the solution to the shortest-route flow problem; the shortest-route flow problem finds all the shortest paths (where the length of a link is defined as the incremental communication delay on the link) and routes the jobs through the shortest path for each O/D pair. In each step, the FD algorithm evaluates from a feasible solution the steepest descent direction and the step size that minimizes the mean job flow time along such a direction to form a new feasible solution. (For details of the FD algorithm, refer to Ref.[FGK73].)

In each step, the Dafermos algorithm finds for each O/D pair the path with maximum incremental delay that has positive job flow rate and the path with minimum incremental delay and obtains another feasible solution in which job flow rates of the two paths are modified so that the incremental delays of both paths should be equal. The incremental

Table 5.1: Storage requirements of algorithms

Algorithm	Storage Requirement
FD	$O(n)$
Dafermos	$O(n^2)$
LBT	$O(n)$

delay of a path is equal to the sum of the incremental communication delays of all links in the path. A column generation technique [LNT73] is used to generate paths for each O/D pair. (For details of the Dafermos algorithm, refer to Ref.[Daf71] and Ref.[LNT73].)

We compare our proposed load balancing algorithm for tree network G (we will call it the LBT algorithm) with these two algorithms in network G'. We examine two performance measures: the storage requirements and the computation time requirements.

5.6.1   Comparison of Storage Requirements

For the LBT algorithm and the FD algorithm, it is necessary to calculate only the elements of the vectors β and x in each iteration. Supposing that the number of elements in β is n ($n = |N|$), we have that the number of elements in x is $2(n-1)$ according to the structure of a tree network. Hence, the amount of storage requirements of these two algorithms are $O(n)$. In the case of the Dafermos algorithm, not only the elements of β and x but also job flow rates in each path for each O/D pair should be kept. The average length of a path for an O/D pair is $n/2$. Therefore, the Dafermos algorithm requires the amount of storage of $O(n^2)$. Table '5.1 summarizes the storage requirements of these three algorithms.

5.6.2   Comparison of Computation Time Requirements

It seems difficult to compare the computation time requirements of these three algorithms in terms of complexity theory, since the structures of these algorithms differ greatly from one another. Here we compare the computation time requirements numerically. We program these three algorithms in FORTRAN and run them in a SPARC workstation. We consider a tree computer network which consists of eleven host computers (nodes) illustrated in Figure 5.2. We select the central server model shown in Figure 5.3 as the node model in the network. In node i model, exponential server 0 is a CPU that processes

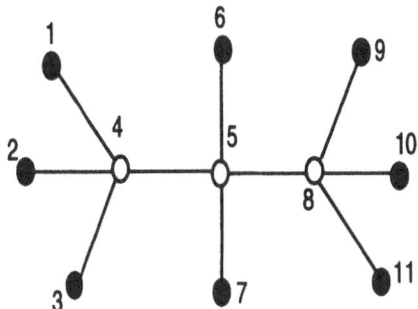

Figure 5.2: An example of the tree computer network

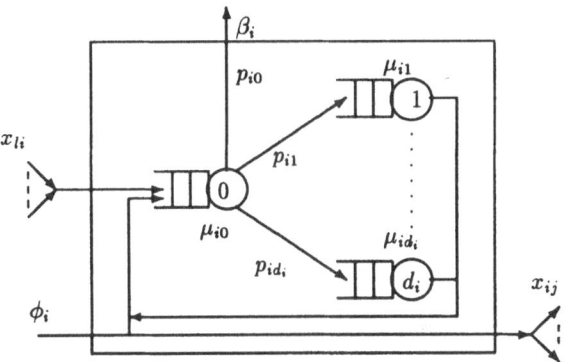

Figure 5.3: An example of node model

Table 5.2: Parameters of node models

Node	Processing rates of servers (jobs/min)					Probabilities of a job leaving CPU					Job arrival rate (jobs/min)
	μ_{i0}	μ_{i1}	μ_{i2}	μ_{i3}	μ_{i4}	p_{i0}	p_{i1}	p_{i2}	p_{i3}	p_{i4}	
1-2	1000	200	200	200	300	0.1	0.2	0.2	0.2	0.3	95
3	200	40	40	40	60	0.1	0.2	0.2	0.2	0.3	19
4-5	1000	200	200	200	300	0.1	0.2	0.2	0.2	0.2	70
6-7	200	40	40	40	60	0.1	0.2	0.2	0.2	0.3	19
8	1000	200	200	200	300	0.1	0.2	0.2	0.2	0.2	70
9-11	200	40	40	40	60	0.1	0.2	0.2	0.2	0.3	19

jobs according to the processor sharing discipline, exponential servers 1, 2, ..., d_i are I/O devices which process jobs according to FCFS. Let p_{i0} and p_{ij}, $j = 1, 2, ..., d_i$, be the probabilities that, after departing from the CPU, a job leaves node i or requests I/O service at device j, $j = 1, 2, ..., d_i$, respectively. The expected node delay of a job in such node model is given as

$$F_i(\beta_i) = \sum_{j=0}^{d_i} \frac{q_{ij}}{\mu_{ij} - q_{ij}\beta_i}, \quad \beta_i < \frac{\mu_{ij}}{q_{ij}}, j = 0, 1, ..., d_i, \quad (5.33)$$

where $q_{i0} = 1/p_{i0}$ and $q_{ij} = p_{ij}/p_{i0}$, and μ_{ij} is the processing rate of server j, $j = 1, 2, ..., d_i$ at node i. Substituting the above equation into eq.(5.4a) yields the incremental node delay function

$$f_i(\beta_i) = \sum_{j=0}^{d_i} \frac{q_{ij}\mu_{ij}}{(\mu_{ij} - q_{ij}\beta_i)^2}, \quad \beta_i < \frac{\mu_{ij}}{q_{ij}}, \ j = 0, 1, ..., d_i \quad (5.34)$$

In this case, the inverse function may not have a closed-form expression; generally, it can be easily obtained numerically by solving a nonlinear equation of a single variable.

Table 5.2 gives the parameters of node models in an example. In this particular example, we get the following mean node delay functions:

$$F_i(\beta_i) = \frac{5}{100 - \beta_i}, \quad \beta_i < 100, \text{ for } i = 1, 2, 4, 5, 8;$$

$$F_i(\beta_i) = \frac{5}{20 - \beta_i}, \quad \beta_i < 20, \text{ for } i = 3, 6, 7, 9, 10, 11.$$

For the link communication model, we have the following consideration. Let the mean communication time for sending a packet from node i to node j be $(1/c_{ij})$ (min.) with

Table 5.3: Computation time of algorithms

Algorithm	Computation time in sec (iteration number) Mean response time in minute		
	$\varepsilon = 10^{-2}$	$\varepsilon = 10^{-3}$	$\varepsilon = 10^{-5}$
FD	9.43 (101) 0.454385538	327.64 (3770) 0.453210699	43970.42 (512466) 0.452855243
Dafermos	26.10 (3) 0.455057846	32.63 (5) 0.453667984	204.64 (99) 0.453120358
LTB	1.81 (3) 0.452941328	2.29 (4) 0.452858451	3.28 (6) 0.452851089

the exponential distribution. Assume that on average there are a packets in a job and there are b packets in a response for a job. Note that there are $ax_{ij} + bx_{ji}$ packets per minute sent from node i to node j and $bx_{ij} + ax_{ji}$ packets per minute sent from node j to node i. We assume that the mean communication delay, $G_{ij}(x_{ij}, x_{ji})$, from node i to node j on link (i, j) $((i, j) \in L)$ has the following form:

$$G_{ij}(x_{ij}, x_{ji}) = \frac{a}{c_{ij} - (ax_{ij} + bx_{ji})} + \frac{b}{c_{ji} - (bx_{ij} + ax_{ji})}, \tag{5.35}$$
$$c_{ij} > ax_{ij} + bx_{ji} \text{ and } c_{ji} > bx_{ij} + ax_{ji}, \ (i, j) \in L.$$

We examine the case where $c_{ij} = c_{ji} = 10000$ packets per minute if both nodes i and j are nonterminal nodes (nodes 4, 5, 8), where $c_{ij} = c_{ji} = 1000$ packets per minute if one of nodes i and j is a terminal node, and where $a = 80$ and $b = 20$.

In the Dafermos algorithm and the LBT algorithm, the stopping rules are based on the acceptance tolerances ε for the relative errors of the mean response time $D(\beta, x)$. In order to obtain the value of the mean response time $D(\beta, x)$ close to that the other two algorithms obtain, the stopping rule for the FD algorithm is based on the acceptance tolerance ε for errors of the flow rates (see page 114 of [FGK73]). The same initial feasible solution is given for these three algorithms. Table 5.3 shows the computation times and number of iterations for obtaining the optimal solution by using these three algorithms in the example. Figure 5.4 shows the mean response time vs. computation time in the example. Note that in Figure 5.4, the start points of computation time for these three algorithms are the CPU time spent for the first iteration.

The numerical results in the above example exhibit that the LBT algorithm has the

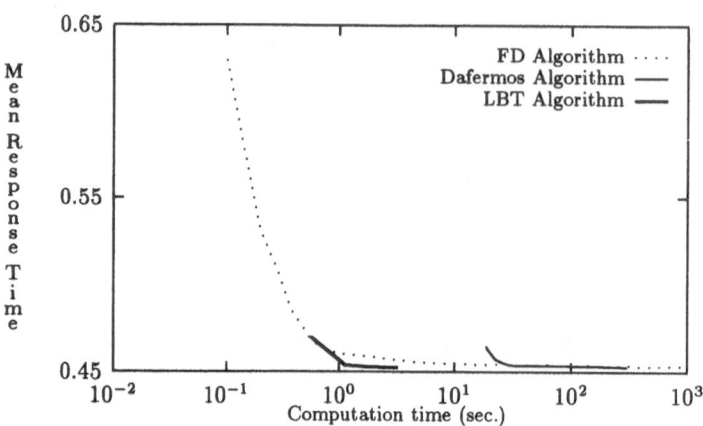

Figure 5.4: Mean response time vs. computation time

fastest convergence among these three algorithms. Let us consider the effect of the size of the acceptance tolerance ε on the computation time requirements. Table 5.3 shows that the computation time and the number of iterations that the FD algorithm requires increase rapidly as the acceptance tolerance ε decreases. We also observe that the Dafermos algorithm spends more than 10 times computation time that the LBT algorithm does to converge to the optimal solution. The above results are expected because the LBT algorithm is straightforward and makes use of the special structure of the problem whereas the others apply to more general problems. Notice that the main process of the sub-algorithm in the LBT algorithm is determining a Lagrange multiplier (in step 2) by applying a golden section search. The LBT algorithm also makes use of the important property in the optimal solution that the job flow rates x_{ij} and x_{ji} ($< i, j > \in E$) cannot be positive both in the same time. Some other numerical experiments wherein we examined different system models were also conducted. They showed results similar to the above example.

5.7   Conclusion

We have studied the problem of statically balancing the load on a set of heterogeneous host computers interconnected by a tree network with two-way traffic.

The optimal load balancing problem was formulated as a nonlinear optimization problem. We have studied the necessary and sufficient conditions for the optimal solution. It was demonstrated that the optimization problem of a particular structure can be solved by a decomposition technique and an algorithm for obtaining the optimal solution is developed.

The proposed algorithm was compared with existing well-known algorithms: the FD algorithm and the Dafermos algorithm with respect to the storage and computation time requirements.

It was proven that for an n-node system, the proposed algorithm and the FD algorithm require only the amount of storage of $O(n)$, whereas the Dafermos algorithm requires the amount of storage of $O(n^2)$. It was shown that the proposed algorithm converges to the optimal solution most quickly, whereas the FD algorithm converges to the optimal solution most slowly. The above results leaded to the conclusion that the proposed algorithm is the most recommendable for optimal load balancing problem in tree network configurations with two-way traffic.

Chapter 6

Uniqueness of Performance Variables in Static Load Balancing

6.1   Introduction

Important performance optimization problems in the communication networks and distributed computer systems may include the problem of how to obtain the optimal routing (or flow assignment) policy for sending data through the communication channels and the problem of how to balance the processing of jobs among processors (i.e. static load balancing). For example, an optimal static routing policy determines the rate at which data flow through each path connecting a given pair of origin and destination points, so as to minimize the overall mean data transmission time. We can regard the combination of such rates as the *solution* of the optimization problem. These optimization problems have close relations to the optimal traffic assignment problems in the transportation science [Daf72, Mag84].

Many authors have discussed the solutions of the static optimization problems for the communication network models or the distributed computer system models which are regarded as open BCMP queueing networks with state-independent arrival and service rates [CG74, FGK73, KH88, KK92a, Kle76, TT84, TT85]. The BCMP queueing network is the queueing network discussed by Baskett *et al.* [BCMP75], to which we can apply the BCMP theorem. In this paper we consider optimal static routing problems in the open BCMP queueing networks. Kelly [Kel79] has indicated that the BCMP theorem can be used to analyze static routing problems.

An optimization problem does not necessarily have a unique solution. For example, it is possible that the combinations of different values of data flow rates for all paths may result in the same minimum overall mean transmission time. If they are not unique, it is necessary to make clear the range and characteristics of the solutions, in particular, when we calculate numerically the optimal solutions and when we intend to analyze the effects of the system parameters on the optimal solutions.

We consider a routing policy whereby the overall mean response time of a job (the total time that a job spends from its arrival into the system until its departure from the system) is minimized. We call that policy the *overall optimal policy*. We show the conditions that the overall optimal solution, i.e., the routing decision of the overall optimal policy, satisfies. We see that the optimal solution may not be unique, and we also obtain the linear relations that characterize the set of the optimal solutions. These optimal solutions must have the same unique utilization factor of each node that is not an infinite-server node. Thus the policy has a unique routing decision if and only if the number of variables whose values express the decision of the policy is not greater than the number of linearly independent equations among the linear relations. These are the main results of this paper.

In Section 6.2, we give some definitions used in this paper and formulate a mathematical model of the queueing network. In Section 6.3 we obtain the overall optimal solution, and discuss the uniqueness of the overall optimal solution. In Section 6.4, we show similar results on the uniqueness of the individually optimal solution. Section 6.5 presents numerical examples. Finally, Section 6.6 concludes this chapter.

6.2   Description of the Model

We consider an open BCMP queueing network model that consists of M service centers, which are either communication channels or processors. Each service center contains either a single-server with the first-come-first-served (FCFS), last-come-first-served-preemptive-resume (LCFS-PR) , or processor-sharing (PS) scheduling policy , or an infinite-server (IS). We assume that the service rate of each single-server is state-independent. We assume that in the network there also exist origin and destination points. An origin or destination point may be regarded as a service center with zero service time. We call the pair of one origin and one destination points an *O-D pair*. The unit entity (like a message packet) that is routed through the network is called a *job* in this paper. Each job arrives at one of the origin points and departs from one of the destination points. The origin and destination points of a job have been determined when the job arrives in the network. Kelly [Kel79] has indicated that the BCMP theorem can be used in the analysis of routing problems even though the route that each job follows is fixed while it is in the network.

Jobs are classified into R different classes. For the sake of simplicity, we assume that jobs do not change their class while passing through the network. A class k job with the O-D pair (o_1, d_1) originates at service center o_1 and destinates for service center d_1 through a series of service centers (we refer to the series as a *path*) and then leaves the system. We classify such paths into a finite number of *path-classes* of job class k O-D pair (o_1, d_1). The arrival process of jobs of each class for each O-D pair forms a Poisson process and is independent of the state of the system. We assume that we can choose the job flow rate of each path class in order to achieve a performance objective, whereas we assume that once the job flow rate of a path class is given, the job flow rate of each path in the path class is also given. That is, the relative flow rate of each path in the same path-class is governed by fixed transfer probabilities between service centers. The solution of a routing problem is characterized by the chosen values of job flow rates of all path classes.

We will use the following notation regarding the network:

D^k Set of O-D pairs for class k jobs.

Π_d^k Set of path classes that class k jobs of O-D pair d flow through.

Π^k Set of all path classes for class k jobs, i.e., $\Pi^k = \bigcup_{d \in D^k} \Pi_d^k$.

$$\gamma_{pd}^{kk'} = \begin{cases} 1 & \text{if } p \in \Pi_d^{k'} \text{ and } k = k', \\ 0 & \text{otherwise.} \end{cases}$$

We will use the following notation regarding arrivals to the network and flow rates:

ϕ_d^k Rate at which class k jobs join O-D pair $d \in D^k$.

Φ System-wide total job arrival rate, i.e., $\Phi = \sum_{k=1}^{R} \sum_{d \in D^k} \phi_d^k$.

x_p^k Rate at which class k jobs flow through path-class p.

δ_{lp} Rate at which jobs that flow through path-class p pass through service center l assuming that $x_p^k = 1$.

λ_l^k Rate at which class k jobs visit service center l, $\lambda_l^k = \sum_{p \in \Pi^k} \delta_{lp} x_p^k$.

We will use the following notation rearding the service and performance values in the BCMP network:

μ_l^k Service rate of class k jobs at service center l. When the service center consists of an FCFS server, μ_l^k must be reduced to μ_l.

I Set of IS service centers.

N Set of service centers except IS centers. ($|N| = M - |I|$)

ρ_l^k $= \lambda_l^k/\mu_l^k$. Utilization factor of service center $l \in N$ for class k jobs.

ρ_l $= \displaystyle\sum_{k=1}^{R} \rho_l^k$. Total utilization factor of service center $l \in N$. (The utilization factor of service center $l \in I$ is always considered zero and is not equal to ρ_l since it has infinite capacity.)

$\hat{T}_l^k$ Mean response time of class k jobs at service center l, i.e., the mean length of the time period that starts when a class k job comes to the service center l and ends when it goes out of the service center.

Δ Overall mean response time of a job, i.e., the mean length of the time period that starts when a job arrives in the system and ends when it leaves the system.

t_p^k $= \partial(\Phi T)/\partial x_p^k$, i.e., class k marginal delay of path-class p, $p \in \Pi^k, k = 1, 2, \ldots, R$.

T_p^k Average class k delay of path-class p, $p \in \Pi^k$, $k = 1, 2, \ldots, R$.

We will use the following notation regarding vectors and matrices:

$\boldsymbol{\rho}$ $[\rho_1, \rho_2, \ldots, \rho_M]^{\mathrm{T}}$ where T means 'transpose'.

$\boldsymbol{\rho}_U$ $\boldsymbol{\rho}|_{\rho_l = 0, l \in I}$. This is the same as $\boldsymbol{\rho}$ except that $\rho_l = 0$ for all $l \in I$. We call this the utilization vector.

$\boldsymbol{\phi}$ $[\phi_1^1, \phi_2^1, \ldots, \phi_1^2, \phi_2^2, \ldots]^{\mathrm{T}}$, i.e., the arrival rate vector.

$\boldsymbol{x}$ $[x_1^1, x_2^1, \ldots, x_1^2, x_2^2, \ldots]^{\mathrm{T}}$, i.e., the path class flow rate vector.

$\boldsymbol{\alpha}$ $[\alpha_1^1, \alpha_2^1, \ldots, \alpha_1^2, \alpha_2^2, \ldots]^{\mathrm{T}}$, i.e., the vector whose elements are α_d^k, $d \in D^k$, $k = 1, 2, \ldots, R$.

$\boldsymbol{A}$ $[A_1^1, A_2^1, \ldots, A_1^2, A_2^2, \ldots]^{\mathrm{T}}$, i.e., the vector whose elements are A_d^k, $d \in D^k$, $k = 1, 2, \ldots, R$.

$\boldsymbol{\Gamma}$ $\begin{bmatrix} \gamma_{11}^{11} & \gamma_{12}^{11} & \cdots & \gamma_{11}^{12} & \gamma_{12}^{12} & \cdots \\ \gamma_{21}^{11} & \gamma_{22}^{11} & \cdots & \gamma_{21}^{12} & \gamma_{22}^{12} & \cdots \\ \vdots & & \ddots & \vdots & & \ddots \\ \gamma_{11}^{21} & \gamma_{12}^{21} & \cdots & \gamma_{11}^{22} & \gamma_{12}^{22} & \cdots \\ \gamma_{21}^{21} & \gamma_{22}^{21} & \cdots & \gamma_{21}^{22} & \gamma_{22}^{22} & \cdots \\ \vdots & & \ddots & \vdots & & \ddots \end{bmatrix}$ i.e., the incident matrix whose (i, j) element is $\gamma_{pd}^{kk'}, p \in \Pi^k$, $k = 1, 2, \ldots, R$, $d \in D^{k'}$, $k' = 1, 2, \ldots, R$, where $i = p + \displaystyle\sum_{\kappa=1}^{k-1} |\Pi^{\kappa}|$ and $j = d + \displaystyle\sum_{\kappa=1}^{k'-1} |D^{\kappa}|$.

$\boldsymbol{x} \cdot \boldsymbol{y}$ $\displaystyle\sum_i x_i y_i$, i.e., the inner product of vectors $\boldsymbol{x} = [x_1, x_2, \ldots]^{\mathrm{T}}$ and $\boldsymbol{y} = [y_1, y_2, \ldots]^{\mathrm{T}}$.

t $[t_1^1, t_2^1, \ldots, t_1^2, t_2^2, \ldots]^T$, i.e., the vector whose elements are $t_p^k, p \in \Pi^k, k = 1, 2, \ldots, R$.

$\mathbf{T}$ $[T_1^1, T_2^1, \ldots, T_1^2, T_2^2, \ldots]^T$, i.e., the vector whose elements are $T_p^k, p \in \Pi^k, k = 1, 2, \ldots, R$.

Let us denote the state of the network by $\mathbf{n} = (\mathbf{n}_1, \mathbf{n}_2, \ldots, \mathbf{n}_M)$ where $\mathbf{n}_l = (n_l^1, n_l^2, \ldots, n_l^R)$ and $n_l = \sum_{k=1}^R n_l^k$ where n_l^k denotes the total number of class k jobs at center l. For an open queueing network [10, 11], the equilibrium probability of the network state $\mathbf{n}$ is obtained as follows:

$$p(\mathbf{n}) = \prod_{l=1}^M \frac{p_l(\mathbf{n}_l)}{G_l}, \tag{6.1}$$

where

$$p_l(\mathbf{n}_l) = \begin{cases} n_l! \prod_{k=1}^R (\rho_l^k)^{n_l^k} / n_l^k! & \text{for } l \in N, \\ \prod_{k=1}^R (\rho_l^k)^{n_l^k} / n_l^k! & \text{for } l \in I, \end{cases}$$

$$G_l = \begin{cases} 1/(1 - \rho_l) & \text{for } l \in N, \\ e^{-\rho_l} & \text{for } l \in I. \end{cases}$$

Let $E[n_l^k]$ be the average number of class k jobs at center l. We have

$$E[n_l^k] = \begin{cases} \dfrac{\rho_l^k}{1 - \rho_l} & \text{for } l \in N, \\ \rho_l^k & \text{for } l \in I. \end{cases} \tag{6.2}$$

By using Little's formula, we have

$$\hat{T}_l^k = \frac{E[n_l^k]}{\lambda_l^k} = \begin{cases} \dfrac{1/\mu_l^k}{1 - \rho_l} & \text{for } l \in N, \\ 1/\mu_l^k & \text{for } l \in I, \end{cases} \tag{6.3}$$

from which the average delay of a class k job that passes through path-class $p \in \Pi^k$ is given by

$$T_p^k = \sum_{l=1}^M \delta_{lp} \hat{T}_l^k = \sum_{l \in N} \delta_{lp} \frac{1/\mu_l^k}{1 - \rho_l} + \sum_{l \in I} \delta_{lp} / \mu_l^k. \tag{6.4}$$

Therefore, the overall mean response time of a job, Δ, can be written as

$$\Delta = \sum_{k=1}^R \sum_{p \in \Pi^k} \frac{x_p^k}{\Phi} T_p^k = \frac{1}{\Phi} \Big[\sum_{l \in N} \frac{\rho_l}{1 - \rho_l} + \sum_{l \in I} \rho_l \Big], \tag{6.5}$$

by noting that $\rho_l = \sum_{k=1}^R \lambda_l^k / \mu_l^k$ and $\lambda_l^k = \sum_{p \in \Pi^k} \delta_{lp} x_p^k$. Note that the following conditions should be satisfied.

$$\sum_{p \in \Pi_d^k} x_p^k = \phi_d^k, \quad d \in D^k, \quad k = 1, 2, \ldots, R, \tag{6.6}$$

$$x_p^k \geq 0, \quad p \in \Pi^k, k = 1, 2, \ldots, R. \tag{6.7}$$

We can express (6.6) as

$$\sum_{k'=1}^{R} \sum_{p \in \Pi^{k'}} \gamma_{pd}^{k'k} x_p^{k'} = \phi_d^k, \ d \in D^k, \ k = 1, 2, \ldots, R,$$

or, equivalently,

$$\Gamma^{\mathrm{T}} x = \phi.$$

Remark We easily see that our model includes those discussed for the static routing problems of communications networks [CG74, FGK73, Kle76]. We also see that our model includes those of the load balancing problems of distributed computer systems such as given by Tantawi and Towsley [TT84, TT85]. From (6.3) it is easy to see that T_l^k is a convex function of $\lambda_l^k, l \in N, \ k = 1, 2, \ldots, R$. It follows that T is also convex with respect to x. It is remarkable that T depends only on ρ.

6.3   Uniqueness of the Overall Optimal Solution

By the overall optimal policy we mean the policy whereby routing is determined so as to minimize the overall mean job response time. The problem of minimizing the overall mean response time is stated as follows:

$$\text{minimize:} \quad \Delta \ = \ \frac{1}{\Phi}\Big[\sum_{l \in N} \frac{\rho_l}{1 - \rho_l} + \sum_{l \in I} \rho_l\Big] \tag{6.8}$$

with respect to x subject to

$$\Gamma^{\mathrm{T}} x \ = \ \phi, \tag{6.9}$$

$$x \ \geq \ 0, \tag{6.10}$$

where $\rho_l = \sum_{k=1}^{R} \lambda_l^k / \mu_l^k$ and $\lambda_l^k = \sum_{p \in \Pi^k} \delta_{lp} x_p^k$. Note that (6.9) and (6.10) are the same as (6.6) and (6.7), respectively. We call the above problem the *overall optimization problem*, and its solution the *overall optimal solution*.

The class k marginal delay of path class $p \in \Pi^k$ is

$$l_p^k(x) - \frac{\partial}{\partial x_p^k}(\Phi \Delta) = \sum_{l \in N} \delta_{lp} \frac{1}{\mu_l^k (1 - \rho_l)^2} + \sum_{l \in I} \delta_{lp} \frac{1}{\mu_l^k}. \tag{6.11}$$

Lemma 6.1 x *is an optimal solution of the problem* (6.8) *if and only if* x *satisfies the following conditions*

$$[t(x) - \Gamma \alpha] \cdot x \ = \ 0, \tag{6.12}$$

$$t(x) - \Gamma \alpha \ \geq \ 0, \tag{6.13}$$

$$\Gamma^{\mathrm{T}} x - \phi \ = \ 0, \tag{6.14}$$

$$x \ \geq \ 0. \tag{6.15}$$

PROOF. Since the objective function (6.8) is convex and the feasible region of its constraints is a convex set, any local solution of the problem is a global solution point. To obtain the optimal solution, we construct the Lagrangian function

$$L(\boldsymbol{x}, \boldsymbol{\alpha}) \;=\; \Phi\Delta + \boldsymbol{\alpha} \cdot (\boldsymbol{\phi} - \boldsymbol{\Gamma}^{\mathrm{T}}\boldsymbol{x}) \tag{6.16}$$

for (6.8). By the Kuhn-Tucker theorem (see, e.g., [Int71]), $\boldsymbol{x}$ is an optimal solution if and only if

$$\frac{\partial L}{\partial \boldsymbol{x}} \;=\; t(\boldsymbol{x}) - \boldsymbol{\Gamma}\boldsymbol{\alpha} \geq 0, \tag{6.17}$$

$$\frac{\partial L}{\partial \boldsymbol{x}} \cdot \boldsymbol{x} \;=\; [t(\boldsymbol{x}) - \boldsymbol{\Gamma}\boldsymbol{\alpha}] \cdot \boldsymbol{x} = 0, \tag{6.18}$$

$$\frac{\partial L}{\partial \boldsymbol{\alpha}} \;=\; \boldsymbol{\phi} - \boldsymbol{\Gamma}^{\mathrm{T}}\boldsymbol{x} = 0, \tag{6.19}$$

$$\boldsymbol{x} \;\geq\; 0, \tag{6.20}$$

where $(\partial L)/(\partial \boldsymbol{x})$ denotes the vector whose elements are $(\partial L)/(\partial x_p^k)$, $p \in \Pi^k$, $k = 1, 2, \ldots, R$. The relations (6.17) - (6.20) are exactly the same as the relations (6.12) - (6.15). $\square$

Lemma 6.2 $\bar{\boldsymbol{x}}$ *is an optimal solution of the problem* (6.8) *if and only if*

$$t(\bar{\boldsymbol{x}}) \cdot (\boldsymbol{x} - \bar{\boldsymbol{x}}) \;\geq\; 0, \quad \text{for all } \boldsymbol{x} \tag{6.21}$$

$$\text{such that } \boldsymbol{\Gamma}^{\mathrm{T}}\boldsymbol{x} \;=\; \boldsymbol{\phi} \text{ and } \boldsymbol{x} \geq 0.$$

PROOF. The problem of finding $\boldsymbol{x}$ such as above can be regarded as the problem of a linear program

$$\begin{aligned}
\text{minimize:} \quad & t(\bar{\boldsymbol{x}}) \cdot \boldsymbol{x} \\
\text{subject to} \quad & \boldsymbol{\Gamma}^{\mathrm{T}}\boldsymbol{x} = \boldsymbol{\phi}, \; \boldsymbol{x} \geq 0
\end{aligned}$$

with $\bar{\boldsymbol{x}}$ fixed. $\boldsymbol{x}$ is an optimal solution of the linear program if and only if $\boldsymbol{x}$ satisfies the Kuhn-Tucker conditions for the Lagrangian

$$L(\boldsymbol{x}, \boldsymbol{\alpha}) \;=\; t(\bar{\boldsymbol{x}}) \cdot \boldsymbol{x} + \boldsymbol{\alpha} \cdot (\boldsymbol{\phi} - \boldsymbol{\Gamma}^{\mathrm{T}}\boldsymbol{x}). \tag{6.22}$$

The Kuhn-Tucker conditions are

$$\frac{\partial L}{\partial \boldsymbol{x}} \;=\; t(\bar{\boldsymbol{x}}) - \boldsymbol{\Gamma}\boldsymbol{\alpha} \geq 0, \tag{6.23}$$

$$\frac{\partial L}{\partial \boldsymbol{x}} \cdot \boldsymbol{x} \;=\; [t(\bar{\boldsymbol{x}}) - \boldsymbol{\Gamma}\boldsymbol{\alpha}] \cdot \boldsymbol{x} = 0, \tag{6.24}$$

$$\frac{\partial L}{\partial \boldsymbol{\alpha}} \;=\; \boldsymbol{\phi} - \boldsymbol{\Gamma}^{\mathrm{T}}\boldsymbol{x} = 0, \tag{6.25}$$

$$\boldsymbol{x} \;\geq\; 0. \tag{6.26}$$

That is, the relation (6.21) (i.e., the statement that $\bar{x}$ is a solution of the above linear program) is equivalent to the set of relations in Lemma 6.1. $\Box$

From eq. (6.8) we see that Δ depends only on the utilization factor of each service center, ρ_l, which results from the path flow rate matrix. It is possible, therefore, that different values of the path flow rate matrix result in the same utilization factor of each service center and the same minimum mean response time. We are uncertain, however, about whether distinct optimal solutions should have the same utilization factor of each service center or not. We first define the concept of monotonicity of vector-valued functions with vector-valued arguments.

Definition   Let $\mathbf{F}(\bullet)$ be a vector-valued function that is defined on a domain $S \subseteq R^n$ and that has values $\mathbf{F}(x)$ in R^n. This function is *monotone* in S if for every pair $x, y \in S$

$$(x - y) \cdot [\mathbf{F}(x) - \mathbf{F}(y)] \geq 0.$$

It is *strictly monotone* if, for every pair $x \in S$ and $y \in S$ with $x \neq y$,

$$(x - y) \cdot [\mathbf{F}(x) - \mathbf{F}(y)] > 0.$$

For the function $t(x)$ we have the following property.

Lemma 6.3 $t(x)$ *is monotone but is not strictly monotone, i.e., for arbitrary* x *and* x' $(x \neq x')$,

$$(x - x') \cdot [t(x) - t(x')] > 0 \quad \text{if} \quad \rho_U \neq \rho'_U, \tag{6.27}$$
$$= 0 \quad \text{if} \quad \rho_U = \rho'_U, \tag{6.28}$$

where ρ_U *and* ρ'_U *are the utilization vectors that* x *and* x' *result in, respectively.*

PROOF. From (6.11) we have

$$
\begin{aligned}
(x - x') \cdot [t(x) - t(x')] &= \sum_{k=1}^{R} \sum_{p \in \Pi^k} (x_p^k - x_p'^k)[t_p^k(x) - t_p^k(x')] \\
&= \sum_{k=1}^{R} \sum_{p \in \Pi^k} \sum_{l \in N} (x_p^k - x_p'^k)\frac{\delta_{lp}}{\mu_l^k}\left[\frac{1}{(1 - \rho_l)^2} - \frac{1}{(1 - \rho_l')^2}\right] \\
&= \sum_{l \in N} \frac{(\rho_l - \rho_l')^2[(1 - \rho_l) + (1 - \rho_l')]}{(1 - \rho_l)^2(1 - \rho_l')^2}. \tag{6.29}
\end{aligned}
$$

In the statistical equilibrium of queueing networks, we have ρ_l, $\rho_l' < 1$, $l \in N$. Therefore we have the relations (6.27) and (6.28). $\Box$

Theorem 6.4 *The overall optimal solution may not be unique. However, the utilization factor of each service center is uniquely determined and is the same for all overall optimal solutions.*

PROOF. The former half of this theorem is clear by noting that T depends only on ρ (see (6.8)). In section 6.5, we present an example of the cases where more than one optimal solution exists.

The latter half is proved as follows. Suppose that the overall optimal policy has two distinct solutions $\hat{x}$ and $\tilde{x}$, which result in the utilization vectors $\hat{\rho}_U$ and $\tilde{\rho}_U$, respectively, and $\hat{\rho}_U \neq \tilde{\rho}_U$. Then we have from Lemma 6.2,

$$t(\hat{x}) \cdot (\tilde{x} - \hat{x}) \geq 0,$$
$$t(\tilde{x}) \cdot (\hat{x} - \tilde{x}) \geq 0.$$

Then we have

$$(\hat{x} - \tilde{x}) \cdot \left[t(\hat{x}) - t(\tilde{x}) \right] \leq 0.$$

From Lemma 6.3 we have

$$(\hat{x} - \tilde{x}) \cdot \left[t(\hat{x}) - t(\tilde{x}) \right] > 0,$$

since $\hat{\rho}_U \neq \tilde{\rho}_U$. This leads to a contradiction. That is, if there exist two distinct optimal solutions, the utilization vectors of both the solutions must be the same. Note that the utilization factor of service center $l \in I$ is considered always zero. Naturally, in that case, $\sum_{l \in I} \rho_l$ must be unique but each of ρ_l, $l \in I$, need not be unique. $\square$

Now let us consider the range of the optimal solutions. From the above, we obtain the following relations that characterize the range of the optimal solutions.

$$\sum_{k=1}^{R} \sum_{p \in \Pi^k} \delta_{lp} \frac{x_p^k}{\mu_l^k} = \rho_l, \quad l \in N, \tag{6.30}$$

$$\sum_{l \in I} \sum_{k=1}^{R} \sum_{p \in \Pi^k} \delta_{lp} \frac{x_p^k}{\mu_l^k} = \sum_{l \in I} \rho_l, \tag{6.31}$$

$$\sum_{p \in \Pi_d^k} x_p^k = \phi_d^k, \quad d \in D^k, \ k = 1, 2, \ldots, R, \tag{6.32}$$

$$x_p^k \geq 0, \quad p \in \Pi^k, k = 1, 2, \ldots, R, \tag{6.33}$$

where the value of each ρ_l is what an optimal solution x results in. From the relations (6.30) - (6.33) we see that optimal path flow rates belong to a convex polyhedron. Then we have the following proposition about the uniqueness of the optimal solutions.

Corollary 6.5 *The overall optimal solution is unique if and only if the total number of elements in x does not exceed the number of linearly independent equations in the set of linear equations (6.30) - - (6.32).*

6.4 Uniqueness of the Individually Optimal Solution

By the individually optimal policy we mean that jobs are scheduled so that each job may feel that its own mean response time is minimum if it knows the mean response time of each path of O-D pair d, $T_p^k(x)$, $p \in \Pi_d^k$. By the *individual optimization problem* we mean the problem of obtaining the routing decision that achieves the objective of the individually optimal policy. We call the solution of the individual optimization problem the *individually optimal solution* or the *equilibrium*. In the equilibrium, no user has any incentive to make a unilateral decision to change his route. Wardrop [War52] considered the policy for a transportation network. We assume that there is a routing decision and that x is the path flow rate matrix which results from the routing decision. The individually optimal policy requires that a class k job of O-D pair d should follow through such a path class p that satisfies

$$T_p^k(x) = \min_{p \in \Pi_d^k} T_p^k(x) \quad \text{for all } d \in D^k, \ k = 1, 2, \dots, R. \tag{6.34}$$

If a routing decision satisfies the above condition we say the routing decision realizes the individually optimal policy.

Definition The path flow rate vector x is said to satisfy the equilibrium conditions for an open BCMP queueing network if the following relations are satisfied for all $d \in D^k$, $k = 1, 2, \dots, R$,

$$T_p^k(x) \ \geq \ A_d^k, \quad x_p^k = 0, \tag{6.35}$$

$$T_p^k(x) \ = \ A_d^k, \quad x_p^k \geq 0, \tag{6.36}$$

$$\sum_{p \in \Pi_d^k} x_p^k \ = \ \phi_d^k, \tag{6.37}$$

$$x_p^k \ \geq \ 0, \quad p \in \Pi_d^k, \tag{6.38}$$

where

$$A_d^k \ = \ \min_{p \in \Pi_d^k} T_p^k(x), \quad d \in D^k, \ k = 1, 2, \dots, R. \tag{6.39}$$

Note that the set of the relations (6.35) - (6.37) is identical with the following set of relations.

$$[T(x) - \Gamma A] \cdot x \ = \ 0, \tag{6.40}$$

$$T(x) - \Gamma A \ \geq \ 0, \tag{6.41}$$

$$\Gamma^T x - \phi \ = \ 0, \tag{6.42}$$

$$x \ \geq \ 0. \tag{6.43}$$

Theorem 6.6 *The individually optimal policy has an equilibrium. That is, there exists an individually optimal solution x which satisfies the relations* (6.40) - (6.43).

PROOF. Define $\tilde{T}(x)$ by

$$\tilde{T}(x) = \frac{1}{\Phi}\Big[\sum_{l \in N} \log_e\Big(\frac{1}{1-\rho_l}\Big) + \sum_{l \in I} \rho_l\Big],$$

where $\rho_l = \sum_{k=1}^{R} \lambda_l^k/\mu_l^k$ and $\lambda_l^k = \sum_{p \in \Pi^k} \delta_{lp} x_p^k$. Note that $\tilde{T}(x)$ is a convex function of x. Then we have by noting (6.4)

$$T_p^k(x) = \frac{\partial}{\partial x_p^k}(\Phi \tilde{T}(x)).$$

Let us consider the following convex nonlinear program:

$$
\begin{array}{ll}
\text{minimize} & \tilde{T}(x) \\
\text{with respect to} & x \\
\text{subject to (6.42) and (6.43).} &
\end{array}
$$

The Kuhn-Tucker conditions are the same as (6.40) - (6.43). Therefore, the program should have an optimal solution which must satisfy relations (6.40) - (6.43). □

We can express the individually optimal solution in the variational inequality form by using the same way as that for the overall optimal solution as follows.

Corollary 6.7 $\bar{x}$ *is an individually optimal solution if and only if*

$$\mathbf{T}(\bar{x}) \cdot (x - \bar{x}) \geq 0, \quad \text{for all } x$$
$$\text{such that } \Gamma^{\mathrm{T}} x = \phi \text{ and } x \geq 0.$$

PROOF. Similar to the proof of Lemma 6.2. □

Lemma 6.8 *Function $\mathbf{T}(x)$ is monotone but is not strictly monotone. That is, for arbitrary x and x' ($x \neq x'$)*

$$(x - x') \cdot [\mathbf{T}(x) - \mathbf{T}(x')] > 0 \quad \text{if} \quad \rho_U \neq \rho_U', \tag{6.44}$$
$$= 0 \quad \text{if} \quad \rho_U = \rho_U' \tag{6.45}$$

where ρ_U and ρ_U' are the utilization vectors that x and x' result in, respectively.

PROOF. This Lemma can be proved by the same way as that for the Lemma 6.3. From (6.4) we have

$$
\begin{aligned}
(x - x') \cdot [\mathbf{T}(x) - \mathbf{T}(x')] &= \sum_{k=1}^{R} \sum_{p \in \Pi^k} (x_p^k - x_p'^k)[T_p^k(x) - T_p^k(x')] \\
&= \sum_{l \in N} \frac{(\rho_l - \rho_l')^2}{(1 - \rho_l)(1 - \rho_l')}.
\end{aligned}
$$

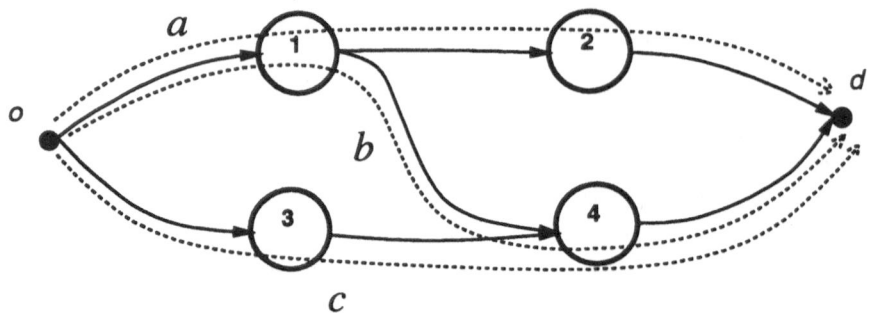

Figure 6.1: A BCMP network model. Each circle denotes a service center.

In the statistical equilibrium, we have ρ_l, $\rho'_l < 1$, $l \in N$. Therefore we have the relations (6.44) and (6.45). $\square$

Theorem 6.9 *In the equilibrium the utilization factor of each service center is unique, but the individually optimal solution may not be unique.*

PROOF. We can prove this theorem in the same way as that for Theorem 6.4. $\square$

The range of the individually optimal solutions is given by the same set of relations as (6.30) - (6.33) but with possibly different values of ρ_l, $l = 1, 2, \ldots, M$.

6.5   Numerical Examples

We consider a network model consisting of four service centers as shown in Figure 6.1. The model has one O-D pair (o, d) and three path classes (a, b, c). Each path class contains only one path. Each service center is modeled as either a PS server or an LCFS-PR server. Then, if the model has R job classes, the number of variables (path flow rates) whose values are determined by a static routing policy amounts to $3R$ and the number of constraints (6.32) is R. Thus, if $R \geq 3$, the overall optimization problem may not have a unique solution, since the number of unique utilization factors of the model may be at most 4. We show two examples under the overall and individually optimal policies where there are three job classes in the following.

$$\begin{array}{llll}
\mu_1^1 = 40.0, & \mu_2^1 = 28.0, & \mu_3^1 = 40.0, & \mu_4^1 = 30.0, \\
\mu_1^2 = 50.0, & \mu_2^2 = 20.0, & \mu_3^2 = 48.0, & \mu_4^2 = 20.0, \\
\mu_1^2 = 60.0, & \mu_2^2 = 10.0, & \mu_3^2 = 60.0, & \mu_4^2 = 30.0, \\
\phi_{(o,d)}^1 = 10.0, & \phi_{(o,d)}^2 = 15.0, & \phi_{(o,d)}^3 = 20.0,
\end{array}$$

Case 1. *The overall optimal policy.* In this case, the overall optimal solution is not unique. The range of optimal solutions is given in terms of x_a^k, x_b^k, x_c^k as follows:

$$3.5603 \cdots \le x_a^1 \le 5.6666 \cdots,$$
$$0.0000 \cdots \le x_b^1 \le 2.1063 \cdots,$$

$$\begin{aligned}
x_c^1 &= 10.0000 \cdots - 1.0000 \cdots x_a^1 - 1.0000 \cdots x_b^1, \\
x_a^2 &= 17.2888 \cdots - 0.6428 \cdots x_a^1, \\
x_b^2 &= 0.6428 \cdots x_a^1 - 2.2888 \cdots, \\
x_c^2 &= 0.0000 \cdots, \\
x_a^3 &= 0.0000 \cdots, \\
x_b^3 &= 8.4999 \cdots - 1.5000 \cdots x_a^1 - 1.5000 \cdots x_b^1, \\
x_c^3 &= 11.5000 \cdots + 1.5000 \cdots x_a^1 + 1.5000 \cdots x_b^1.
\end{aligned}$$

We show an example of a pair of distinct points in the range. The underlined numerical figures show those which are uniquely determined.

One is as follows.

$$\begin{array}{lll}
x_a^1 = 3.5603 \cdots, & x_b^1 = 2.1063 \cdots, & x_c^1 = 4.3333 \cdots, \\
x_a^2 = 15.0000 \cdots, & x_b^2 = 0.0000 \cdots, & x_c^2 = \underline{0.0000} \cdots, \\
x_a^3 = \underline{0.0000} \cdots, & x_b^3 = 0.0000 \cdots, & x_c^3 = 20.0000 \cdots.
\end{array}$$

This yields

$$\rho_1 = \underline{0.4416} \cdots, \; \rho_2 = \underline{0.8771} \cdots, \; \rho_3 = \underline{0.4416} \cdots, \; \rho_4 = \underline{0.8813} \cdots,$$
$$\Delta = \underline{0.35885756} \cdots.$$

The other is as follows.

$$\begin{array}{lll}
x_a^1 = 3.5603 \cdots, & x_b^1 = 1.4396 \cdots, & x_c^1 = 5.0000 \cdots, \\
x_a^2 = 15.0000 \cdots, & x_b^2 = 0.0000 \cdots, & x_c^2 = \underline{0.0000} \cdots, \\
x_a^3 = \underline{0.0000} \cdots, & x_b^3 = 1.0000 \cdots, & x_c^3 = 18.9999 \cdots.
\end{array}$$

This yields

$$\rho_1 = \underline{0.4416} \cdots, \; \rho_2 = \underline{0.8771} \cdots, \; \rho_3 = \underline{0.4416} \cdots, \; \rho_4 = \underline{0.8813} \cdots,$$
$$\Delta = \underline{0.35885756} \cdots.$$

Case 2. *The individually optimal policy.* In this case, the optimal solution of the individual optimization problem is not unique either. The range of optimal solutions is given in terms of x_a^k, x_b^k, x_c^k as follows:

$$3.5000\cdots \le x_a^1 \le 5.6666\cdots,$$
$$0.0000\cdots \le x_b^1 \le 2.1666\cdots,$$
$$x_c^1 = 10.0000\cdots - 1.0000\cdots x_a^1 - 1.0000\cdots x_b^1,$$
$$x_a^2 = 17.2500\cdots - 0.6428\cdots x_a^1,$$
$$x_b^2 = 0.6428\cdots x_a^1 - 2.2500\cdots,$$
$$x_c^2 = 0.0000\cdots,$$
$$x_a^3 = 0.0000\cdots,$$
$$x_b^3 = 8.4999\cdots - 1.5000\cdots x_a^1 - 1.5000\cdots x_b^1,$$
$$x_c^3 = 11.5000\cdots + 1.5000\cdots x_a^1 + 1.5000\cdots x_b^1.$$

We show an example of a pair of distinct points in the range. The underlined numerical figures show those which are uniquely determined.

One is as follows.

$$x_a^1 = 3.5000\cdots, \quad x_b^1 = 1.3237\cdots, \quad x_c^1 = 5.1762\cdots,$$
$$x_a^2 = 15.0000\cdots, \quad x_b^2 = 0.0000\cdots, \quad x_c^2 = \underline{0.0000}\cdots,$$
$$x_a^3 = \underline{0.0000}\cdots, \quad x_b^3 = 1.2644\cdots, \quad x_c^3 = 18.7355\cdots.$$

This yields

$$\rho_1 = \underline{0.4416}\cdots, \; \rho_2 = \underline{0.8750}\cdots, \; \rho_3 = \underline{0.4416}\cdots, \; \rho_4 = \underline{0.8833}\cdots,$$
$$\Delta = \underline{0.35896706}\cdots.$$

The other is as follows.

$$x_a^1 = 3.5000\cdots, \quad x_b^1 = 2.1666\cdots, \quad x_c^1 = 4.3333\cdots,$$
$$x_a^2 = 15.0000\cdots, \quad x_b^2 = 0.0000\cdots, \quad x_c^2 = \underline{0.0000}\cdots,$$
$$x_a^3 = \underline{0.0000}\cdots, \quad x_b^3 = 0.0000\cdots, \quad x_c^3 = 20.0000\cdots.$$

This yields

$$\rho_1 = \underline{0.4416}\cdots, \; \rho_2 = \underline{0.8750}\cdots, \; \rho_3 = \underline{0.4416}\cdots, \; \rho_4 = \underline{0.8833}\cdots,$$
$$\Delta = \underline{0.35896706}\cdots.$$

6.6   Concluding Remarks

In the parametric analysis of the effect of system parameters, the utilization factors of service centers are unique meaningful performance variables. In contrast, the values of such variables as the utilization factor for each class may not be uniquely determined and may not be suitable for parametric analysis. We see similar difficulties when we obtain the values of performance variables by numerical calculation.

Even when the assumptions of the open BCMP queueing model hold approximately, we may have the conjecture that near optimal solutions must have nearly identical utilization factors of service centers although the solutions themselves may be much different from one another.

Chapter 7

A Survey of Dynamic Load Balancing

Static policies use only the system statistical information in making load balancing decisions, and therefore have the advantages in mathematical analysis and in implementation because of their simplicity. They do not, however, adapt to fluctuations of workload. Under a situation where the system workload is statistically balanced, some computers may be heavily loaded at a given instant (hence suffering from performance degradation), while others are idle or lightly loaded. On the other hand, dynamic policies [ELZ86a, WM85, Zho88, GDI93, MTS90, SK90] attempt to balance the workload dynamically corresponding to the current system state and are therefore thought to be able to further improve system performance. Dynamic policies are, however, much more complex than static policies. Studies concerning on dynamic load balancing have been usually limited to specific models that assume either that all the nodes in the system are identical or that overheads involved in load balancing are negligible [MTS89, MTS90, ELZ86a, ELZ86b, LM82, VD90]. In order to take account of the current system state in load balancing decisions, much information is needed, e.g., load state of each node in the system, information on each job, and the congestion information of the network. The schemes for information collection , job selection for transfer, and job allocation are also important and affect the performance of a load balancing policy. In this chapter the performance metrics for dynamic load balancing, the key components of a load balancing policy, and a brief review of dynamic load balancing policies are described.

One problem involved in dynamic load balancing is how to determine a criterion for load index selection. Many indices have been explicitly or implicitly used to express the load existing on a node at a given time. Examples of such indices include the utilization of the CPU, the length of the ready queue (in UNIX terminology, the *load average*), the stretch factor (defined as the ratio between the execution time of a job on a loaded machine and its execution time on the same node when it is empty), and more complicated functions of these simple variables. Ferrari and Zhou [FZ86] proposed an efficient load index based on a mean-value equation for closed multichain queueing network models with population-independent service rates and Processor-Sharing (PS), Last-Come-First-Served-Preemptive-Resume (LCFSPR), Infinite-Server (IS), or single-server First-Come-First-Served (FCFS) service centers. They showed that for CPU-bound jobs, the CPU queue length accurately predicts the response time, whereas the disk queue lengths have negligible effect. They [FZ88] also studied the effects on performance of the choice of load index, the averaging interval, the load information exchange period, and the characteristics of the workload. They found that load indices based on resource queue lengths perform better than those based on resource utilization, and the use of an exponential smoothing method yields further improvements over that of instantaneous queue lengths. That is, averaging the queue lengths over a short interval can produce significant improvements over the instantaneous values, as well as stability against the load oscillation.

When the resource requirements of jobs (i.e., CPU, memory, and file I/O requirements) can be predictable like UNIX [DI89], one can define more effective load indices. Gowswami, Devarakonda, and Iyer [GDI93] introduced the CPU load imposed by a job as the ratio of its predicted CPU time over its predicted execution time (i.e., CPU time plus the elapsed file I/O time), and the total CPU load at a node as the sum of the CPU loads of all the jobs running on that node. They showed the effectiveness of the load index.

A dynamic load balancing policy may be either *preemptive* or *nonpreemptive*. A nonpreemptive load balancing policy assigns a newly arriving job to what appears at that moment to be the best node. Once the job execution begins, it is not moved even though its run-time characteristics, or the run-time characteristics of any other jobs, later change after assigning the job in such a way as to cause the nodes to become much unbalanced. Many authors studied this policy [ELZ86a, ELZ86b, Zho88, ZHKS95, ZKS94]. Nonpreemptive job transfers are also referred to as *job placement* or *job assignment*. A preemptive load balancing policy [PM83, BS85, DO87, DO91], on the other hand, allows load balancing to occur whenever the imbalance appears in the workloads among the nodes. If a job that should be migrated to a new node is in the course of execution, its execution will be continued at the new node. The increased adaptability of preemptive load balancing could make it more effective than nonpreemptive load balancing [BS85]. Since in most systems the service demands of jobs are not known *a priori*, with initial assignment jobs are assigned to nodes in ignorance of these demands. An initial distribution of jobs across nodes that appears balanced will therefore become unbalanced as shorter jobs complete and leave behind an uneven distribution of longer jobs. Migration allows such imbalances to be corrected. To migrate a job in execution, however, is much complex and is accompanied with much overhead caused by gathering and transferring the state of the job

[KL88, PM83, TLC85], resulting in performance degradation.

Eager, Lazowska, and Zahorjan [ELZ88] explored the question of whether migration can offer significant performance benefits, or whether initial placement alone is sufficient. They identified the potential performance benefits of migration through the application of simple analytic and simulation models. They showed that there are likely no conditions under which migration could yield major performance improvements over those offered by nonmigratory load balancing. Under some extreme conditions, migration can offer modest additional performance improvements. These extreme conditions are characterized by high variability in both job service demands and the workload generation process. They pointed out that the benefits of migration are not limited by its cost, but rather by the inherent effectiveness of non-migratory load balancing.

This chapter focuses only on nonpreemptive load balancing policies. A nonpreemptive load balancing policy typically has three components: (1) a transfer policy that determines a node is in a suitable state to participate in a job transfer; (2) a location policy that determines to which node a job selected for transfer should be sent; and (3) an information policy that determines the amount of load information made available to the location policy and what load information should be collected and how this information is obtained. A transfer policy typically requires information only on the state of the local node and no exchange of state information among the nodes in deciding whether to transfer a job. It determines whether a job should be transferred either to it or from it, by identifying the node as a suitable sender of a job, a suitable receiver, or neither. A large number of the transfer policies proposed are *threshold* policies [ELZ86a, ELZ86b, ZF87, SK90, ZKS94, ZG94, KS94, ZHKS95]. Thresholds are expressed in units of load. A location policy at a node determines the allocation of a job and takes the action of the transfer if the job is determined to be sent.

An information policy may be based on a *time-driven* or *event-driven* way. In a time-driven approach, a node periodically announces its load information to other nodes or issues a request-for-bid message to other nodes to collect their load information. Periodic policies do not adapt their activity to the system state. Overheads due to periodic information announcement or collection at high system loads continue to increase the system load and thus worsen the situation. In an event-driven approach, on the other hand, a node does not announce its load information or issue a request-for-bid message for negotiation until its load changes. The information on the load state or the request-for-bid message at a node can be broadcasted to all other nodes, or only to a subset of node or a single node. Since overhead and delay due to state information manipulation have strong effects on performance of a dynamic load balancing policy and cannot usually be negligible, many studies effort to evaluate the effects of the amounts of the state information on the performance of load balancing and to try to minimize the overheads caused by information collection [ELZ86b, ZG94, KS94, ZKS94, BMDK94].

Load balancing policies can be classified as *centralized* or *decentralized*. In centralized policies [LY95, LR92, Zho88], the global load information is collected at a single node, called the *central scheduler*. The other nodes, called *local nodes*, send their load states to the central scheduler. All load balancing decisions are made at the central scheduler based on the collected state information. In decentralized policies [Sta84, ELZ86b, HS90, Zho88, ZKS94, KS94], on the other hand, each node broadcasts its load state to the other nodes or sends a request-for-bid message to collect the load information of the other nodes, and updates its locally maintained load tables. Each node makes its own load balancing decision to determine whether to send jobs to the other nodes or to request jobs from the other nodes. Centralized load balancing approaches may impose less overhead for manipulating state information, but have low reliability. Failure of the central scheduler will make the load balancing inoperable. Furthermore, centralized approaches are not appropriate for large-scale systems, since the central scheduler itself may become the bottleneck. Hereafter in this chapter decentralized load balancing policies will be mainly discussed.

Decentralized load balancing policies can be broadly characterized as *sender-initiated*, *receiver-initiated*, and *symmetrically-initiated*. Senders generate jobs to be processed, and receivers process these jobs. A node may be both a sender and/or receiver. Sender-initiated policies [BF81, ELZ86b, RS84, MTS90, ZF87, HS90, SK90, KS94, ZKS94] allow the congested nodes (senders) to search for lightly loaded nodes (receivers) where the jobs can be moved to. Receiver-initiated policies [LM82, ELZ86b, MTS90, HS90] allow the receivers to search for senders where the jobs can be transferred from. When a new job originates at a node, and when the load at that node exceeds a threshold T, the transfer policy decides the node to be a *sender*. If the load at a node falls below T, the transfer policy allows the node to be a *receiver* for a remote job. Symmetrically-initiated policies [HS90, BMDK94, Kru88, KL88, SK90, KS94] are a mix of sender-initiated and receiver-initiated policies. They seek to find suitable receivers when senders wish to send jobs, and to find suitable senders when receivers wish to acquire jobs. When a new job originates at a node, and the load at that node exceeds a threshold T_2, the node becomes a sender, while when a job leaves from a node, and the load at that node falls below a threshold T_1 ($T_1 \geq T_2$), the node becomes a receiver. In the following, a brief review of dynamic load balancing policies is described.

Eager, Lazowska, and Zahorjan [ELZ86b] examined typical sender-initiated and receiver-initiated policies, in which nodes were probed randomly until a complementary node was found or until a *probe limit* was reached. They found that, using a simple queueing model and assuming that the overhead of load balancing is negligible, the type of location policy that performs best depends on the workload. Such policies are representative of a large class of global scheduling algorithms suitable for medium and large-scale distributed systems, where determining the state of every other node in the system is prohibitively expensive. At low to moderate offered loads, sender-initiated searching was found to result in better performance than receiver-initiated searching, since at such loads, potential senders can find underloaded nodes to which they transfer jobs. Under receiver-initiated searching, the overloaded node must wait for an underloaded node to contact it, which may require a significant amount of time. Conversely, when the offered load from arriving jobs is high, receiver-initiated searching was found to result in better performance than sender-initiated searching, since nodes that become underloaded are almost immediately able to find overloaded nodes from which to transfer jobs. With sender-initiated searching under these conditions, such nodes may remain underloaded for long periods of time before they are found.

Rotithor and Pyo [HS90] examined two load balancing algorithms which combine sender-initiated (they called shedding) and receiver-initiated (they called requesting) policies in a VAX based distributed system. One is *random policy*, in which a decision maker (a node attempting to send a job or receive a job) in state l (lightly-loaded) or h (heavily-loaded) chooses another processing element (PE) at random to make a scheduling decision. Another is *fixed policy*, in which a decision maker in state l or h chooses a predetermined (fixed) neighbor to make a scheduling decision. The decision maker will make a job shedding decision if it is in state h, while make a job requesting decision if it is in state l. Results showed that the combination policies outperform either Random Requesting only or Shedding only policy with respect to mean response time.

It has been found that the symmetrically-initiated policies involve a larger number of load balancing messages at heavy load levels. This is due to its multiple probing nature. Higher negotiation failures also result from the concentration of the probing on the time scale. Benmohammed-Mahieddine, Dew, and Kara [BMDK94] proposed a so-called *Periodic Symmetrically-Initiated* (PSI) algorithm that emulates a *gas diffusion* process in its negotiation policy. It uses periodic polling of a single remote and random node. It has the same probing policies as the above mentioned ones, but aims at fixing (reducing) the L_p parameter (probe limit) at one, and at the same time fixing the frequency of algorithm invocation through a timer parameter P_t. This would result in a spreading of the probing messages over time and limit the overhead at high system loads in comparison to the traditional symmetrically-initiated algorithm where the algorithm is invoked each time a job departs or starts. For every timer period the node load is checked against the threshold T. If the load exceeds the threshold, a request is sent to a random node

($L_p = 1$), the node replies with an ACCEPT message if it is underloaded, otherwise it ignores the request. The requesting node transfers a job from its transferable jobs queue as a response to an ACCEPT message, or ignores the request if it is no longer overloaded. If the load is below the threshold, a request to receive a job is made to a random node. The chosen node will respond by sending a job from its transferable jobs queue, or just ignores the message if it is also underloaded. If the load is normal, no load balancing is attempted.

Shivaratri and Krueger [SK90, KS94] proposed two global adaptive scheduling policies. The first of the policies is *symmetrically-initiated* and is suitable for systems that have the capacity both to transfer newly arriving jobs and to migrate partly-executed jobs. The second of the policies is *sender-initiated* and is suitable for systems where only newly arriving jobs can be transferred. These policies adapt to global characteristics of the system workload by using as hints information that is normally discarded by static policies. The symmetrically-initiated policy utilizes the information gathered by polling and classify the nodes in the system as either Sender/overloaded, Receiver/underloaded, or OK (i.e., nodes having manageable load). The knowledge concerning the state of nodes is maintained by a data structure at each node, which comprises a sender list, a receiver list, and an OK list. These lists are maintained using an efficient scheme in which list manipulative actions, such as moving a node from one list to another, or finding the list to which a node belongs, impose a small and constant overhead irrespective of the number of nodes in the system. The transfer policy is a threshold policy where decisions are based on CPU queue length. The transfer policy is triggered when a new job originates or when a job departs. The transfer policy makes use of two threshold values to classify the nodes: a lower threshold (LT) and an upper threshold (UT). A node is said to be a sender if its queue length > UT, a receiver if its queue length < LT, and OK if LT $\leq$ node's queue length $\leq$ UT. The sender-initiated component is triggered at a node when it becomes a sender. The sender polls the node at the head of the receiver list to determine whether it is still a receiver. The polled node removes the sender node ID from the list it is presently in, puts it at the head of its sender list, and informs the sender whether it is a receiver, sender, or OK node based on its current status. On receipt of this reply, the sender transfers the new task if the polled node has indicated that it is a receiver. Otherwise, the polled node's ID is removed from the receivers list and put at the head of the OK list or at the head of sender list based on its reply. Then the sender polls the node at the head of the receivers list. The polling process stops if a suitable receiver is found for the job that newly arrived, if the number of polls reaches a PollLimit (a parameter of the algorithm), or if the receivers list at the sender node becomes empty. If polling fails to find a receiver, the job is processed locally, though it can later migrate as a result of receiver-initiated load balancing. The receiver-initiated component is triggered at a node when the node becomes a receiver. The receiver polls the selected node to determine whether it is a sender. On receipt of the message, the polled node, if it is a sender, transfers a job to the polling node and informs it of its state after the job transfer. If the polled node is not a sender, it removes the receiver node ID from the list it is presently in, puts it at the head of its receivers list, and informs the receiver whether it (the polled node) is a receiver or OK. On receipt of the reply, the receiver node removes the polled node ID from whatever list it is presently in and puts it at the head of the appropriate list based on its reply. The polling process stops if a sender is found, if the receiver is no longer a receiver, or if the number of polls reaches a static PollLimit. The sender-initiated policy is quite similar to the symmetrically-initiated policy and has only few differences. In addition to the data structure of the symmetrically-initiated policy, each node maintains a *state vector* (SV). The state vector is used by each node to keep track of which list (sender, receiver, or OK) it belongs to at all the other nodes in the system. When a sender probes a selected node, the sender's state vector is updated to indicate that the sender now belongs to the sender list at the selected node. Also, the probed node updates its state vector based on the reply it sent to the sender node to indicate which list it will belong to at the sender. The receiver-initiated component of the policy is strictly informational. A new receiver uses its state vector to determine which nodes believe it is not a receiver, and informs only those nodes.

Collecting system state information by issuing a request-for-bid (bidding) message has been accepted as an efficient approach and has been used by many researchers. Stankovic and Sidhu [SS84] proposed a bidding policy which utilizes the McCulloch-Pitts evaluation procedure (MPEP). This procedure is based on a decision cell which has a number of excitators and inhibitors as inputs, and a single value output. The output is the sum of the excitators, or zero if any of the inhibitors are set. Their policy works in the following way. 1). All jobs arriving at a node are periodically evaluated using an MPEP to decide whether to transmit a bid request for a particular job. The input to the MPEP includes both job and network characteristics. 2). If the output from the MPEP is above some threshold, then the job is performing well in the current node. Otherwise, if the output is non-zero, the job should be moved because it performs poorly. If the MPEP returns zero, as a result of setting one of the inhibitors, the job may not be moved. 3). When at least one job in a node needs to be moved, the node broadcasts the so-called *request-for-bid* message to all other nodes within a distance $i, i = 1, 2, \cdots, n$, where n is the diameter of the network. The bid contains all of the characteristics of the potential migration. Thus all receivers can run the MPEP to check whether the job would run well if it were moved. 4). After some predefined time of period t, all responses to the bid are adjusted by the cost of transferring the job. The best bid is then considered to be a potential destination node. One additional test for the job is performed before the transfer is taken. If the job is still performing poorly, then the job is transferred to the best bidder. The simulation results under various system conditions showed that bidding is quite effective even with simple parameters.

The level of state information involved in load balancing decisions is also a key factor for a dynamic policy. Maintaining too much state information and/or keeping state information too tightly up-to-date results in a high overhead. It therefore needs a tradeoff between the level of state information and the overhead. Selecting an efficient scheme to collect state information is also important for a load balancing policy. Eager, Lazowska and Zahorjan [ELZ86a] examined the effects of the level of information in scheduling decisions on performance of a load balancing policy and studied three policies each of which uses different level of information: 1). *Random policy* which uses no information of other nodes. Under this policy, a destination node is selected at random. Since newly arriving jobs and transferred jobs are treated equally, a transferred job may be transferred again if it arrives at a node with queue length exceeding the threshold value. To avoid instability, a limit is imposed on the number of transfers a job may experience. 2). *Threshold policy* which uses a small amount of information about potential destination nodes. With this policy a node is selected at random and then its queue length is probed to determine if the queue length is below the threshold. If so, the job is transferred. If not, another node is probed up to some specified maximum number of probes. A job is processed locally if the probing limit is exceeded. 3). *Shortest queue policy* which requires additional system state information and attempts to make the best choice given this information. Under this policy, a given number of nodes are chosen at random and their queue lengths are probed. The node with the shortest queue length is chosen to receive the job, if its queue is less than a given threshold. Otherwise, the job is processed locally. A surprising result is that extremely simple load balancing policies such as the Random policy and the Threshold policy perform quite well and more complex policies, such as the Shortest queue policy that utilizes more information cannot improve performance much further.

Zhang, Kameda, and Shimizu [ZKS94] examined three bidding policies for a *hetero-geneous* distributed system. These policies use different amounts of state information in their job scheduling decisions. By the term *heterogeneous*, it means that all nodes in the system have the same function, but may have different processing capacities. A key feature of these policies is that they take account of not only the queue length of each node but also the differences in capacity between the nodes when making the scheduling decisions. The three policies have the same transfer and location policies, but different information policies. The three information policies, named BR, LB, and LBT, are based on different bidding approaches. In BR, a node attempting to transfer a job broadcasts a request-for-bid message to all other nodes to collect system state information. In LB, L_b distinct nodes, rather than all other nodes, are selected at random and a request-for-bid message is sent to these nodes. In LBT, the similar technique to LB is used but more information is employed. A time window T_w is introduced. When a node sends a job to node i at an instant, it records the event. Then for a time interval T_w, this node will not bid node i. Results show that the simple policy, LB, utilizing small amounts of system state information (e.g., for 3 or 4 nodes) yields dramatic performance improvements over the case of no load balancing and perform much better than the Shortest algorithm of Eager [ELZ86a], an efficient algorithm in homogeneous systems which takes account of only the queue length of each node. LBT yields negligible further performance improvements even though it uses more information than does LB.

Kremien, Kramer and Magee [KKM93] proposed a flexible load balancing algorithm, which utilizes the idea that simple algorithms with small amounts of state information perform quite well and adds a function for scalability. It partitions a system into sets of nodes, called *domains*. A node determines which other nodes to include in its domain based on their load state: overloaded, underloaded, or under a medium load. Domain membership is symmetrical in two ways: (1). Two nodes are *candidates* for load balancing – and hence for inclusion in each other's domain – if one is overloaded while the other is underloaded. (2). If (underloaded) node B is in (overloaded) node A's domain, then node A is also in node B's domain. It is said that $candidate_{A,B}$ is true if nodes A and B are of opposite states, and false otherwise. The members of a domain are chosen using a biased random selection approach, and the membership of a domain is changed by either *message receipt* or *domain refresh*. When node A receives a state message from node B, node A reacts as follows. If B is not part of A's domain (D_A), and if B's load state $(state_B)$ is the opposite of A's, then B is a new candidate $(candidate_{A,B})$. A inserts an entry for B and confirms to B that A is a candidate. The entry is a tuple containing the sending node's ID, the time when the entry was updated, and the sending node's load state. If B is a candidate that is already part of A's domain, then A retains it but terminates the state exchange; that is, A gives no further response to B. Noncandidates (nodes with the same or medium state) are discarded if they are currently part of the domain, or ignored if they are new. Because A includes and updates only candidates, its domain contains only candidate nodes. On initialization (the domain is initially empty), node A first informs any current domain members of its new state. It then discards nodes that are no longer candidates, randomly selects new nodes from the rest of the system, and sends each of them a message with its name and state. A's domain contains only candidate nodes, so if A's load state changes, then the node empties its domain. However, if A's new state is either overloaded or underloaded, the responding messages will quickly reestablish mutual domain membership; if A changes to a medium load, its domain remains empty. The algorithm was examined using simulation and compared with a *Probe* algorithm, Eager's Random algorithm [ELZ86a], which randomly selects a remote location for a job and a *Periodic* algorithm which randomly and periodically selects all new nodes for a domain even if the domain already had candidates. Results showed that their algorithm outperforms the other two algorithms.

Zhu and Goscinski [ZG94] proposed two algorithms, OTSB and MTSB, which reduce network traffic. A load index similar to that in Ferrari and Zhou [FZ88] is used to measure the load of a node and two tables are introduced to store the information of negotiation between the nodes. The *candidate table* stores the information of the received bids, while the *offer table* stores the information of the nodes to which the node has offered bids. The information of each bid is, rather than discarding it, stored in both the initiator and the received nodes. This can reduce much of the network traffic. When an overloaded node broadcasts a request-for-bid message, only underloaded nodes respond to it and store the initiator on their offer table. If the initiator is on their candidate table, it is deleted since it is no longer underloaded. The initiator stores the bidders on its candidate table and starts negotiation with the fastest bidder immediately. When the candidate table is not empty, an overloaded node, rather than broadcasting a request-for-bid message, selects on bid from its candidate table and starts negotiation immediately. In OTSB, only one threshold is employed to distinguish an overloaded node from underloaded one. In MTSB, on the other hand, multi-thresholds are employed, for example, two thresholds, T_{over} and T_{under}, $(T_{under} \leq T_{over})$, are used to divide the nodes into three groups: overloaded if the load is greater than or equal to T_{over}, fully loaded if the load is in $[T_{under}, T_{over})$, and underloaded if the load is less than T_{under}. The fully-loaded nodes have enough amounts of resources to accomplish their current work, but cannot accept jobs from other nodes. Due to the introduction of a multi-threshold, the burden for the fully-loaded nodes to send bids is removed, which also reduces the possibility of network congestion.

Under a situation where resource requirements are predictable, more efficient scheduling mechanisms can be used. Goswami, Devarakonda, and Iyer [GID89, GDI93] examined four load balancing heuristics, based on the prediction method [DI89] wherein the CPU, memory, and I/O requirements of a job can be predicted prior to its execution using a statistical pattern-recognition method in Unix systems. Two of the heuristics, MINQ and MINRT, are centralized, and the other two, DMINQ and FDMINQ, are distributed. In MINQ, a new job is assigned to the node with the least CPU load. For each node i, the scheduler maintains a list of active jobs and their predicted CPU and I/O requirements.

When a new job X arrives the scheduler computes CPU load at each node i, and the predictor predicts its CPU and I/O requirements. Job X is sent to node k with the smallest CPU load value and entry containing information of job X is added to node k's list. In MINRT, each node maintains a list of active jobs like MINQ. The difference between the two policies is the job assignment for a job. Since when round-robin CPU scheduling is used, selecting the node with the smallest CPU load does not guarantee the best response time for the job, and therefore additional factors are needed. MINRT considers three factors, the number of jobs in the queue, their CPU requirements and the CPU requirement of the job being scheduled. Based on these factors, MINRT computes the expected response time a job X will receive at each node. The expected response time of a new job X on node i is the sum of the CPU requirements of the jobs in the node, if they are less than X's CPU requirement and an amount equal to job X's CPU requirement for each of the remaining jobs. For a job that newly arrives at a node, the estimated response time of the job r_i is computed as

$$r_i = \sum_{j=1}^{N_i} I \times CPUREQ_j + (1 - I) \times CPUREQ_X$$

$$I = \begin{cases} 1 & CPUREQ_j < CPUREQ_X \\ 0 & otherwise \end{cases}$$

where N_i is the number of jobs in node i, $CPUREQ_j$ is the CPU time required by job j. The node with the lowest r_i value is selected to execute job X and then job X is added to the appropriate node's list. In DMINQ, scheduling is similar to MINQ's but in a decentralized way. Each node periodically broadcasts its $CPULOAD$ and the resource usages of the terminated jobs. The scheduler at each node makes its own scheduling decisions based on its partial knowledge of the system. When a job X arrives at a node, the scheduler checks the $CPULOAD$ of this node against a preset threshold T. If the $CPULOAD$ is below the threshold, then the job will be processed locally. Otherwise, the table is searched for the node with the lowest $CPULOAD$ and the job is sent. In FDMINQ, instead of using a fixed threshold mechanism based on the load of a node as in DMINQ, a filtering mechanism based on prediction is used. Predicted resource requirements are used to identify and filter out short jobs (i.e., jobs requiring little CPU time). Regardless of the load on the node, all short processes are executed locally. This policy reduces the burden on the scheduler and significantly improves the response times for short jobs.

Some studies explored on the comparison of various load balancing policies. Zhou [Zho88] studied dynamic load balancing in homogeneous distributed systems using a simulation model driven by job traces collected from a production system. He considered explicitly the costs of load information exchange and job transfers. Seven load balancing algorithms were studied. He showed that, under moderate system loads, the system performance can be improved substantially through load balancing. Algorithms using periodic and aperiodic information policies yield comparable performance, and for the periodic load information policies, the centralized load balancing policies have much less overhead than the distributed load balancing policies. He also showed that receiver-initiated policies perform better than sender-initiated policies. He showed that system performance improves as the number of hosts in the system increases, but when the number of hosts goes beyond few tens of hosts, little further improvement results.

Although it is difficult to analyze dynamic models as described in the beginning of the section, there are a few studies on analysis models with some limitations. Mirchandaney, Towsley and Stankovic [MTS89] studied the performance characteristics of simple load balancing algorithms using Matrix-Geometric solution technique for homogeneous distributed systems. Nonnegligible delays in transferring jobs and in gathering remote state information are taken into account. They examined three algorithms, Forward (sender-initiated), Reserve (receiver-initiated), and Symmetric (symmetrically-initiated). By developing queueing theoretic models, they derived the important performance metrics and studied the effects of these delays on the performance of the algorithms. Mirchandaney, Towsley, and Stankovic [MTS90] also studied the performance characteristics of load balancing algorithms for heterogeneous distributed systems. They took account of delays in transferring jobs from one node to another and analyzed the effects of these delays on the performance of two threshold-based algorithms, Forward and Reserve. They focused on two types of heterogeneous system: In the first instance, the nodes in the system are identical with regard to their processing capacities and speeds, but the rates at which external jobs arrive at nodes may be different. In the second instance, nodes, although functionally identical, may process jobs at different rates. The arrival rates of jobs at different nodes may also differ. Results confirm that simple Forward and Reserve algorithms are quite effective in heterogeneous systems.

Although the communication network plays a key role in load balancing, the delay associated with the network often is either neglected or considered to be constant because of difficulties in analysis. This omission is particularly serious when the number of nodes in the network is large, due to the communication overhead associated with gathering global state information and communicating decisions. Schaar, Efe, Delcambre, and Bhuyan [SEDB91] introduced an analytical model for a homogeneous system, and proposed a load balancing policy that use the operational characteristics of the network in a fundamental way. They used a formula for computing the response time based on the basic formula in the Mean Value Analysis algorithm (for example, see [LZGS84]). To compute the network delay, they employed CSMA/CD style protocol model developed by Lam [Lam80] assuming that the frame length is a random variable from an exponential distribution. Schaar, Efe, and Delcambre [SED91] predicted the response time of jobs for heterogeneous systems using an analytical model. They took account of the network delay time for job transfer and packing time for preparing transfer for a job, but regardless of the overheads of load information collection. To adopt their model to heterogeneous systems, they match one type of processor to a class of jobs. In a heterogeneous system, the nodes may all be different and therefore there are as many classes of jobs as there are nodes. Even for a system model with two types of nodes there are 22 equations must be solved.

There are also some papers worth reading. For example, Kremien and Kramer [KK92b] presents a method for qualitative and quantitative analysis of load balancing algorithms, using a number of well-known examples as illustration. Casavant and Kuhl [CK88] describes a taxonomy of scheduling schemes for distributed systems. Smith [Smi88] discusses a survey of process migration schemes. Singhal and Shivaratri [SS94] provides a comprehensive review of key issues in load balancing.

Chapter 8

A Comparison of Static and Dynamic Load Balancing

By intuition we anticipate that dynamic load balancing policies outperform static ones, but dynamic policies have to react to the current system state. This makes dynamic policies necessarily more complex than static ones. One concern is that the overhead caused by such complexity may negate the benefits of dynamic load balancing. Other concerns are the effect of occasionally poor load-balancing decisions and the potential for instability in dynamic load balancing because of the inherent inaccuracy of system state-information. Even though it is known that dynamic policies have advantages over static ones in homogeneous systems, it is questionable whether it is also true in heterogeneous systems. The *homogeneous* systems under consideration denote systems where all the nodes in the system are identical in both function and processing capacity. The *heterogeneous* systems under consideration denote systems where all the nodes in the system have the same functions but might have different processing capacities. In this chapter the above concerns are addressed and the performance of static and dynamic policies in a heterogeneous system is compared. The overheads and the delays in both job transfer and state-information exchange are taken into account.

The dynamic policies under consideration take into account not only the queue lengths of the nodes but also the differences in capacity between the nodes in load-balancing decisions. For comparison, Zhou's DISTED policy [Zho88] was selected. This policy takes account of only the queue length of each node in load-balancing decisions and has been proven to be quite efficient in homogeneous systems. The static policies under consideration are *probabilistic* policies, which require only aggregate information of jobs instead of individual jobs [TT84, TT85, KK92a, KK90b, ZKS92].

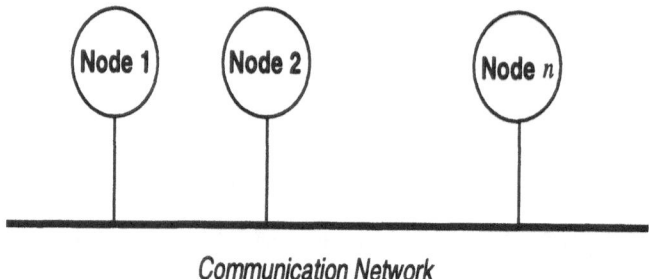

Communication Network

Figure 8.1: System model used in the simulation

8.1 System Model

A distributed system model consisting of N heterogeneous nodes connected by a single-channel communication network (e.g., an Ethernet) is considered as shown in Figure 8.1. The set of nodes is also denoted by N, and the nodes are assumed to have the same function but might process jobs at different speeds. Each node is modeled as a central-server model which consists of a CPU and infinite I/O devices as shown in Figure 8.2. The symbols p_0 and p_1 respectively denote the probabilities that a job, after departing from the CPU, finishes and requests I/O service. Therefore, p_1/p_0 denotes the average number of I/O requests per job. The average service time at I/O devices is denoted by t_{IO}. It is assumed that the CPU at each node processes jobs according to a round-robin discipline and that the size of time quantum is considerably small compared with the mean job service time.

External jobs arrive at each node according to a Poisson process with a mean interarrival time of a_i (sec). Arrival jobs at node i (referred to as the origin node) may be either processed at node i or transferred through the network to another node j (referred to as the destination node) for remote processing. It is assumed that a job being executed is not eligible for transfer for the reasons depicted in the previous section. The job service time at the CPU of node i is assumed to be generally distributed with a mean s_{0_i} (sec). The *load level* is defined as the ratio of the sum of the external job arrival rates at all nodes to the sum of the job service rates of all nodes: ρ ($\rho = (\sum_{i\in N} 1/a_i)/(\sum_{i\in N} 1/s_i)$). It is assumed that the communications network contains only one server that transfers jobs based on the round-robin discipline. Because there is no correlation between job size and job service time, the transmission time (not including waiting time) of a job is assumed to be exponentially distributed with a mean t_c (sec) regardless of its service time.

Two kinds of overheads involved in dynamic load balancing are taken into account.

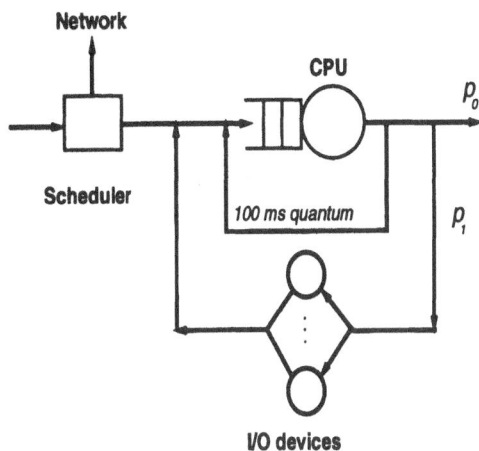

Figure 8.2: Node model

The first is that for measuring the load states at the nodes and for exchanging the load states among the nodes. The second is that associated with transferred jobs at both the sending and the receiving nodes. CPU time and network bandwidth are consumed for these purposes. The exchange of state information and the transfer of job are assumed to have preemptive priority over job execution. Note that, in static load balancing, a job transfer incurs CPU time overhead as well as network delay.

8.2 Load Balancing Policies: Static and Dynamic

8.2.1 Static Load Balancing Policies

Static policies can, for example, be implemented by simply transferring all jobs originating at node X to nodes Y and Z with probabilities p and $1-p$, respectively. The probabilities wherewith jobs arriving at a node should be transferred to the other nodes are determined by using the methods developed in [TT85, KK92a, LK94a, LK94b, LK93]. The basic definitions and assumptions for the static policies are reviewed in the following for the self-containment.

The job processing rate at node i is denoted by β_i and the job transfer rate from node i to node j is denoted by x_{ij}. The network traffic is denoted by λ and is defined as $\lambda = \sum_{i,j \in N, i \neq j} x_{ij}$. The mean node delay of a job processed at node i is denoted by $F_i(\beta_i)$. The mean job transfer delay is denoted by $G(\lambda)$ and is assumed to be independent of the origin and the destination nodes. It is assumed that a job can be transferred once and that the communications network satisfies the *triangle inequality* property; that is, there is no node k sending jobs to node i and receiving jobs from another node j. For the static model, the following relations (refer, e.g., to [GM80, Kle76]) can be obtained:

$$F_i(\beta_i) = \frac{s_{0_i}}{1 - s_{0_i}\beta_i} + \frac{p_1}{p_0}t_{IO}, \qquad (8.1)$$

$$G(\lambda) = \frac{t_c}{1 - t_c\lambda}. \qquad (8.2)$$

We can also write $F_i(\beta_i)$ and $G(\lambda)$ as

$$F_i(\beta_i) = s_{0_i}(N_{0_i} + 1) + \frac{p_1}{p_0}t_{IO}, \qquad (8.3)$$

$$G(\lambda) = \frac{t_c}{1 - \rho_c}, \qquad (8.4)$$

where N_{0_i} denotes the mean number of jobs at the CPU of node i, and ρ_c denotes the mean utilization of the communication channel. The *marginal node delay* function denoted by $f_i(\beta_i)$ and the *marginal job transfer delay* function denoted by $g(\lambda)$ are defined as follows:

$$f_i(\beta_i) = \frac{d}{d\beta_i}\beta_i F_i(\beta_i), \qquad (8.5)$$

$$g(\lambda) = \frac{d}{d\lambda}\lambda G(\lambda). \qquad (8.6)$$

$f_i(\beta_i)$ and $g(\lambda)$ can also be writen as

$$f_i(\beta_i) = s_{0_i}(N_{0_i} + 1)^2 + \frac{p_1}{p_0}t_{IO}, \qquad (8.7)$$

$$g(\lambda) = \frac{t_c}{(1 - \rho_c)^2}. \qquad (8.8)$$

Static Overall Optimal Policy (SOO)

The Static Overall Optimal (SOO) policy determines load balancing so as to minimize the average response time of jobs. That is, its goal is system-wide optimization. It can be formulated as a nonlinear optimization problem and its optimal solution can be obtained by using the Kuhn-Tucker theorem [Int71]. According to the results of Tantawi and Towsley [TT85], the optimal solution of SOO, β_i ($i \in N$), satisfies the following relations:

$$f_i(\beta_i) \geq \alpha + g(\lambda), \qquad \beta_i = 0, \qquad (8.9)$$

$$f_i(\beta_i) = \alpha + g(\lambda), \qquad 0 < \beta_i < 1/a_i, \qquad (8.10)$$

$$\alpha \leq f_i(\beta_i) \leq \alpha + g(\lambda), \quad \beta_i = 1/a_i, \qquad (8.11)$$

$$\alpha = f_i(\beta_i), \qquad \beta_i > 1/a_i, \qquad (8.12)$$

subject to $\sum_{i \in N} \beta_i = \sum_{i \in N} 1/a_i$, where α is the Lagrange multiplier.

The optimal solution can be calculated by using the algorithm developed by Kim and Kameda [KK92a]. The solution can be used to obtain the transfer rate x_{ij} for each node i (see [TT85]) and then to obtain the probabilities wherewith the SOO policy is realized. For example, the probability wherewith jobs arriving at node i will be transferred to node j is $x_{ij}a_i$.

Static Individually Optimal Policy (SIO)

The Static Individually Optimal (SIO) policy determines load balancing so that each job (precisely, the user of the job) may feel that its own mean response time is minimum if it knows the mean node delay at each node and the mean job transfer delay. It appears that this policy is closely related to a completely decentralized scheme in that each job itself determines, on the basis of the information on the mean node and job transfer delays, which node should process it.

In Zhang, Kameda, and Shimizu [ZKS92], the SIO policy is defined to be an optimal solution β_i $(i \in N)$ that satisfies the following relations:

$$
\begin{align}
F_i(\beta_i) &\geq R + G(\lambda), & \beta_i &= 0, & (8.13) \\
F_i(\beta_i) &= R + G(\lambda), & 0 < \beta_i &< 1/a_i, & (8.14) \\
R &\leq F_i(\beta_i) \leq R + G(\lambda), & \beta_i &= 1/a_i, & (8.15) \\
R &= F_i(\beta_i), & \beta_i &> 1/a_i, & (8.16)
\end{align}
$$

subject to $\sum_{i \in N} \beta_i = \sum_{i \in N} 1/a_i$, where $R = \min_i F_i(\beta_i)$.

A striking parallelism between the solutions of the two policies can be observed from relations (8.9) – (8.12) in the SOO and relations (8.13) – (8.16) in the SIO. The algorithms for calculating their optimal solutions of the two policies are also strikingly parallel. By replacing $f_i(\beta_i)$, $g(\lambda)$, and α in the algorithm of SOO by $F_i(\beta)$, $G(\lambda)$, and R, respectively, an algorithm that calculates the optimal solution of SIO is obtained. Then the probabilities wherewith the SIO policy can be realized.

8.2.2 Dynamic Load Balancing Policies

The information policy of the dynamic policies under consideration is a *periodic* policy. That is, each node broadcasts information on its own load state to all the other nodes every P time units. It is assumed that each node knows the precise value of its own load at any given instant, but the estimated loads of other nodes may be inaccurate (out of date). Each node has a load table that contains the mean service times and the estimated loads of all the nodes. A periodic load balancing policy works as follows. When a job arrives at a node, the transfer policy component is activated and determines whether the job is eligible for transfer. If the job is eligible for transfer, the location policy component is activated to determine, using the system state information collected periodically, where the job should be allocated.

In the following, the transfer and the location policies are described in details.

Adaptive Marginal Virtual Delay Policy (AMVD)

The Adaptive Marginal Virtual Delay (AMVD) policy attempts to balance the workload dynamically to minimize the system-wide mean job response time. To implement AMVD, a method that makes use, with some modification, of the SOO techniques is employed. Replacing the mean job numbers N_{0_i} and the mean utilization of the network ρ_c in relations (8.7) and (8.8) by n_{0_i}, the number of jobs at the CPU at node i at a given instant, and ρ'_c, the utilization factor of the network, respectively, yield relations (8.17) and (8.18). These relations are given in terms of f_i and g respectively are used as the estimates the *marginal virtual node delay* at node i and the *marginal virtual job transfer delay* :

$$f_i \;=\; s_i(n_{0_i}+1)^2, \tag{8.17}$$

$$g \;=\; \frac{t_c}{(1-\rho'_c)^2}, \tag{8.18}$$

In SOO, the marginal node delays among the nodes are balanced when SOO is realized (See relations (8.9) - (8.12)). The AMVD policy attempts to balance dynamically the marginal virtual delays among the nodes in order to realize the system-wide optimization.

Note that I/O requirements are not taken into consideration in relation (8.17). This does not pose a problem for load balancing because what we need is not the absolute marginal virtual delay, but rather a consistent comparative measure that can be used to distinguish between the nodes. To estimate ρ'_c, it is assumed that there exists a device, either software or hardware, capable of measuring the utilization of the network. Implementing such a hardware device does not appear to be difficult [Kru88, LM85].

• *Transfer policy*: This is a *threshold* policy by which each node uses only local state information and determines whether an arriving job is eligible for transfer. A job arriving at a node becomes eligible for transfer when the number of jobs in the node is greater than some threshold (T_i). Otherwise the job will be processed there. The nodes in heterogeneous systems may have different capacities and therefore have different viewpoints on the system workload. Under such a situation, it is appropriate to set different thresholds for different nodes.

The determination of the thresholds of the nodes reflects the objective of the policy. The threshold of the fastest node k, T_k, is chosen as the basis. For node i ($i \in N, i \neq k$), by letting $f_i = f_k$ and by substituting Eq. (8.17) into this relation, we have $n_{0_i} = \sqrt{s_{0_k}/s_{0_i}}\,(n_{0_k}+1) - 1$. We replace n_{0_i} and n_{0_k} in this relation by T_i and T_k, respectively, so that T_i and T_k satisfy this relation. T_i is determined using this relation, but it should be nonnegative and an integer. Therefore, we have $s_{0_k}/s_{0_i} \leq 1$ and $0 \leq T_i = \sqrt{s_{0_k}/s_{0_i}}(T_k + 1) - 1 \leq T_k$. When the threshold of a node is determined to be zero, all the arriving jobs become eligible for transfer.

• *Location policy.* For an arriving job that is eligible for transfer, the *best* potential destination node is sought as the one with the shortest marginal virtual delay (lightest node), and then the allocation of the job is determined.

(1) *Search for the node with the lightest load.* For a job arriving at node i, each potential destination node j ($j \in N$, $j \neq i$) is compared in turn with node i as follows:

If $f_i > f_j + g$, or, by substituting relations (8.17) and (8.18) into this relation, if

$$n_{0_i} > n_{ij}, \tag{8.19}$$

where $n_{ij} = \sqrt{(s_{0_j}/s_{0_i})(n_{0_j} + 1)^2 + t_c/s_{0_i}(1 - \rho_c')^2} - 1$, then node i is said to be more heavily loaded than node j. Otherwise, if $f_i \leq f_j + g$ or

$$n_{0_i} \leq n_{ij}, \tag{8.20}$$

then node i is said not to be heavily loaded, compared to node j. Let $\delta_{ij} = n_{0_i} - n_{ij}$ and $\delta_i = \max_j \delta_{ij}$. If δ_i is greater than zero, then node j is the node with the lightest load.

(2) *Job allocation.* If $w^c \delta_i > \Delta$ when the job has been transferred c times, then the job will be transferred to node j; otherwise, it will be processed at node i. Here w $(0 < w \leq 1)$ is a *weighting factor* used to prevent a job from being transferred continuously. When $w = 1$ any transferred jobs are treated as jobs that newly arrive from the external world. When $w < 1$, on the other hand, a transferred job will be transferred again only if it finds a node better than it found last time. Δ $(\Delta > 0)$ is a *bias* used to protect the system from potential instability caused by the occasionally poor load-balancing decisions due to the inherent inaccuracy and rapidly changing nature of system state information. Without such a bias ($\Delta = 0$), the load-balancing policy has to react to small load distinctions between nodes. This may put the system into an unstable state. At the extreme, a form of *processor thrashing*, in which all the nodes are spending all their time transferring jobs, can occur.

Adaptive Virtual Delay Policy (AVD)

The Adaptive Virtual Delay (AVD) policy attempts to balance the *virtual delay (un-finished work)* among the nodes in order to minimize the response time of each job. The *virtual node delay* at node i denoted by F_i and the *virtual job transfer delay* denoted by G are estimated as follows:

$$F_i = s_{0_i}(n_{0_i} + 1), \tag{8.21}$$

$$G = \frac{t_c}{1 - \rho_c'}. \tag{8.22}$$

The virtual node (job transfer) delay is the amount of time that a job has to spend in a node (communication channel) if the job enters the node (communication channel) at a given instant. Note also that I/O requirements are not taken into consideration here for the same reason as in AMVD. In the following, the transfer and the location policy components of AVD are described in detail.

• *Transfer policy.* This is, except for the method of deciding the threshold for each node, the same as that of AMVD. That is, the threshold of the fastest node k, T_k, is chosen as the basis and then the thresholds of all the other nodes are determined. For node i ($i \in N, i \neq k$), by letting $F_i = F_k$ and by substituting Eq. (8.21) into this relation, we have $n_{0_i} = (s_{0_k}/s_{0_i})(n_{0_k} + 1) - 1$. Replacing n_{0_i} and n_{0_k} in this relation by T_i and T_k, respectively, lead T_i and T_k to satisfy this relation. T_i is determined using this relation, but it should be nonnegative and an integer. Therefore, we have $s_{0_k}/s_{0_i} \leq 1$ and $0 \leq T_i = (s_{0_k}/s_{0_i})(T_k + 1) - 1 \leq T_k$.

• *Location policy.* Procedures similar to those of AMVD are used here.

(1) *Search for the node with the lightest load.* For a job arriving at node i, each potential destination node j ($j \in N$, $j \neq i$) is compared in turn with node i as follows:

If $F_i > F_j + G$, or, by substituting relations (8.21) and (8.22) into this relation, if

$$n_{0_i} > n_{ij}, \tag{8.23}$$

where $n_{ij} = (s_{0_j}/s_{0_i})(n_{0_j} + 1) + t_c/s_{0_i}(1 - \rho_c') - 1$, then node i is said to be more heavily loaded than node j. Otherwise, if $F_i \leq F_j + G$ or

$$n_{0_i} \leq n_{ij}, \tag{8.24}$$

then node i is said not to be heavily loaded, compared to node j. Let $\delta_{ij} = n_{0_i} - n_{ij}$ and $\delta_i = \max_j \delta_{ij}$. If δ_i is greater than zero, then node j is the node with the lightest load.

(2). *Job allocation.* If $w^c \delta_i > \Delta$ when the job has been transferred c times, the job will be transferred to node j; otherwise, it will be processed at node i. As in AMVD, w ($0 < w \leq 1$) is a *weighting factor* used to prevent a job from being transferred continuously, and Δ ($\Delta > 0$) is a *bias* used to protect the system from potential instability.

Homogeneous Policy (HM)

For comparison, a homogeneous policy (HM) (called DISTED in [Zho88]) is used. The HM policy attempts to equalize only the queue length (the number of jobs in service and waiting for service) of each node and does not consider the virtual delays of the nodes or the congestion in the network. The thresholds of all the nodes in HM are identical.

Table 8.1: Mean service and interarrival times of jobs (load level $\rho = 0.1$).

Node	mean service time(sec)	mean interarrival time(sec)
1	5.0	500.0
2	5.0	400.0
3	10.0	400.0
4~6	10.0	300.0
7, 8	30.0	300.0
9 ~11	30.0	120.0
12~16	30.0	210.9
17~20	50.0	210.9
21~30	50.0	200.0

8.3 Simulation Results

The dynamic and static policies under consideration were examined on a distributed system model consisting of 30 nodes connected by a single-channel communication network as shown in Figure 8.1. The values of the system parameters in the simulation were chosen with regard to systems that are in use or will be in use in the near future. Each node was assumed to use a round-robin CPU scheduling discipline with a 100-millisecond time-slice. The job service time was chosen from a two-stage hyperexponential distribution with a standard deviation of $3s_i$, a figure based on measurement results. The mean service and interarrival times of jobs at each node are listed in Table 8.1. The nodes in the load table at each node were ordered in descending order of processing capacity. Since I/O devices are not the points of contention, it is simply assumed that I/O operations were evenly spread throughout the execution of each job and that each disk I/O request took 30 milliseconds. The number of I/O requests for each job was chosen from a normal distribution with a mean of 12 and a standard deviation of 10 and was assumed to be greater than 0.

In estimating the network utilization ρ'_c, single measurement of the network device can result in oscillatory behavior. For example, low utilization over one measurement interval makes job transfer appear more advantageous, causing more to occur during the next interval. Therefore, the network utilization was estimated as the mean of the most recent 10 measurements, taken at regular intervals. Each interval was set to be 20 seconds, a figure based on simulation results. An upper limit of $1 - 10^{-7}$ on the estimate of ρ'_c was set in order to avoid division by zero in Eqs. (8.18) and (8.22).

The job transmission time, t_c, was chosen from an exponential distribution with a mean of 1 second. It is observed that the data transmission rate for a 10-Mbps local area

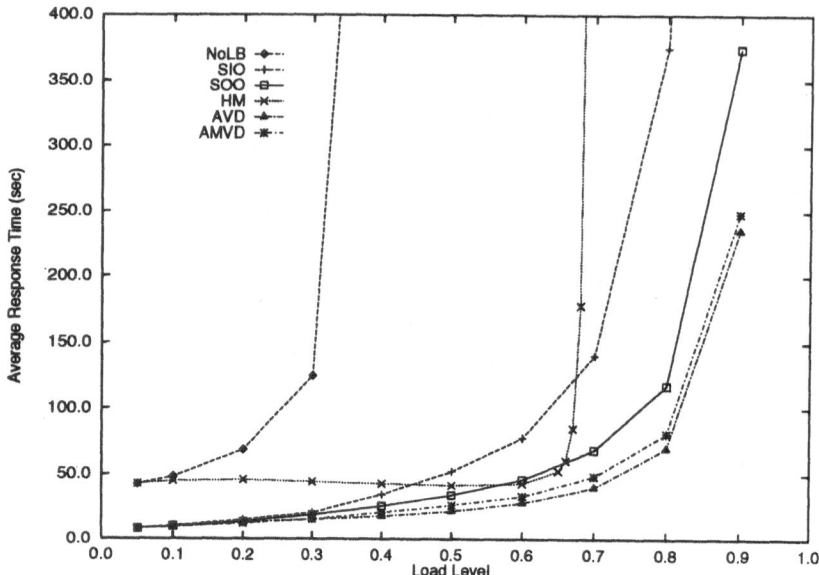

Figure 8.3: Average response time vs. load level (OV=5%)

network (e.g., Ethernet) is about 1 Mbps and that the average file size of UNIX is about 100 KBytes. Transferring a 100-KByte job would therefore take about 1 second if there were no jobs queued for transfer. The delay for state information exchange is thought to be much shorter than that for job transfer and therefore was chosen to be a fixed value of 5 milliseconds (500 Bytes).

The simulation results were obtained with 95% confidence intervals which were within 5% of the sample mean. The average response time of the boundary case where no load balancing is attempted is shown in the figures by "NoLB." To make the comparison between the dynamic and the static policies accurate, a large number of simulation runs for AMVD, AVD, and HM with different values of adjustable parameter values (e.g., T_i, Δ, P) were conducted, and the results used for comparison with those of static policies shown in the figures were selected from the best combination of those parameter values.

Figure 8.3 shows the average response time versus load level for the policies under consideration. The threshold of node 1, T_1, was selected to be the basis and was set to be 3 in AMVD and 5 in AVD. The thresholds of all the nodes in HM were set to be 2. The biases for job transfer, Δ, in AMVD, AVD, and HM were respectively set to be 0.4, 0.7, and 1. The overhead for sending or receiving a job, OV, was set to be 5/100 of the

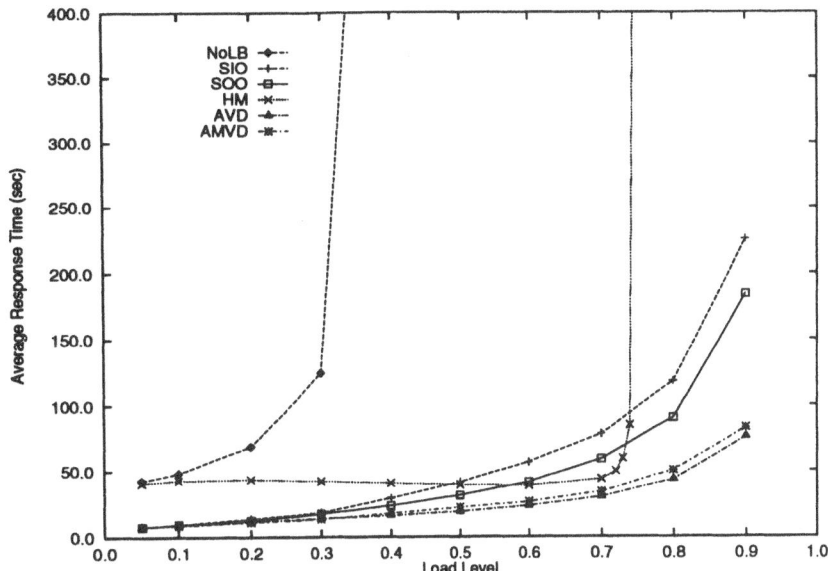

Figure 8.4: Average response time vs. load level (OV=0%)

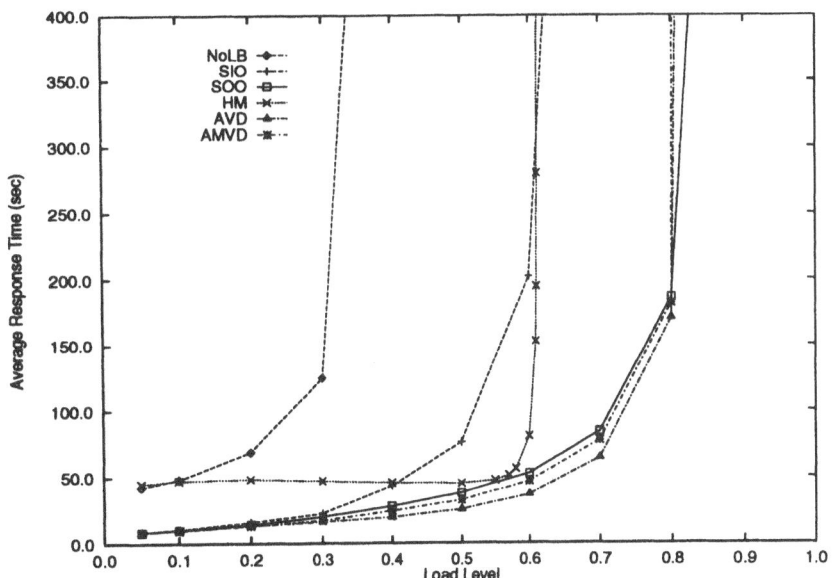

Figure 8.5: Average response time vs. load level (OV=10%)

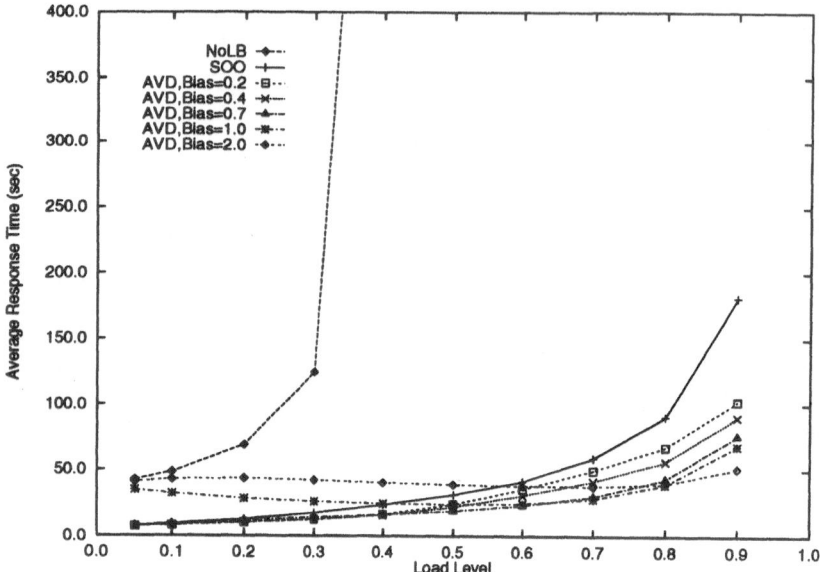

Figure 8.6: Average response time vs. load level for various biases(OV=0%)

mean job service time at a node. The overhead for broadcasting the load state at a node is much smaller than that for job transfer [ZZWD93] and was set to be 5/100 of that for job transfer. The exchange period for system state information, P, was 2 seconds; that is, every 2 seconds each node broadcasts its load state to other nodes. The weighting factor for job transfer, w, in AMVD, AVD, and HM was set to be 0.9. Results showed that by introducing w the frequency of job transfer is significantly reduced.

Figure 8.3 shows that both the dynamic policies (AMVD, AVD, and HM) and the static policies (SOO and SIO) provide substantial performance improvements over the situation where no load balancing is attempted. A key observation from this figure is that, contrary to our intuition, at light and moderate system loads the performance provided by SOO and SIO is close to that provided by AMVD and AVD. Furthermore, at heavy system loads the performance provided by SOO is not much inferior to that provided by either AMVD or AVD. This figure also shows that HM does not appear to be effective in heterogeneous systems, performing much worse than SOO and SIO over a wide range of load levels.

Figures 8.3 and 8.5 show the average response time of the policies versus the load level for two cases: one in which there are no overheads for both job transfer and system state-information exchange, and the other in which the overhead for job transfer is 10/100 of the mean job service time. The other parameters were fixed as in Figure 8.3. The overhead

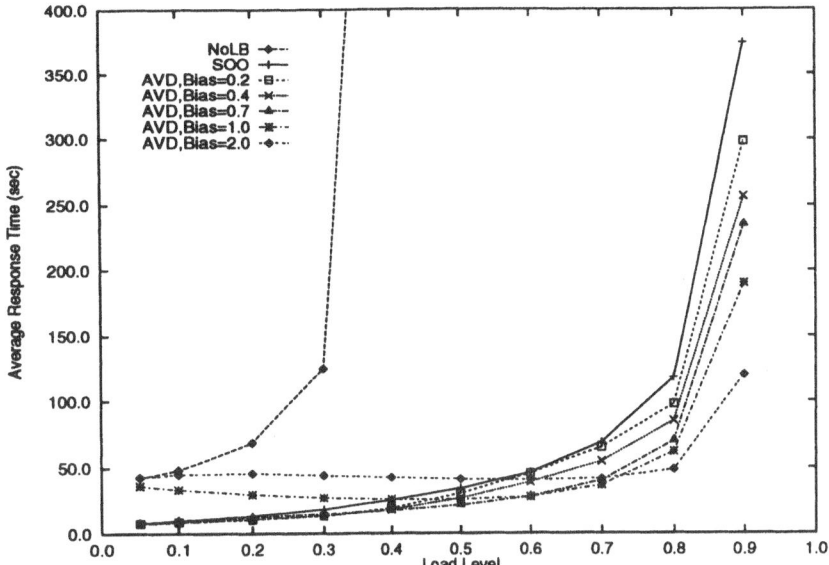

Figure 8.7: Average response time vs. load level for various biases $(OV=5\%)$

for job transfer, although nonnegligible, can be expected to be quite low compared with the mean job service time. The overhead of job transfer is quite low, for example, in the systems that include file servers and database servers, where only a descriptor will be shipped.

Figures 8.3 and 8.5, in addition to Figure 8.3, show that the average response time is insensitive to the overheads when system loads are light or moderate (e.g., load level is below 0.6 in Figure 8.5) but increases rapidly when system loads are heavy (e.g., load level exceeds 0.7 in Figure 8.5). This implies that the performance benefits can be gained at light and moderate system loads even when the overheads are nonnegligibly high, whereas at heavy system loads high overheads rapidly lead to instability. From these figures, it was surprising to see that AMVD and AVD perform slightly better than SOO and SIO at light and moderate system loads. It can be also observed that SOO may have chances to behave better than both AMVD and AVD at very heavy system loads (see Figure 8.5). These results indicate that static policies are preferable when system loads are light or moderate, or the overheads are very high.

Figures 8.6, 8.7, and 8.8 show the average response time versus load level for various biases for job transfer under AVD. The other parameters were fixed as in Figure 8.3. These figures show that the average response time depends strongly on the bias for job

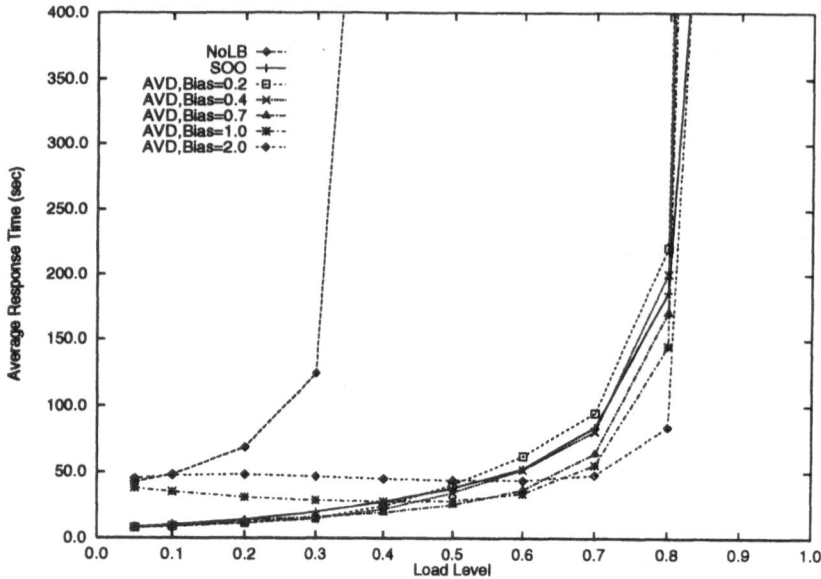

Figure 8.8: Average response time vs. load level for various biases (OV=10%)

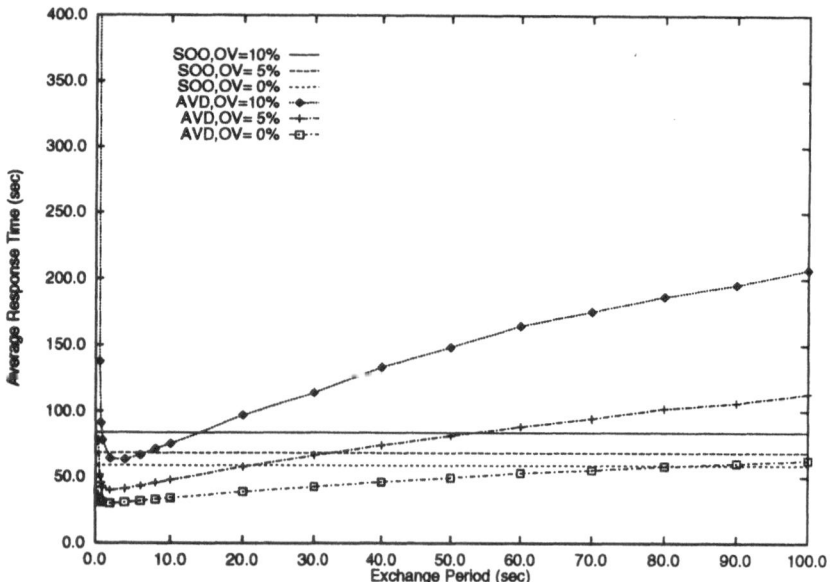

Figure 8.9: Average response time vs. exchange period

transfer and that a bias of less than 1 results in short average response time over a wide range of load levels (except very heavy system loads). They also show that at light and moderate system loads an improper bias (e.g., greater than 1 here) results in average response time much worse than that provided by SOO and even worse than that when no load balancing is attempted (e.g., load level is 0.05 in Figure 8.8). The average response time of dynamic policies can be improved much further by tuning the bias to the load level. To do this, however, it needs to monitor the load level, choosing a small bias at light and moderate system loads and choosing a large bias at heavy system loads.

Figure 8.9 shows the average response time versus the exchange period of system state information at a fixed load level of 0.7 for AVD. The results in AMVD were similar to those in AVD and are not shown here. "OV" in the figure denotes the ratio of the overhead for job transfer to the mean job service time at a node. The other parameters were fixed as in Figure 8.3. This figure shows again that SOO is less sensitive to the overheads and may perform better than AVD. It is observed that the magnitude of the exchange period has a strong influence on the average response time in AVD. When P is too short (e.g., 0.4 seconds in Figure 8.9), the overhead for state information exchange becomes so high that the performance suffers, although the system state information on which the load-balancing decisions are based is kept up-to-date. On the other hand, if P is too long (e.g., 100 seconds in Figure 8.9), the system state information is so out-of-date that frequent mistakes are made in load-balancing decisions. That is, jobs are often sent to nodes with loads equal to or higher than that of the local node.

8.4 Discussion

The dynamic policies were expected to perform much better than the static policies, but the simulation results showed otherwise. The results show that both the dynamic policies (AMVD, AVD, and HM) and the static policies (SOO and SIO) improve performance to levels far above those attained without load balancing, and that SOO is not much inferior to either AMVD or AVD despite its simplicity. There might be two main reasons for such results. The first is that static policies do not incur overhead in exchanging system state information, and the second is the simplicity of static load balancing. In static load balancing, the allocation of a job cannot be changed. In dynamic load balancing, however, when a job arrives at a node from either the external world or the network, the scheduler of the node attempts to allocate the job to the *best* destination node. Because of the inherent inaccuracy of system state information, it is inevitable that poor scheduling decisions may be made. In such a situation, a job may be kept being transferred continuously, wasting CPU capacity and degrading performance.

The reason for HM performing worse than the other policies seems clear. At light and moderate system loads, many nodes in the system are lightly loaded and load balancing is rarely attempted in HM. This clearly fails to produce good performance in heterogeneous systems. On the other hand, load-balancing decisions in AMVD, AVD, SOO, and SIO are determined so that the nodes with small capacities send almost all their arriving jobs to the nodes with large capacities. When system loads are heavy, HM performs load balancing without considering the congestion conditions of the system and hence rapidly leads to system instability. This suggests that in heterogeneous systems the differences in capacity between the nodes should be taken into account in scheduling decisions.

It needs a set of the values for various system parameters to implement a dynamic policy. These parameters usually have strong effects on system performance. The results show that some key system parameters (e.g., the exchange period of state information and the bias for job transfer) have strong effects on the system performance, and can easily be adjusted. Introducing some mechanisms by which tuning the values of these parameters based on the system conditions, that is, adapting the values of the system parameters to the current system state, system performance can be significantly improved.

Bibliography

[AK91]   B. W. Abeysundary and A. E. Kamal. High-speed local area networks and their performance: A survey. *ACM Computing Surveys*, 23(2):221–264, June 1991.

[Alb83]   A. Albanese. Star networks with collision-avoidance. *Bell Syst. Tech. J.*, 62:631–638, March 1983.

[BCMP75] Forest Baskett, K. Mani Chandy, Richard R. Muntz, and Fernando G. Palacios. Open, closed, and mixed networks of queues with different classes of customers. *Journal of the Association for Computing Machinery*, 22(2):248–260, April 1975.

[BF81]   R.M. Bryant and R.A. Finkel. A stable distributed scheduling algorithm. In *Proc. 2nd Int. Conf. Dist. Computing Syst.*, pages 314–323, April 1981.

[BMDK94] K. Benmohammed-Mahieddine, P.M. Dew, and M. Kara. A periodic sysmmetrically-initiated load balancing algorithm for distributed systems. In *Proc. 14th Int. Conf. Dist. Computing Syst.*, pages 616–623, Poznan, Poland, June 1994.

[BS85]   Amnon Barak and Amnon Shiloh. A distributed load-balancing policy for a multicomputer. *Software-Practice and Experience*, 15(9):901–913, September 1985.

[CG74]   D. G. Cantor and M. Gerla. Optimal routing in a packet switched computer network. *IEEE Transactions on Computers*, 10:1062–1068, 1974.

[CK88]   T. L. Casavant and J. G. Kuhl. A taxonomy of scheduling in general-purpose distributed computing systems. *IEEE Trans. on Software Engineering*, SE-14(2):141–154, February 1988.

[Cou77]   P. J. Courtois. *Decomposability: Queueing and Computer System Applications*. Academic Press, New York, 1977.

[Daf71]   S. C. Dafermos. An extended traffic assignment model with applications to two-way traffic. *Transportation Science*, 5:366–389, 1971.

[Daf72]   S. C. Dafermos. Traffic assignment problem for multiclass user transportation networks. *Transportation Science*, 6:73–87, 1972.

[DI89]   M.V. Devarakonda and R.K. Iyer. Predictability of process resource usage: A measurement-based study on UNIX. *IEEE Trans. Softw. Eng.*, 15(12):1579–1586, December 1989.

[DO87]   F. Douglis and J. Ousterhout. Process migration in the Sprite operating system. In *Proc. 7th Int. Conf. Dist. Computing Syst.*, pages 18–25, 1987.

[DO91]   F. Douglis and J. Ousterhout. Transparent process migration: Design alternatives and the Sprite implementation. *Software – Practice and Experience*, 21(8):757–785, 1991.

[DP78]   A. M. Despain and D. A. Patterson. X-tree: A tree structured multi-processor computer architecture. In *Proc. IEEE&ACM 5th Symp. on Comput. Arch.*, pages 144–151, April 1978.

[DS69]   S. C. Dafermos and F. T. Sparrow. The traffic assignment problem for a general network. *Journal of Research of National Bureau of Standards-B*, 73B(2):91–118, 1969.

[dSeSG84] E. de Souza e Silva and M. Gerla. Load balancing in distributed sytems with multiple classes and site constraints. In E. Gelenbe, editor, *PERFORMANCE '84*, pages 17–33. Elsevier Science Publishers B. V. (North-Holland), 1984.

[ELZ86a]   D. L. Eager, E. D. Lazowska, and J. Zahorjan. Adaptive load sharing in homogeneous distributed systems. *IEEE Trans. on Software Engineering*, SE-12(5):662–675, May 1986.

[ELZ86b]   D. L. Eager, E. D. Lazowska, and J. Zahorjan. A comparison of receiver-initiated and sender-initiated adaptive load sharing. *Performance Evaluation*, 6(1):53–68, March 1986.

[ELZ88]   D.L. Eager, E.D. Lazowska, and J. Zahorjan. The limited performance benefits of migrating active processes for load sharing. In *Proc. 12th ACM Symp. Oper. Syst. Principles*, pages 63–72, May 1988.

[FGK73]   L. Fratta, M. Gerla, and L. Kleinrock. The flow deviation method: An approach to store-and-forward communication network design. *Networks*, 3:97–133, 1973.

[FZ86]   D. Ferrari and S. Zhou. A load index for dynamic load balancing. In *Proc. 1986 Fall Joint Comput. Conf.*, pages 684–690, November 1986.

[FZ88]   D. Ferrari and S. Zhou. An empirical investigation of load indices for load balancing applications. In P. J. Courtois and G. Latouche, editors, *Performance '87*, pages 515–528. Elsevier Science Publishers B. V. (North-Holland), 1988.

[GDI93]   K.K. Goswami, M. Devarakonda, and R.K. Iyer. Prediction-based dynamic load-sharing heuristics. *IEEE Trans. Para. Dist. Syst.*, 4(6):638–648, June 1993.

[GF88]   M. Gerla and L. Fratta. Tree structured fiber optics MANs. *IEEE J. Selected Areas Commun.*, 6(6):934–942, July 1988.

[GID89]   K.K. Goswami, R.K. Iyer, and M. Devarakonda. Load sharing based on task resource prediction. In *Proc. 22nd Annu. Hawaii Int. Conf. Syst. Sci.*, pages 921–927, 1989.

[GM80]   E. Gelenbe and I. Mitrani. *Analysis and Synthesis of Computer Systems.* Academic Press, New York, 1980.

[HS90]   H.G.Rotithor and S.S.Pyo. Decentralized decision making in adaptive task sharing. In *2nd IEEE Symp. Parallel and Distr. Processing*, pages 34–41, December 1990.

[Int71]   M. D. Intriligator. *Mathematical Optimization and Economic Theory.* Prentice-Hall, Inc., Englewood Cliffs, N.J., 1971.

[Kam87]   A. E. Kamal. Star local area networks: a performance study. *IEEE Trans. on Computers*, c-36(4):483–499, April 1987.

[Kel79]   F.P. Kelly. *Reversibility and Stochastic Networks.* John Wiley & Sons, Ltd., New York, 1979.

[KH88]   H. Kameda and A. Hazeyama. Individual vs. overall optimization for static load balancing in distributed computer systems. Computer science report, The University of Electro-Communications, Tokyo, Japan, 1988.

[KK90a]   Chonggun Kim and Hisao Kameda. Optimal static load balancing of multi-
           class jobs in a distributed computer system. In *10th IEEE International
           Conference on Distributed Computing Systems*, pages 562–569, Paris, France,
           June 1990.

[KK90b]   Chonggun Kim and Hisao Kameda. Optimal static load balancing of multi-
           class jobs in a distributed computer system. *The Transactions of the IEICE*,
           E73(7):1207–1214, July 1990.

[KK92a]   C. Kim and H. Kameda. An algorithm for optimal static load balancing in
           distributed computer systems. *IEEE Trans. on Computers*, 41(3):381–384,
           March 1992.

[KK92b]   O. Kremien and J. Kramer. Methodical analysis of adaptive load sharing
           algorithms. *IEEE Trans. Para. Dist. Syst.*, 3(6):747–760, November 1992.

[KKM93]   O. Kremien, J. Kramer, and J. Magee. Scalable, adaptive load sharing for
           distributed systems. *IEEE Parallel and Distributed Technology*, pages 62–70,
           August 1993.

[KL88]   P. Krueger and M. Livny. A comparison of preemptive and non-preemptive
           load distributing. In *Proc. 8th Int. Conf. Dist. Computing Syst.*, pages 123–
           130, June 1988.

[Kle76]   L. Kleinrock. *Queueing Systems, Vol. 2: Computer Applications*. John Wiley
           & Sons, Inc., New York, 1976.

[Kru88]   P. Krueger. *Distributed scheduling for a changing environment*. PhD thesis,
           University of Wisconsin–Madison, 1988.

[KS94]   P. Krueger and N.G. Shivaratri. Adaptive location policies for global schedul-
           ing. *IEEE Trans. Softw. Eng.*, 20(6):432–444, June 1994.

[Kum89]   Anurag Kumar. Adaptive load control of the central processor in a distributed
           system with a star topology. *IEEE Trans. on Computers*, 38(11):1502–1512,
           November 1989.

[KZ95]   Hisao Kameda and Yongbing Zhang. Uniqueness of the solution for optimal
           static routing in open bcmp queueing networks. *Mathl. Comput. Modelling*,
           22:119–130, 1995.

[Lam80]   S.S. Lam. A carrier sense multiple access protocol for local networks. *Com-
           puter Networks*, 4:21–32, 1980.

[LB83]   E. S. Lee and P. I. P. Boulton. The principles and performance of hubnet: A 50 mbits/s glass fiber local area network. *IEEE J. Select. Areas Commun.*, SAC-1:711–720, November 1983.

[LK93]   J. Li and H. Kameda. Optimal load balancing in tree networks with two-way traffic. *Computer Networks and ISDN Systems*, 25(12):1335–1348, 1993.

[LK94a]   J. Li and H. Kameda. A decomposition algorithm for optimal static load balancing in tree hierarchy network configurations. *IEEE Trans. on Parallel and Distributed Systems*, 5(5):540–548, 12 1994.

[LK94b]   J. Li and H. Kameda. Optimal load balancing in star network configurations with two-way traffic. *Journal of Parallel and Distributed Computing*, 23(3):364–375, 12 1994.

[LM82]   M. Livny and M. Melman. Load balancing in homogeneous broadcast distributed systems. In *Proc. ACM Computer Network Performance Symposium*, pages 47–55, April 1982.

[LM85]   M. Livny and U. Manber. Distributed computation via active messages. *IEEE Trans. Comput.*, 34(12):1185–1190, 12 1985.

[LNT73]   T. Leventhal, G. Nemhauser, and L. Trotter, Jr. A column generation algorithm for optimal traffic assignment. *Transportation Science*, 7:168–176, 1973.

[LR92]   H.C. Lin and C.S. Raghavendra. A dynamic load-balancing policy with a central job dispatcher (LBC). *IEEE Trans. Softw. Eng.*, 18(2):148–158, February 1992.

[LVM79]   J. G. Lee, W.G. Vogt, and M.H. Mickle. Optimal decomposition of large-scale networks. *IEEE Trans. on Systems, Man, and Cybernetics*, SMC-9(7):369–375, July 1979.

[LY95]   Y. Lan and T. Yu. A dynamic central scheduler load balancing mechanism. In *Proc. of IEEE 14th Annual Int. Pheonix Conf. Comput. and Commun.*, pages 734–740, 1995.

[LZGS84] E.D. Lazowska, J. Zahorjan, G.S. Graham, and K.C. Sevcik. *Quantitative System Performance*. Prentice-Hall, Inc., 1984.

[Mag84]   T. L. Magnanti. Models and algorithms for predicting urban traffic equilib-
             ria. In M. Florian, editor, *Transportation Planning Models*, pages 153–185.
             Elsevier Science Publishers B. V. (North-Holland), 1984.

[MAHE88] M. K. Mehmet-Ali, J. F. Hayes, and A. K. Elhakeem. Traffic analysis of a
             local area network with a star topology. *IEEE Trans. on Communications*,
             36(6):703–712, June 1988.

[MTS89]   R. Mirchandaney, D. Towsley, and J. A. Stankovic. Analysis of the effects
             of delays on load sharing. *IEEE Trans. on Computers*, 38(11):1513–1525,
             November 1989.

[MTS90]   R. Mirchandaney, D. Towsley, and J.A. Stankovic. Adaptive load sharing in
             heterogeneous distributed systems. *J. Parallel and Distributed Computing*,
             9:331–346, 1990.

[Ota92]   T. Ota. Coupled star network: A new configuration for optical local area
             network. *IEICE Trans. on Commun.*, E75-B(2):67–75, February 1992.

[PM83]   M.L. Powell and B.P. Miller. Process migration in DEMOS/MP. In *Proc. 9th
             Symp. on Operating System Principles*, pages 110–119, 1983.

[RM78]   E. G. Rawson and R. M. Metcalfe. Fibernet: Multimode optical fibers for local
             computer networks. *IEEE Trans. on Communications*, COM-26(7):983–990,
             July 1978.

[RS84]   K. Ramamritham and J.A. Stankovic. Dynamic task scheduling in hard real-
             time distributed systems. *IEEE Software*, pages 65–75, July 1984.

[Sch83]   R. V. Schmidt *et al.* Fibernet II: A fiber optic Ethernet. *IEEE J. Select.
             Areas Commun.*, SAC-1:702–711, November 1983.

[SED91]   M. Schaar, K. Efe, and L. Delcambre. An analytical model for predicting
             performance in a heterogeneous system. In *Proc. 3rd IEEE Symp. Parallel
             and Dist. Process.*, pages 334–341, December 1991.

[SEDB91] M. Schaar, K. Efe, L. Delcambre, and L.N. Bhuyan. Load balancing with
             network cooperation. In *Proc. 11th Int. Conf. Dist. Computing Syst.*, pages
             328–335, Arlington, Texas, May 1991.

[SK90]   N.G. Shivaratri and P. Krueger. Two adaptive location policies for global
             scheduling algorithms. In *Proc. 10th Int. Conf. Dist. Computing Syst.*, pages
             502–509, Paris, France, May 1990.

[SM89]   T. Suda and S. Morris. Tree LANs with collision avoidance: Station and switch protocols. *Computer Networks and ISDN systems*, 17(2):101–110, July 1989.

[Smi88]   J.M. Smith. A survey of process migration mechanisms. *ACM Operating Syst. Review*, 22(3):28–40, 1988.

[SS84]   J.A. Stankovic and I.S. Sidhu. An adaptive bidding algorithm for processes, clusters and distributed groups. In *Proc. 4th Int. Conf. Dist. Computing Syst.*, pages 49–59, 1984.

[SS94]   M. Singhal and N.G. Shivaratri. *Advanced Concepts in Operating Systems.* McGraw-Hill, Inc., New York, 1994.

[Sta84]   J. A. Stankovic. A perspective on distributed computer systems. *IEEE Trans. on Computers*, c-33(12):1102–1115, December. 1984.

[Tam84]   T. Tamura *et al.* Optical cascade star network - a new configuration for passive distribution system with optical collision detection. *IEEE J. Lightwave Technol.*, LT-2(1):61–66, 1984.

[TLC85]   M.M. Theimer, K.A. Lantz, and D.R. Cheriton. Preemptable remote execution facilities for the V-system. In *Proc. 10th Symp. on Operating System Principles*, pages 2–12, 1985.

[TT84]   A. N. Tantawi and D. Towsley. A general model for optimal static load balancing in star network configurations. In E. Gelenbe, editor, *PERFORMANCE '84*, pages 277–291. Elsevier Science Publishers B. V. (North-Holland), 1984.

[TT85]   A. N. Tantawi and D. Towsley. Optimal static load balancing in distributed computer systems. *Journal of the ACM*, 32(2):445–465, April 1985.

[VD90]   R. Venkatesh and G.R. Dattatreya. Adaptive optimal load balancing of loosely coupled processors. In *1990 Int. Conf. Parallel Process.*, volume Vol.I, pages 22–25, Chicago, August 1990.

[War52]   J. G. Wardrop. Some theoretical aspects of road traffic research. In *Proc. Inst. Civ. Eng.*, pages 325–378. Part 2, 1, 1952.

[WM85]   Y. T. Wang and R. J. T. Morris. Load sharing in distributed systems. *IEEE Trans. on Computers*, c-34(3):204–217, March 1985.

[Yem83]   Y. Yemini. Tinkernet: Or, is there life between LANs and PBXs ? In *Proc. of the IEEE International Communications Conference*, pages 1501–1505, 1983.

[ZF87]   S. Zhou and D. Ferrari. A measurement study of load balancing performance. In *Proc. 7th Int. Conf. Dist. Computing Syst.*, pages 490–497, Berlin, West Germany, September 1987.

[ZG94]   W. Zhu and A. Goscinski. Low cost load balancing algorithms. In *Proc. 27th Annual Simulation Symposium, IEEE Computer Society Press*, pages 226–235, 1994.

[ZHKS95]   Y. Zhang, K. Hakozaki, H. Kameda, and K. Shimizu. A performance comparison of adaptive and static load balancing in heterogeneous distributed systems. In *Proc. 28th Annual Simulation Symposium, IEEE Computer Society Press*, pages 332–340, Phoenix, A.Z., 1995.

[Zho88]   S. Zhou. A trace-driven simulation study of dynamic load balancing. *IEEE Trans. Softw. Eng.*, 14(9):1327–1341, September 1988.

[ZKS91]   Y. Zhang, H. Kameda, and K. Shimizu. Parametric analysis of optimal static load balancing in star network configurations. *IEICE Transactions*, J74-D-I(9):644–655, 1991.

[ZKS92]   Y. Zhang, H. Kameda, and K. Shimizu. Parametric analysis of optimal static load balancing in distributed computer systems. *Journal of Information Processing*, 14(4):433–441, 1992.

[ZKS94]   Y. Zhang, H. Kameda, and K. Shimizu. Adaptive bidding job scheduling in heterogeneous distributed systems. In *Proc. 1994 Int. Workshop Modeling, Analysis, and Simulation of Computer and Telecommunication Systems, IEEE Computer Society Press*, pages 250–254, Durham, N.C., 1994.

[ZZWD93]   S. Zhou, X. Zheng, J. Wang, and P. Delisle. Utopia: A load sharing facility for large, heterogeneous distributed computer systems. *Software–Practice and Experience*, 23(12):1305–1336, December 1993.

Index